Lecture Notes in Networks and Systems

Volume 13

The series "Lecture Notes in Networks and Systems" publishes the latest developments in Networks and Systems—quickly, informally and with high quality. Original research reported in proceedings and post-proceedings represents the core of LNNS.

Volumes published in LNNS embrace all aspects and subfields of, as well as new challenges in, Networks and Systems.

The series contains proceedings and edited volumes in systems and networks, spanning the areas of Cyber-Physical Systems, Autonomous Systems, Sensor Networks, Control Systems, Energy Systems, Automotive Systems, Biological Systems, Vehicular Networking and Connected Vehicles, Aerospace Systems, Automation, Manufacturing, Smart Grids, Nonlinear Systems, Power Systems, Robotics, Social Systems, Economic Systems and other. Of particular value to both the contributors and the readership are the short publication timeframe and the world-wide distribution and exposure which enable both a wide and rapid dissemination of research output.

The series covers the theory, applications, and perspectives on the state of the art and future developments relevant to systems and networks, decision making, control, complex processes and related areas, as embedded in the fields of interdisciplinary and applied sciences, engineering, computer science, physics, economics, social, and life sciences, as well as the paradigms and methodologies behind them.

More information about this series at http://www.springer.com/series/15179

Ignacio Chang · José Baca
Héctor A. Moreno · Isela G. Carrera
Manuel N. Cardona
Editors

Advances in Automation and Robotics Research in Latin America

Proceedings of the 1st Latin American Congress on Automation and Robotics, Panama City, Panama 2017

Springer

Editors
Ignacio Chang
Facultad de Ingeniería Eléctrica (FIE)
Universidad Tecnológica de Panamá
Panama
Panama

Isela G. Carrera
Facultad de Ingeniería Mecánica y Eléctrica, Unidad Norte
Universidad Autónoma de Coahuila
Monclova, Coahuila
Mexico

José Baca
Computer Science Department, College of Information Science & Technology
University of Nebraska at Omaha
Omaha, NE
USA

Manuel N. Cardona
Universidad Don Bosco
Soyapango, San Salvador
El Salvador

Héctor A. Moreno
Facultad de Ingeniería Mecánica y Eléctrica, Unidad Norte
Universidad Autónoma de Coahuila
Monclova, Coahuila
Mexico

ISSN 2367-3370 ISSN 2367-3389 (electronic)
Lecture Notes in Networks and Systems
ISBN 978-3-319-54376-5 ISBN 978-3-319-54377-2 (eBook)
DOI 10.1007/978-3-319-54377-2

Library of Congress Control Number: 2017932795

Printed on acid-free paper

This Springer imprint is published by Springer Nature
The registered company is Springer International Publishing AG
The registered company address is: Gewerbestrasse 11, 6330 Cham, Switzerland

Preface

LACAR 2017 is the first Latin American Congress on Automation and Robotics that promotes an open forum where researches, scientists, and engineers from academia and private industries come together to present current and exciting research applications, and discuss future challenges in this fascinating field. It aims to generate activities in research, development, and application on automation and robotics by exchanging knowledge, experiences, and the synergy of research groups from different places from Latin American.

This year, the congress is held at the Universidad Tecnologica de Panama (UTP), Panama City. The congress highlights different research areas such as control algorithms, systems automation, perception, mobile robotics, computer vision, educational robotics, robotics modeling and simulation, and robotics and mechanism design. Main workshops discuss topics related to automation and robotics research in Latin American, the collaboration between research groups, development of low-cost robotic systems and disaster risk management.

We would like to extend our sincere gratitude and appreciation for all the hard work and dedication provided by the members of the organizing committee, program committee, volunteers, and staff serving LACAR 2017. Also, we thank the authors and reviewers who have contributed to this event by preparing papers of high quality. We would like to express our grateful thanks to the publisher and editorial staff of Springer for accepting and supporting the publication of this proceedings volume. We believe the LACAR 2017 proceedings are a valuable source of reference for future research and development that considers the sustainable economics and social development of Latin American countries.

February 2017

Ignacio Chang
José Baca
Héctor A. Moreno
Isela G. Carrera
Manuel N. Cardona

Contents

Robot Position Optimization Based on Joint Mean Torque Norm

Aníbal Alexandre Campos Bonilla, Leandro da Silva, and Luis Eduardo Garcia Gonzalez(✉)

UDESC, Joinville, SC, Brazil
alexandre.campos@udesc.br, dasilvaleandro21@gmail.com, luisgarciagonzalez.10@gmail.com

Abstract. Optimal plant-floor robot positioning is a non-trivial task for robotic manufacturing. An unsuitable robot positioning may cost time and energy, which after thousands of cycles may be significant. The present article presents an approach to choose the robot base position through a numerical optimization tool. The problem may be defined as follows: given a set of constraints for the base placement of a serial robot, determine the base placement which requires minimum norm of the averaged absolute joint torques through a given set of trajectories in a process. In order to obtain the joint torques, the dynamic model of the robot is used, by means of the recursive Newton-Euler approach. Optimization process is based on a numerical software function. Additionally, trajectories are determined using polynomial interpolation for joints motion, assuring motor smoothness suitable for pick and place tasks.

1 Introduction

Industrial robots, also called manipulators, are widely used due to their flexibility to change function by software. The type of automation which makes use of robots is called soft automation, as opposed to hard automation, which cannot be reconfigured by software, and manual labor, which is very expensive for large batches [9].

When using a manipulator to perform a given task, there is always a need to carefully consider the limitations imposed by the work environment. All these factors may change if the robot placement is altered. This means that if the robot base changes, all joint motors experience a different torque regime, which in its turn may reduce the robot lifespan, due to the increased effort in each actuator and mechanical components wear. Reduction in actuator torques also leads to reduced energy consumption [5].

Considering an industrial repetitive task such as pick and place, even a small amount of reduction in actuator torque may, after thousands of cycles, be high enough to justify the procedure of choosing an adequate positioning for the robot base. This is even more so, given the one-time nature of the calculation, which is made offline.

I. Chang et al. (eds.), *Advances in Automation and Robotics Research in Latin America*, Lecture Notes in Networks and Systems 13, DOI 10.1007/978-3-319-54377-2_1

For this article, the function $f(\mathbf{x})$, which returns a norm of the absolute torque average in the task path, is associated with a given base position x, and z, granted that the trajectories, and the processes are held constant, the only difference being the base position and orientation. The base position may, therefore, be chosen as the one with the minimum return value of the function $f(\mathbf{x})$ associated.

As an example of this method, the IRB6620LX robot from ABB is presented. This robot is a serial robot with 6 joints, mounted on a gantry (which is the link 0). It is a PRR robot (prismatic, revolute, revolute) robot, with the last 3 joints being revolute, and forming a spherical wrist.

The task to be executed is a set of three processes (A, B and C) the robot needs to perform, carrying a part. Process A comprises "sliding" the part alongside the wall of the room, process B comprises "sliding" the object along the roof, and process C comprises resting (or waiting) for 3 s in the roof. The part is fed by a conveyor belt (1) with a given speed and orientation, and must be placed on a final conveyor belt (2) with a given speed and orientation, as indicated by Fig. 1 [6].

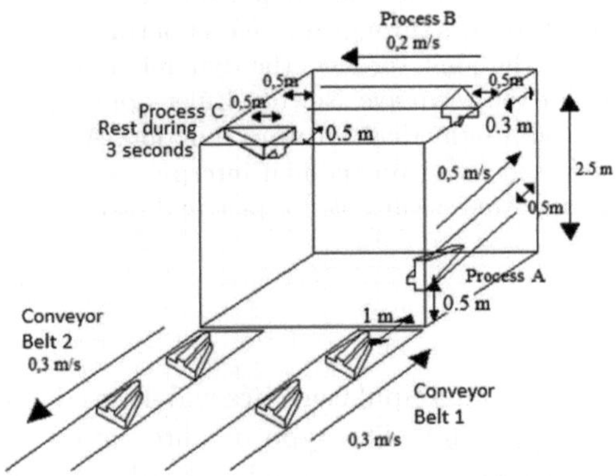

Fig. 1. Proposed processes to be executed by the robotic arm. Room size is $2.5 \times 2.5 \times 2.5\,m$

2 Forward Kinematics

The link coordinate frames may be assigned to the links using Denavit-Hartenberg algorithm [10]. Figure 2 shows each frame, including the room frame S_r and the robot base S_b frame. Robot's Denavit-Hartenberg parameters are given by Table 1.

Using the notation ${}^{a}T_{b}$ for the homogeneous transformation matrix representing frame b on the frame a, and $c(\theta)$, $s(\theta)$ for the sine and cosine of θ, the matrices ${}^{i-1}T_{i}$ represent the forward kinematics of IRB6620LX robot, for $1 < i < 6$, (Eqs. 1 and 2).

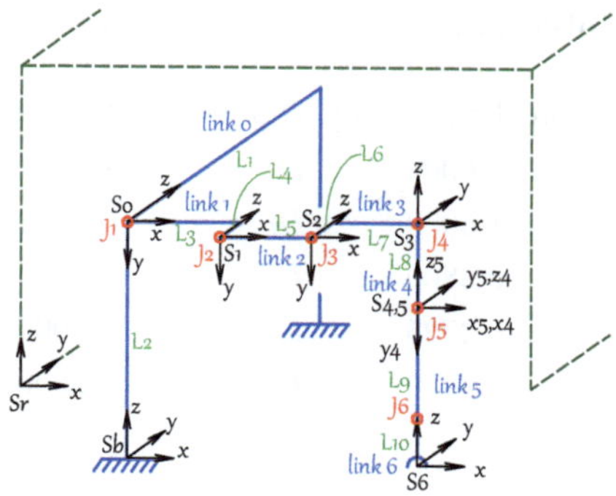

Fig. 2. Coordinate frames associated with each link

Table 1. Denavit-Hartenberg parameters for IRB6620LX

Joint	θ	d	a	α
1	0	$(d_1 - L_4)$	L_3	0
2	$-\theta_2$	0	L_5	0
3	$-\theta_3$	L_6	L_7	$\pi/2$
4	θ_4	$-L_8$	0	$-\pi/2$
5	$-\theta_5$	0	0	$\pi/2$
6	θ_6	$(-L_9 - L_{10})$	0	0

$$ {}^{i-1}T_i = \begin{bmatrix} c_{\theta_i} & -s_{\theta_i} c_{\alpha_i} & s_{\theta_i} s_{\alpha_i} & a c_{\theta_i} \\ s_{\theta_i} & c_{\theta_i} c_{\alpha_i} & -c_{\theta_i} s_{\alpha_i} & a s_{\theta_i} \\ 0 & s_{\alpha_i} & c_{\alpha_i} & d \\ 0 & 0 & 0 & 1 \end{bmatrix} \tag{1} $$

$$ {}^0T_6 = {}^0T_1 \cdot {}^1T_2 \cdot {}^2T_3 \cdot {}^3T_4 \cdot {}^4T_5 \cdot {}^5T_6 \tag{2} $$

Using the datasheet of the robot and the CAD models supplied by the manufacturer [1], the dimensional parameters are obtained and presented in Table 2.

Room and base frames (S_r and S_b) can be represented by the homogeneous transformation matrices 4 and 5 respectively. Vector ${}^s\boldsymbol{b}_b$ represents the position and orientation of base frame relative to the room frame, using the Euler angles ZYZ [10].

$$ {}^r\boldsymbol{b}_b = \left[x\ y\ z\ \phi\ \theta\ \psi\right]^T \tag{3} $$

$$ {}^rT_b = \begin{bmatrix} c_\phi c_\theta c_\psi - s_\phi s_\psi & -c_\phi s_\psi c_\theta - s_\phi c_\psi & c_\phi s_\theta & x_b \\ s_\phi c_\theta c_\psi + c_\phi s_\psi & c_\phi c_\psi - s_\phi s_\psi c_\theta & s_\phi s_\theta & y_b \\ -c_\psi s_\theta & s_\psi s_\theta & c_\theta & z_b \\ 0 & 0 & 0 & 1 \end{bmatrix} \tag{4} $$

Table 2. Link lengths for IRB6620LX

Dimension	Length $[m]$	Dimension	Length $[m]$
L_1	4.0000	L_6	L_4
L_2	1.5000	L_7	0.2000
L_3	0.3800	L_8	0.8870
L_4	0.2725	L_9	0.2000
L_5	0.9750	L_{10}	0.2000

$$^{b}T_0 = \begin{bmatrix} 1 & 0 & 0 & 0 \\ 0 & 0 & 1 & 0 \\ 0 & -1 & 0 & L_2 \\ 0 & 0 & 0 & 1 \end{bmatrix} \tag{5}$$

3 Inverse Kinematics

Considering the fact that this robot possesses a spherical wrist, the inverse kinematics may be uncoupled at the wrist, yielding two, less complex, inverse kinematics problems [11]. The first step is to determine the wrist position P_w based on the tool-tip position and orientation. Given the tool length d_6, the tool-tip position and orientation $^0T_6$, the wrist position can be determined as in Eq. (6).

$$\begin{aligned} ^0P_w &= {}^0P_e - d_6 \cdot a_e & \qquad ^0P_e &= [^0T_{6(11)}, {}^0T_{6(12)}, {}^0T_{6(13)}]^T \\ a_e &= [^0T_{6(13)}, {}^0T_{6(23)}, {}^0T_{6(33)}]^T & d_6 &= L_9 + L_{10} \end{aligned} \tag{6}$$

Using the value of ^0P_w, the value of the first 3 joint variables (d_1, θ_2 and θ_3) can be determined as in Eq. (7). Figure 3 shows the scheme used to calculate the inverse kinematics, as well as auxiliary variables r, α, β, ω and γ. The two

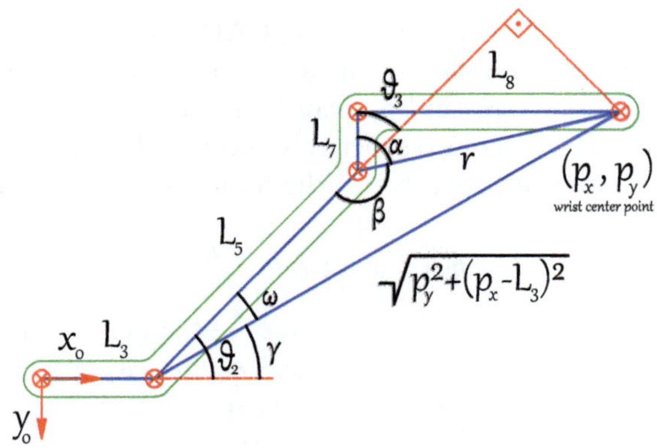

Fig. 3. Scheme used to derive the inverse kinematics

possible configurations for the arm are elbow up, and elbow down, represented by the plus-minus sign of $sin(\beta)$. Elbow up was chosen in this case (plus sign).

$$^0P_w = [p_x,\ p_y,\ p_z]^T \tag{7}$$

$$\begin{aligned} d_1 &= p_z \\ r &= \sqrt{L_8^2 + L_7^2} \\ \alpha &= atan(L_8,\ L_7) \\ cos(\beta) &= \frac{r^2 + L_5^2 - p_y^2 - (p_x - L_3)^2}{2 \cdot r \cdot L_5} \\ sin(\beta) &= \pm\sqrt{1 - cos(\beta)^2} \end{aligned}$$

$$\begin{aligned} \beta &= atan(sin(\beta),\ cos(\beta)) \\ \theta_3 &= \beta + \alpha - \pi \\ \gamma &= atan(-p_y,\ p_x - L_3) \\ \omega &= atan(r \cdot sin(\alpha - \theta_3), \\ &\qquad r \cdot cos(\alpha - \theta_3) + L_5) \\ \theta_2 &= \gamma + \omega; \end{aligned}$$

Applying the numeric values of the first three joints variables (d_1, θ_2 and θ_3) to the homogeneous matrices, and multiplying them, yields the matrix $^0T_3$, which is used to obtain $^0R_3$. The matrix $^0R_3$ is used to obtain $^3R_6$.

$$^0T_3(d_1, \theta_2, \theta_3) = {}^0T_1(d_1) \cdot {}^1T_2(\theta_2) \cdot {}^2T_3(\theta_3) \tag{8}$$

$$^0R_{3(i,j)} = {}^0T_{3(i,j)} \quad 1 \le i, j \le 3 \tag{9}$$

$$^3R_6 = {}^0R_3^T \cdot {}^0R_6 \tag{10}$$

Finally, the matrix $^3R_6$ can be used to determine the last 3 joint angles. These are the euler angles, and can be calculated as indicated as in Eq. (11)

$$\begin{aligned} \theta_4 &= atan({}^3R_{6(2,3)}, {}^3R_{6(1,3)}) \\ \theta_5 &= -atan(\sqrt{{}^3R_{6(1,3)}^2 + {}^3R_{6(2,3)}^2},\ {}^3R_{6(3,3)}) \\ \theta_6 &= atan({}^3R_{6(3,2)}, -{}^3R_{6(3,1)}) \end{aligned} \tag{11}$$

4 Differential Kinematics

The differential kinematics gives the relationship between the joint velocities and the end-effector linear and angular velocities [11]. It can be obtained by differentiating the kinematic expression. The first order partial derivatives of the forward kinematic equations yields an entity called *jacobian* matrix, which can also be obtained in a recursive way. This matrix possesses a dimension of $(6 \times n)$ for a serial robot with n DOF (degrees-of-freedom). The *jacobian* matrix J can be written as two $(3 \times n)$ matrices, as in Eq. (12)

$$J = \left[J_P,\ J_O\right]^T \tag{12}$$

The first matrix J_P indicates the relationship between the joint angular velocity $\dot{\boldsymbol{q}}$ and the end-effector linear velocity $\dot{\boldsymbol{p}}_{\boldsymbol{e}}$. The second matrix J_O relates the joint angular velocity $\dot{\boldsymbol{q}}$ with the end-effector angular velocity $\boldsymbol{\omega_e}$. Both of these matrices are dependent on the joint position vector $\boldsymbol{q}$, as can be seen in Eq. (13).

$$\dot{\boldsymbol{p}}_{\boldsymbol{e}} = J_P(\boldsymbol{q}) \cdot \dot{\boldsymbol{q}} \qquad \boldsymbol{\omega_e} = J_O(\boldsymbol{q}) \cdot \dot{\boldsymbol{q}} \tag{13}$$

Matrices J_P and J_O are dependent on the kind of joint (prismatic or revolute) and can be obtained recursively using the homogeneous transformation matrices from the forward kinematics. Using the vector $\boldsymbol{v_e}$ to represent the end-effector linear and angular velocities, and the vector $\boldsymbol{q}$ to represent the joint variables, the recursive differential kinematics of the robot becomes as indicated in Eq. (14). The inverse diferential kinematics is given by Eq. (16).

$$J_P = \left[jp_1\; jp_2\; jp_3\; jp_4\; jp_5\; jp_6\right] \qquad J_O = \left[jo_1\; jo_2\; jo_3\; jo_4\; jo_5\; jo_6\right] \tag{14}$$

$$\begin{aligned} jp_1 &= {}^0z_0 \\ jp_{i+1} &= {}^0z_i \times (\,{}^0p_e - {}^0p_i), \;\; 1 \le i \le 5 \\ jo_1 &= [0,0,0]^T \\ jo_{i+1} &= {}^0z_i, \;\; 1 \le i \le 5 \end{aligned} \qquad \begin{aligned} {}^0z_0 &= [0,0,1]^T \\ {}^0z_i &= {}^0T_{i(1:3,3)}, \;\; 1 \le i \le 5 \\ {}^0p_0 &= [0,0,0]^T \\ {}^0p_i &= {}^0T_{i(1:3,4)}, \;\; 1 \le i \le 5 \\ {}^0p_e &= {}^0T_{6(1:3,4)} \end{aligned}$$

$$\boldsymbol{q} = [d1,\; \theta_2,\; \theta_3,\; \theta_4,\; \theta_5,\; \theta_6]^T \tag{15}$$

$$\begin{aligned} \dot{\boldsymbol{q}} &= d\boldsymbol{q}/dt \\ \boldsymbol{v_e} &= \left[\dot{\boldsymbol{p}}_{\boldsymbol{e}}, \boldsymbol{\omega_e}\right]^T \end{aligned} \qquad \begin{aligned} \boldsymbol{v_e} &= J(\boldsymbol{q}) \cdot \dot{\boldsymbol{q}} \\ \dot{\boldsymbol{q}} &= J(\boldsymbol{q})^{-1} \cdot \boldsymbol{v_e} \end{aligned} \tag{16}$$

5 Trajectory

In this section a path is created based on a given set of *waypoints*, using a 3rd degree polynomial to interpolate each joint movement relative to time, between each two consecutive points. Considering that the velocities are specified at the *waypoints*, the curve obtained by this process is continuous at the first order derivatives, but not at the second order derivatives.

The problem of finding such a curve between two points, can be defined as follows: "given an initial point ${}^0P_i$, an initial orientation ${}^0R_i$, the initial linear and angular velocities $\boldsymbol{v}_i$, and the initial time t_i, as well as the final point ${}^0P_f$, the final orientation ${}^0R_f$, the final velocity $\boldsymbol{v}_f$ on the tool-tip and the final time t_f, find the expression for the joint vector $\boldsymbol{q}$ in the generic time t, for $t_i \le t \le t_f$".

For a single joint, this problem can be described as follows:

$$\begin{aligned} &\mathit{Find:} && q_n(t) = a_0 + a_1 t + a_2 t^2 + a_3 t^3 \\ &\mathit{Given:} && {}^0P_i,\; {}^0R_i,\; \boldsymbol{v}_i,\; t_i,\; {}^0P_f,\; {}^0R_f,\; \boldsymbol{v}_f,\; t_f. \end{aligned}$$

The curve obtained as the solution of the above problem yields a polynomial expression for q_n between two points in time. This is called *joint space trajectory*, because the trajectory is generated in the robot joint space. It possesses the following characteristics compared to a trajectory generated in the operational space [11]:

1. The trajectory generated is not very demanding from a computational viewpoint;
2. Joint positions and velocities are continuous functions of time;
3. Undesirable effects, such as non-smooth trajectories, are minimized.

The first step to solve the described problem, is to convert the given positions and velocities from the operational space to the joint space. This can be done using the inverse kinematics and the differential inverse kinematics, which will result in the four vectors $\boldsymbol{q_i}$, $\dot{\boldsymbol{q}}_{\boldsymbol{i}}$, $\boldsymbol{q_f}$, and $\dot{\boldsymbol{q}}_{\boldsymbol{f}}$. Using the initial and final time, and the 3rd degree polynomials, as well as their derivatives, the interpolating function is found. Mathematically, for one joint:

$$
\begin{aligned}
q_i(t_i) &= a_0 + a_1 t_i + a_2 t_i^2 + a_3 t_i^3 & \qquad q_f(t_f) &= a_0 + a_1 t_f + a_2 t_f^2 + a_3 t_f^3 \\
\dot{q}_i(t_i) &= a_1 + 2a_2 t_i + 3a_3 t_i^2 & \qquad \dot{q}_f(t_f) &= a_1 + 2a_2 t_f + 3a_3 t_f^2
\end{aligned}
$$

Solving this set of equations for each joint, the 6 interpolating polynomials can be found, as:

$$
\boldsymbol{q}(t) = \left[q_1(t),\ q_2(t),\ \ldots,\ q_n(t)\right]^T,\quad n = 6
$$

If the velocities were not provided at the *waypoints*, a curve called *spline* could be used to yield continuity at the first and second order derivatives [4].

6 Dynamics

Robot dynamics model allows for control strategies and motion planning techniques to be tested without a need to use a physically available system [11]. There are two basic problems related to the dynamics of a manipulator [4]:

1. Given the trajectory vectors $\boldsymbol{q}(t)$, $\dot{\boldsymbol{q}}(t)$, $\ddot{\boldsymbol{q}}(t)$, find the torque vector $\boldsymbol{\tau}(t)$;
2. Given the torques applied to the joints $\boldsymbol{\tau}(t)$, find the resulting motion of the robot $\boldsymbol{q}(t)$, $\dot{\boldsymbol{q}}(t)$, $\ddot{\boldsymbol{q}}(t)$.

Since the interest at this point is to find the torques for a given trajectory, the first problem needs to be addressed.

There are two basic approaches to model the dynamics of a manipulator: the lagrangian formulation, and the Newton-Euler formulation. While the first one is an *energy-based* approach and conceptually simpler, the second one is a *force-balance* recursive approach and is computationally more efficient, since it exploits the structure of the manipulator open kinematic chain [4,9,11]. Thus, the Newton-Euler approach is used.

Newton-Euler approach consists of using the Newton-Euler equations to find the force and torque acting on the center of mass of each link. The Newton-Euler equations for the link i are given in Eq. (17).

$$\boldsymbol{F_i} = m_i \dot{\boldsymbol{v}}_{ci} \qquad \boldsymbol{N_i} = {}^{ci}I\dot{\boldsymbol{\omega}}_{\boldsymbol{i}} + \boldsymbol{\omega_i} \times {}^{ci}I\boldsymbol{\omega_i} \tag{17}$$

In order to calculate the Eq. (17), the mass (m_i) and the inertia tensor at the C.G. (center of gravity) (${}^{ci}I$) of each link (i) needs to be specified, as well as the linear acceleration at the center of mass ($\dot{\boldsymbol{v}}_{ci}$,) and the angular velocity and acceleration ($\boldsymbol{\omega_i}$ and $\dot{\boldsymbol{\omega}}_{\boldsymbol{i}}$) of the link need to be calculated.

The masses of each link were obtained from the toolbox *ARTE* [7], and given in Eq. (18). The links are assumed to be cylinders to estimate the inertia tensors. This assumption imply the algorithm is stable relative to uncertainty in the C.G. position. However, if the C.G. of each joint change the torques may change, which may be analyzed in further studies.

$$\begin{aligned} m_1 &= 327\,\text{kg} & m_3 &= 81\,\text{kg} & m_5 &= 7\,\text{kg} \\ m_2 &= 90\,\text{kg} & m_4 &= 49\,\text{kg} & m_6 &= 2\,\text{kg} \\ & & & & m_{part} &= 1\,\text{kg} \end{aligned} \tag{18}$$

For a solid cylinder with height h and radius r, the inertia tensor at the C.G. can be calculated as in Eq. 19.

$${}^{ci}I = (m_i/12) \cdot \begin{bmatrix} 3r^2 + h^2 & 0 & 0 \\ 0 & 3r^2 + h^2 & 0 \\ 0 & 0 & 6r^2 \end{bmatrix} \tag{19}$$

The heights and radius for each link are estimated as in Eq. (20).

$$\begin{aligned} r_1 &= 0.220\ m, & h_1 &= L_3 & r_4 &= 0.123\ m, & h_4 &= L_8 \\ r_2 &= 0.126\ m, & h_2 &= L_5 & r_5 &= 0.100\ m, & h_5 &= L_9 \\ r_3 &= 0.161\ m, & h_3 &= L_7 & r_6 &= 0.100\ m, & h_6 &= L_{10} \end{aligned} \tag{20}$$

There are basically two steps to calculate the force and torque in Eq. (17). The first one is to determine the values of $\dot{\boldsymbol{v}}_{ci}$, $\boldsymbol{\omega_i}$, $\dot{\boldsymbol{\omega}}_{\boldsymbol{i}}$, $\boldsymbol{F_i}$ and $\boldsymbol{N_i}$, in the so called *outward iterations*, i.e., calculate the linear velocities and angular velocities and accelerations, as well as the forces and torques due to the link velocities and accelerations from the base to the tool-tip. The second step is to determine the resultant force and torque ($\boldsymbol{f_i}$ and $\boldsymbol{n_i}$) at the C.G. of each link in the *inward iterations*, i.e., from the tool-tip to the base. For the first link, the outward iteration is simpler, since it is a prismatic joint, as shown in Eq. (21).

$$\begin{aligned} \boldsymbol{\omega_1} &= [0,0,0]^T & \boldsymbol{F_1} &= m_1 \cdot \dot{\boldsymbol{v}}_{c1} \\ \dot{\boldsymbol{\omega}}_{\boldsymbol{1}} &= [0,0,0]^T & \boldsymbol{N_1} &= {}^{c1}I\dot{\boldsymbol{\omega}}_{\boldsymbol{1}} + \boldsymbol{\omega_1} \times {}^{c1}I\boldsymbol{\omega_1} \\ \dot{\boldsymbol{v}}_{c1} &= \ddot{q}_1(t_n) \cdot z_1 \end{aligned} \tag{21}$$

The outward iterations for the other 5 revolute joints are calculated as Eq. (22).

$$
\begin{aligned}
&\boldsymbol{for}\ \ \boldsymbol{i=1..5} \\
&\qquad \boldsymbol{z}_{i+1} = {}^{i+1}R_{i\ (1:3,3)} \\
&\qquad \boldsymbol{\omega}_{i+1} = {}^{i+1}R_i\boldsymbol{\omega}_i + \dot{q}_{i+1}\boldsymbol{z}_{i+1} \\
&\qquad \dot{\boldsymbol{\omega}}_{i+1} = {}^{i+1}R_i\dot{\boldsymbol{\omega}}_i + {}^{i+1}R_i\boldsymbol{\omega}_i \times \dot{q}_{i+1}\boldsymbol{z}_{i+1} + \ddot{q}_{i+1}\boldsymbol{z}_{i+1} \\
&\qquad {}^{i}P_{i+1} = {}^{i}T_{i+1\ (1:3,4)} \\
&\qquad \dot{\boldsymbol{v}}_{i+1} = {}^{i+1}R_i(\dot{\boldsymbol{\omega}}_i \times^i P_{i+1} + \boldsymbol{\omega}_i \times (\boldsymbol{\omega}_i \times^i P_{i+1}) + \dot{\boldsymbol{v}}_1) \\
&\qquad \dot{\boldsymbol{v}}_{ci+1} = \dot{\boldsymbol{\omega}}_{i+1} \times P_{ci+1} + \boldsymbol{\omega}_{i+1} \times (\boldsymbol{\omega}_{i+1} \times P_{ci+1}) + \dot{v}_2 \\
&\qquad \boldsymbol{F}_{i+1} = m_{i+1} \cdot \dot{\boldsymbol{v}}_{c\ i+1} \\
&\qquad \boldsymbol{N}_{i+1} = {}^{i+1\ c}I \cdot \dot{\boldsymbol{\omega}}_{i+1} + \boldsymbol{\omega}_{i+1} \times {}^{i+1\ c}I \cdot \boldsymbol{\omega}_{i+1} \\
&\boldsymbol{endfor}
\end{aligned}
\tag{22}
$$

Part weight is included by adding an external force pointing towards the ground at the tool-tip. This force is designated by F_{ext}. There is no momentum N_{ext} applied to the tool-tip. The inward iterations are given as Eq. (23). Supposing a given trajectory is defined as a 3rd degree polynomial between times t_i and t_f, the torque vector $\boldsymbol{\tau}(t)$ is calculated iteratively for a set of times $(t_i, t_{i+1}, ..., t_n, ..., t_f)$ using the described outward and inward iterations at each given time t_n.

$$
\begin{aligned}
&\boldsymbol{F}_{ext} = {}^{6}R_r \cdot m_{part} \cdot g \cdot (-\boldsymbol{z}_r) \\
&\boldsymbol{N}_{ext} = 0 \\
&\boldsymbol{f}_6 = \boldsymbol{F}_{ext} + \boldsymbol{F}_6 \\
&\boldsymbol{n}_6 = \boldsymbol{N}_6 + \boldsymbol{P}c_6 \times \boldsymbol{F}_6 + \boldsymbol{N}_{ext} \\
&\tau_6 = \boldsymbol{n}_6^T \cdot \boldsymbol{z}_6 \\
&\boldsymbol{for}\ \ \boldsymbol{i=5:-1:2} \\
&\qquad \boldsymbol{f}_i = {}^{i}R_{i+1} \cdot \boldsymbol{f}_{i+1} + \boldsymbol{F}_i \\
&\qquad \boldsymbol{n}_i = \boldsymbol{N}_i +^i R_{i+1}\boldsymbol{n}_{i+1} + \\
&\qquad\qquad \boldsymbol{P}_{ci} \times \boldsymbol{F}_i + {}^{i}\boldsymbol{P}_{i+1} \times^i R_{i+1}\boldsymbol{f}_{i+1} \\
&\qquad \tau_i = \boldsymbol{n}_i^T \cdot \boldsymbol{z}_i \\
&\boldsymbol{endfor} \\
&\boldsymbol{f}_1 = {}^{1}R_2 \cdot \boldsymbol{f}_2 + \boldsymbol{F}_1 \\
&\boldsymbol{n}_1 = \boldsymbol{N}_1 +^1 R_2\boldsymbol{n}_2 + \boldsymbol{P}_{c1} \times \boldsymbol{F}_1 \\
&\qquad + {}^{1}\boldsymbol{P}_2 \times {}^{1}R_2\boldsymbol{f}_2 \\
&\tau_1 = \boldsymbol{f}_1^T \cdot \boldsymbol{z}_1
\end{aligned}
\tag{23}
$$

7 Optimization

Optimization is the process of finding the minimum or maximum value of a function, usually not using an explicit approach [2]. The numerical-computational analysis software *Scilab* offers a toolbox, *fmincon*, for optimization. The *fmincon*() optimization function present in this toolbox is applied to minimize or maximize a multivariable function [12]. Function *fmincon*() takes an "objective function" which is the one to be minimized, an input parameter "$\boldsymbol{x}$" and the problem constraints, if they exist, and returns the point at which the objective function yields its minimum value. Such that

$$\begin{array}{llll} min_{\boldsymbol{x}} & f(\boldsymbol{x}) & & c_{eq}(\boldsymbol{x}) = 0 \\ subject \quad to & c(\boldsymbol{x}) \leq \quad 0 & & lb \leq \quad \boldsymbol{x} \quad \leq \quad ub \end{array} \tag{24}$$

Given the expression of torque as a function of the base position, presented in the previous sections, the goal is to find the position where the norm of the average absolute torques is minimal, subject to the restrictions showed above. The objective function $f(\boldsymbol{x})$ takes the base position and returns the norm of the average absolute torques, ignoring the first joint (because it is prismatic). [3] Since the relationship established between the base position and the torque at the joints is not linear, the optimization function $fmincon()$ present in the scilab toolbox is used [8].

The function $f(\boldsymbol{x})$ can be mathematically defined as:

$$f(\boldsymbol{x}) = \sqrt{\sum_{i=2}^{6} \tau_{i\_mean}^2(\boldsymbol{x})}, \qquad where \qquad \tau_{i\_mean}(\boldsymbol{x}) = \frac{\sum_{i=1}^{n} |\tau_i(\boldsymbol{x})|}{n} \tag{25}$$

where n is the number of all the path points. Vector $\boldsymbol{x} = [x, z]$ is the base position in the room frame.

The first constraint is derived from the inverse kinematics. The value of $cos(\beta)$ defines the existence of a solution for the inverse kinematics problem. Physically this means the absolute value of $cos(\beta)$ will be greater than 1 when the target is too distant for the end-effector to reach. So the value of $cos(\beta)$ must be less than or equal to 1 to assure the existence of a solution for the inverse kinematics problem. Thus:

$$cos(\beta) = \frac{r^2 + L_5^2 - p_y^2 - (p_x - L_3)^2}{2 \cdot r \cdot L_5}, \qquad cos(\beta) \leq 1 \tag{26}$$

The second constraint, is the singularity position. Given by the minimum *jacobian's* determinant which is a function of base position, where the minimum *jacobian* can't be zero, or in practice, less than a given tolerance $\varepsilon = 0.001$. In physical terms, when $det(J) = 0$ the robot loses one degree of freedom. The restriction then becomes as in Eq. (27). The last constraints are the room boundaries (Eq. 28), where x and z are the components of vector $\boldsymbol{x}$.

$$J_{min} = min(det(J_i)), \qquad where \qquad J_{min} \geq \varepsilon \tag{27}$$

$$x_{lb} \quad \leq \quad x \quad \leq \quad x_{ub} \quad and \quad z_{lb} \quad \leq \quad z \quad \leq \quad z_{ub} \tag{28}$$

8 Results

Figure 4 shows the base positions in which the constraints are met, the detail of the region where the minimum torque was obtained and the norm of average

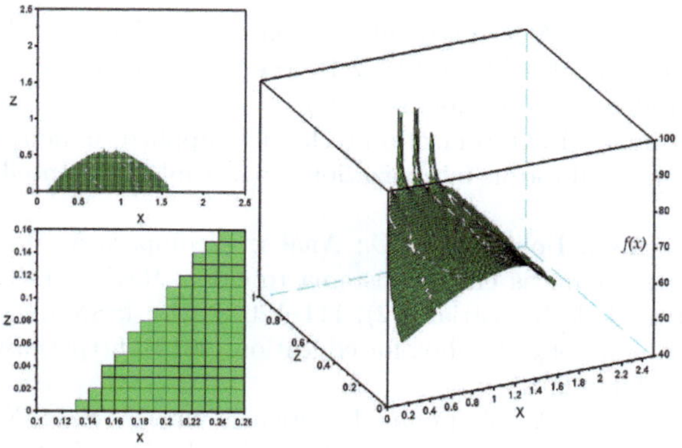

Fig. 4. *Upper left:* Regions in which the constraints are met. *Lower left:* Detail of minimum torque region. *Right:* Norm of average absolute torques in function of x and z base position

absolute torques in function of x and z positioning. The y coordinate is kept at -1. The minimum torque position was obtained as:

$$\boldsymbol{x} = [0.13117320, 0] \ \ m, \qquad f(\boldsymbol{x}) = 57.4965256 \ \ N.m \tag{29}$$

9 Conclusion

The appropriate selection of the robot base coordinates can reduce the torque required by the robot joints. This paper demonstrated the importance of optimization analysis, and a method for finding the base position which yields the minimum value of the norm of average absolute torques among the revolute joints. Using constraints in the optimization problem it is possible to reduce the time necessary to perform the algorithm by restricting the region of analysis.

The assumption that the C.G.'s are at the center of each link is a simplification implying the algorithm is stable relative to uncertainty in the C.G. positioning. However torques will change if the C.G.'s of each joint change. This stability may be analyzed in further studies. The results confirm that the torque profile of the robot is sensitive to the base position. Thus, this type of analysis brings great benefits not only in operation of the robot, but also in the dynamic requirements and joint wear, factors that are generally associated with costs, in the industry.

References

1. ABB: IRB 6620lx industrial robot on linear axis (2010)
2. Chou, H., Sadler, J.: Optimal location of robot trajectories for minimization of actuator torque. Mech. Mach. Theory **28**(1), 145–158 (1993)

3. Ciupitu, L., Ivanescu, A.N.: Optimal location of an industrial robot used in forge applications. In: Annals of DAAAM & Proceedings, pp. 453–455 (2010)
4. Craig, J.: Introduction to Robotics (2005)
5. Garg, D.P., Kumar, M.: Optimization techniques applied to multiple manipulators for path planning and torque minimization. Eng. Appl. Artif. Intell. **15**(3), 241–252 (2002)
6. Gembarowski, A.G., Bonilla, A.A.C.: Análise e comparação da geração de trajetórias lineares e cúbicas em um sistema robótico. Revista E-Tech: Tecnologias para Competitividade Industrial **8**(2), 111–126 (2015). ISSN:1983-1838
7. Gil, A.: ARTE: a robotics toolbox for education (2012). http://www.arvc.umh.es/arte. Accessed 3 Jan 2017
8. Kamrani, B., Berbyuk, V., Wäppling, D., Stickelmann, U., Feng, X.: Optimal robot placement using response surface method. Int. J. Adv. Manufact. Technol. **44**(1–2), 201–210 (2009)
9. Schilling, R.J.: Fundamentals of Robotics: Analysis and Control. Simon & Schuster Trade, New York (1996)
10. Siciliano, B., Khatib, O.: Springer Handbook of Robotics. Springer, Heidelberg (2008)
11. Siciliano, B., Sciavicco, L., Villani, L., Oriolo, G.: Robotics: Modelling Planning and Control. Springer, London (2010)
12. Srinivasan, M.: Trajectory optimization, a brief introduction. In: Powerpoint Presentation at Dynamic Walking Conference (2010). Citeseer

On the Constrained Walking of the NAO Humanoid Robot

Jesus E. Fierro[1], J. Alfonso Pamanes[1(✉)], Hector A. Moreno[2], and Victor Nunez[3]

[1] Instituto Tecnológico de la Laguna, Torreón, Mexico
japamanesg@correo.itlalaguna.edu.mx
[2] Universidad Autónoma de Coahuila, Monclova, Mexico
[3] Universidad Politécnica de Sinaloa, Mazatlán, Mexico

Abstract. The legs of the NAO humanoid robot have only 11 DOF; consequently, the motion of the pelvis and the free foot cannot be entirely specified in order to fully controlling the zero moment point (ZMP) in the contact area of the support foot. Thus, constraints on motion of both the pelvis and free foot must be appropriately taken into account in the motion planning process in order to specify suitable stable gaits. Specifically, the take-off and landing of the free foot have to be achieved in such a way that undesirable forces in the contact point with the floor be avoided. In this paper, a procedure is proposed to planning suitable walks in such a way that contact between the foot and the floor produces a stable behavior of the robot. To the knowledge of the authors, this problem has not be previously studied in the literature. The proposed method is applied in a study case to show its efficacy to get a stable gait.

1 Introduction

The research on humanoid robots has been expanding since the creation of the WABOT-1 in 1973 in Waseda University [1] and then moving forward to the year 2000 with the creation of the ASIMO robot by Honda [2]. Further advances led us to the All-new ASIMO [3] in 2011 and comparable humanoid robots like the HRP-4C [4] and the Robonaut 2 [5]. Then, a few years later arrives the humanoid Nao robot [6], in which this paper takes a closer look.

In regards of motion planning for biped robots there has been several approaches, like footstep planning [7,8] and walking patterns. The study in [9] is based on walking of humans and the efficiency that brings emulating the extended knee walk in our natural pattern. Other way of approaching the walking control problem is by generating trajectories based on the inverted pendulum model [10,11] or by using fuzzy logic systems as seen in [12] which brings a faster response for the online walking mode. Previous particular works on the Nao robot were presented in [13,14] where the motion planning is based on the inverse pendulum approach. On the other hand, a gait planning focused on energy efficiency was proposed in [15]. A similar kinematic model of the robot was applied in [16]. The methods proposed in all these works give interesting schemes of walking for

I. Chang et al. (eds.), *Advances in Automation and Robotics Research in Latin America*,
Lecture Notes in Networks and Systems 13, DOI 10.1007/978-3-319-54377-2_2

that humanoid. However, constraints on motion of the pelvis and the free foot of the robot were not explicitly taken into account and therefore the ZMP cannot be fully controlled; as consequence, a stable walking could not be guaranteed by applying such approaches. In this paper, a procedure is proposed to define a suitable walking in such a way that full contact between the foot and the floor be achieved. As result, a feasible stable behavior of the robot could be obtained for constrained motions of the pelvis and the free foot during the walking. A section is included to present the constraints that exist in motion of the pelvis and the free foot. The proposed method is finally applied in a study case that show its efficacy in getting a suitable walking.

2 Humanoid Robot NAO

Made by the Aldebaran Robotics Company, the Nao is a programmable humanoid robot with 25 DOF: five for each arm and one for each hand, two for the head, and eleven for the legs. It has 0.57 m of height and 4.5 kg of weight. The robot is shown in Fig. 1 and the kinematic scheme corresponding to the legs is appreciated in Fig. 2. Each leg has six rotational joints, but the motion of the joints attached to the pelvis are constrained in such a way that one actuator produces symmetrical rotations on them, as shown in Fig. 3.

To describe the kinematic chain of the Nao, we apply the modified Denavit-Hartenberg notation [17]. One orthonormal reference frame is attached to each link. The links are numbered from *0* to *12* starting with the fixed link (support

Fig. 1. The Nao humanoid robot from the Aldebaran Company.

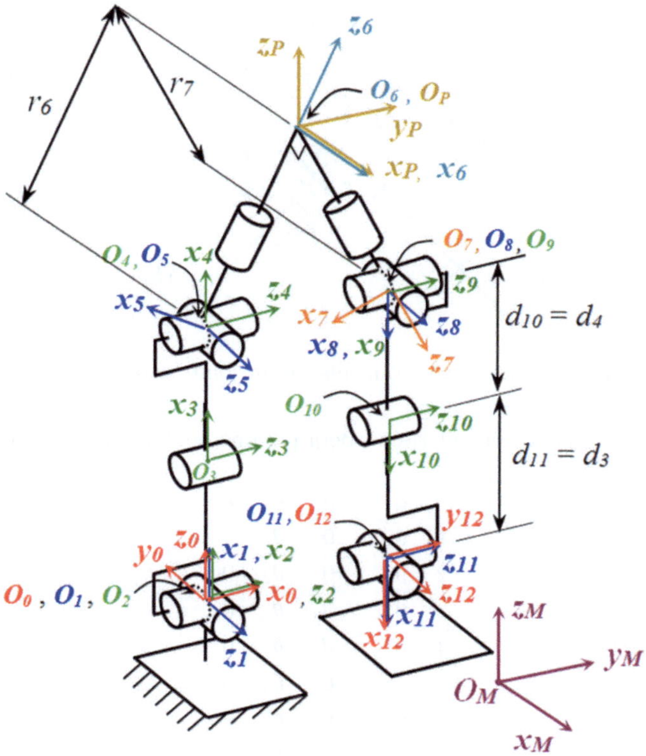

Fig. 2. Kinematic scheme of legs of the Nao robot

foot) and ending on the moving foot (free foot), as shown in Fig. 2. In this figure, we can also appreciate the orthonormal frame $x_M - y_M - z_M$ attached to the floor (*world frame*).

For the kinematic analysis of the walking pattern, we establish two sets of reference frames in the same kinematic chain: one for the fixed right foot (*frf*), and the corresponding one of the fixed left foot (*flf*). The kinematic chain showed in Fig. 2 is that one corresponding to *frf*. In both cases *frf* and *flf*, the numeration of frames begins with zero, assigned to the fixed foot. The modified Denavit-Hartenberg (MDH) parameters of the kinematic chain for the legs are given in Table 1. These parameters are the same for both chains *frf* and *flf*. The values for the non-null distances in Table 1 are $d_3 = d_{11} = 0.103\ m, d_4 = d_{10} = 0.10\ m$ and $r_6 = r_7 = 0.0712\ m$.

In the power transmission scheme of Fig. 3 we can appreciate how a single motor provides motions to joints 6 and 7 attached to the pelvis of the kinematic chain. Thus, the following constrain is imposed by the power transmission:

$$\theta_7 = \theta_6 \tag{1}$$

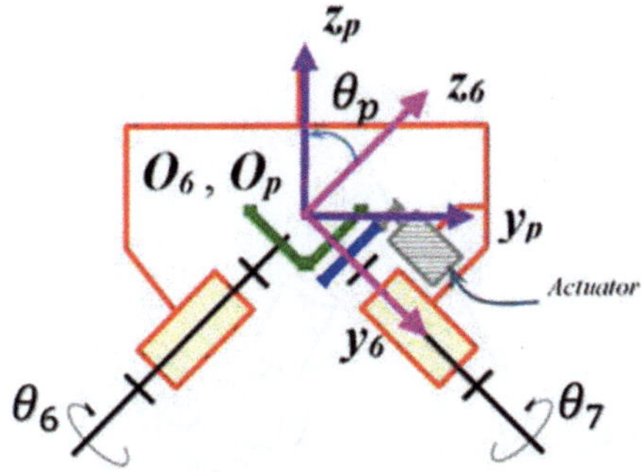

Fig. 3. Scheme of the power transmission to joints attached to the pelvis

Table 1. Modified Denavit-Hartenberg parameters for the legs of the Nao

	α	d	θ	r
1	90°	0	θ_1	0
2	90°	0	θ_2	0
3	0°	d_3	θ_3	0
4	0°	d_4	θ_4	0
5	−90°	0	θ_5	0
6	90°	0	θ_6	r_6
7	−90°	0	θ_7	r_7
8	90°	0	θ_8	0
9	−90°	0	θ_9	0
10	0°	d_{10}	θ_{10}	0
11	0°	d_{11}	θ_{11}	0
12	90°	0	θ_{12}	0

This equation holds while the robot is moving and, evidently, restrain the mobility of the kinematic chain. One orthonormal reference frame (frame $x_p - y_p - z_p$) attached to the pelvis is shown in Fig. 3. This frame is rotated by a fixed angle θ_p with respect to the x_6 axis of frame 6. Note that $\theta_p = 45°$ for *frf* and $\theta_p = -45°$ for *flf*.

3 Elemental Walking Pattern

A general walking pattern is defined by specifying the poses (position and orientation) of the frames attached to the pelvis (frame $x_p - y_p - z_p$, Fig. 2) and the free foot (frame $x_{12} - y_{12} - z_{12}$, Fig. 2) with respect to the world frame as functions of time. In previous works [18] we have used closed cycloidal functions to define both motions. However, as stablished in Sect. 1, other approaches have been applied to specify the gait of humanoid robots. A particular case of walking for a humanoid is obtained when the frame $x_p - y_p - z_p$ remains parallel to the frame $x_m - y_m - z_m$ and the sole of the free foot remains parallel to the plane $x_M - y_M$ of the world's frame during the walk. This kind of walking will be termed *elemental walking* in this paper. In such a walking, the displacement of the torso is only translational while the motion of the free foot is general but constrained to have rotations only about z_m. The desired orientation of the pelvis and the free foot will be specified with respect to the world's frame by using the Bryant angles λ, μ, ν, which corresponds respectively to successive rotations about the axes x, y, z attached to the corresponding link beginning with this frame matched with the world's frame. Thus, in elemental walking the three Bryant angles of the pelvis are nulls, and therefore, the rotation matrix of the pelvis with respect to the world's frame is:

$$ {}^{M}_{p}R^{*} = \begin{bmatrix} t_{11} & t_{12} & t_{13} \\ t_{21} & t_{22} & t_{23} \\ t_{31} & t_{32} & t_{33} \end{bmatrix} = \begin{bmatrix} 1 & 0 & 0 \\ 0 & 1 & 0 \\ 0 & 0 & 1 \end{bmatrix} \tag{2} $$

For rotation of the free foot in elemental walking, we only specify the orientation of the vector x_{12} of the foot frame. To hold the conditions of an elemental walking, such a vector must be normal to the floor surface (plane $x_M - y_M$). Hence, the rotation matrix for the free foot with respect to the world's frame is:

$$ {}^{M}_{p}R^{*} = \begin{bmatrix} m_{11} & m_{12} & m_{13} \\ m_{21} & m_{22} & m_{23} \\ m_{31} & m_{32} & m_{33} \end{bmatrix} = \begin{bmatrix} 0 & m_{12} & m_{13} \\ 0 & m_{22} & m_{23} \\ -1 & 0 & 0 \end{bmatrix} \tag{3} $$

The terms m_{12}, m_{13}, m_{22} and m_{23} in this matrix are defined by the values specified for λ, μ and ν. In a previous analysis of the elemental walking [19] it was proved that for a specified pose of the pelvis and a specified position of the free foot, the terms m_{12}, m_{13}, m_{22} and m_{23} must hold the following equations:

$$ -c_7 s_8 m_{13} + (c_P c_8 + s_P s_7 s_8) m_{23} = 0 \tag{4} $$

$$ \begin{aligned} &[c_7^2 c_8^2 + s_7^2] m_{13}^2 + 2[c_7 c_8 (c_P s_8 - s_P s_7 c_8) + s_7 c_7 s_P] m_{13} m_{23} \\ &+ [(c_P s_8 - s_P s_7 c_8)^2 + s_P^2 c_7^2] m_{23}^2 = 1 \end{aligned} \tag{5} $$

where $s_i \equiv sin\theta_i, c_i \equiv cos\theta_i (i = 7, 8), s_p \equiv sin\theta_p$ and $c_p \equiv cos\theta_p$. The angles θ_7 and θ_8 are the joint variables corresponding to joints 7 and 8, respectively, of the kinematic chain of Fig. 2.

For an elemental walking the Bryant angles of the free foot are constrained in such a way that m_{12}, m_{13}, m_{22} and m_{23} hold Eqs. (4) and (5). These equations are defined in [19] as the *conditions for an elemental walking*. During a walk, these conditions are accomplished if the path of the point O_p of the pelvis remains on the sagittal plane of the robot, and the point O_{12} of the free foot remains on a plane parallel to the sagittal plane. Otherwise, the sole of the free foot will not be parallel to the plane of the floor and a stable contact of the foot with the floor could not be achieved.

4 General Walking

As shown in the precedent Section, for an elemental walking of the Nao robot it is possible to obtain poses of the support foot such that a full contact of the sole with the floor is achieved. Nevertheless, under elemental walking, the point O_p of the pelvis is constrained to remain on the sagittal plane. This condition generally do not allow keeping the ZMP in the support surface of the robot during the single support phase. Thus, in order to fully controlling the position of the ZMP, a component of motion of the point O_p must be on the frontal plane. Such a component, nevertheless, avoids the elemental walking; consequently, the conditions for elemental walking must be disregarded if we want improve the position of the ZMP in the sole of the support foot. We propose to retain the full contact condition of the support foot (Eq. 3), but admitting rotations and displacements of the pelvis on the frontal plane. Therefore, the positions and orientations of the free foot and the positions of the pelvis will be fully specified. The rotations of the pelvis will be admitted in order to improve the mobility of the free leg to suitably controlling the ZMP. To specify the orientation of the free foot we apply the Bryant angles $\lambda_{12}, \mu_{12}, \nu_{12}$ that define the orientation of frame 12 with respect to the world's frame. The Bryant angles λ_p, μ_p and ν_p of the orientation of the pelvis are determined by the joint variables θ_4, θ_5 and θ_6 of the support leg. Because of λ_p, μ_p and ν_p are not specified, then θ_4, θ_5 and θ_6 will be considered as the independent variables of an optimization method. Indeed, to solve the inverse kinematic problem by using only 11 DOF for the partially specified motion of the robot we propose a numerical procedure based on a non-linear optimization method. The phases of this procedure are:

1. To specify the Cartesian coordinates x_p, y_p, z_p of the point O_p in pelvis, and $x_{O12}, y_{O12}, z_{O12}$ of the point O_{12} of frame attached to the free foot, and the Bryant angles $\lambda_{12}, \mu_{12}, \nu_{12}$ that define the orientation of frame 12 with respect to the world's frame.
2. To propose initial values of the joint variables θ_4, θ_5 and θ_6.
3. By starting with the proposed initial values of θ_4, θ_5 and θ_6, minimize the following objective function:

$$f = (\theta_7^* - \theta_7)^2 \tag{6}$$

   The joint variable θ_7^* in Eq. (6) is the value of the 7th joint variable of the free leg that ideally should have the Nao robot considered as one of 12 DOF and, therefore, without the constraint of Eq. (1). On the other hand, θ_7 is the current value of that joint variable in the optimization process. Such a value is obtained by considering the real architecture of 11 DOF and by taking into account the constraint of Eq. (1).

Note that by following the proposed procedure at each path point, when the convergence be attained the objective function must be zero and the specified pose of the free foot will be obtained. In this case, the specified pose of the free foot and the specified position of the pelvis are fully accomplished.

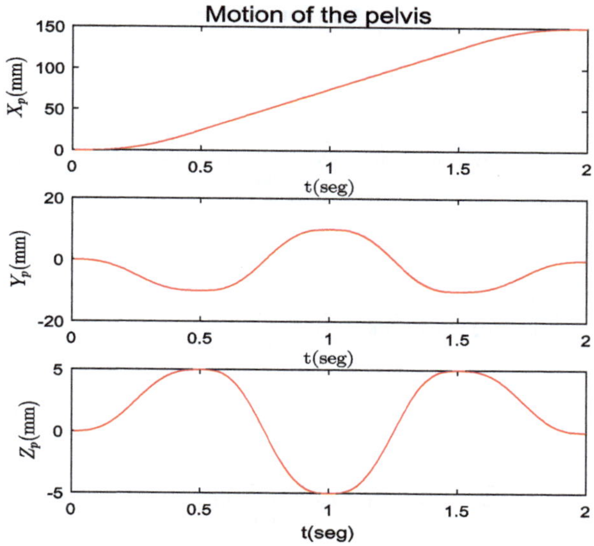

Fig. 4. Motion of the pelvis

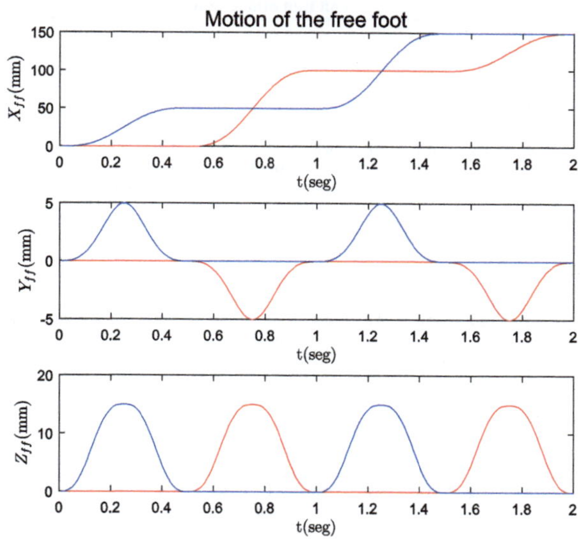

Fig. 5. Position of feet

The orientation of the pelvis will be such that cooperate as much as possible with the free leg in order to get the desired pose of the free foot. At each iteration of the optimization process, for any time during the walking, by using the current set of values of the independent variables θ_4, θ_5 and θ_6 the other joint variables of the robot are computed by applying the inverse kinematic model

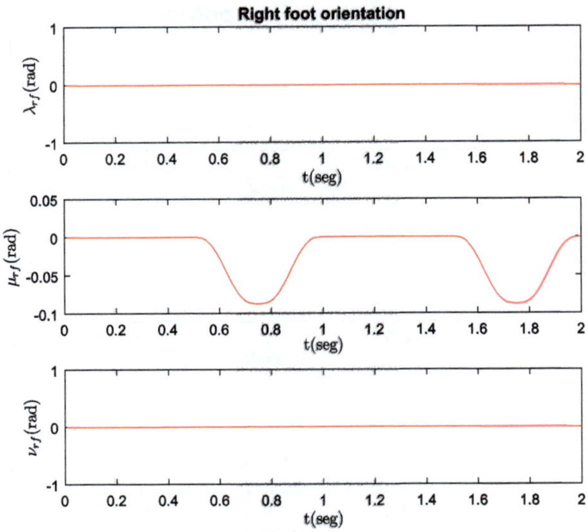

Fig. 6. Orientation right foot

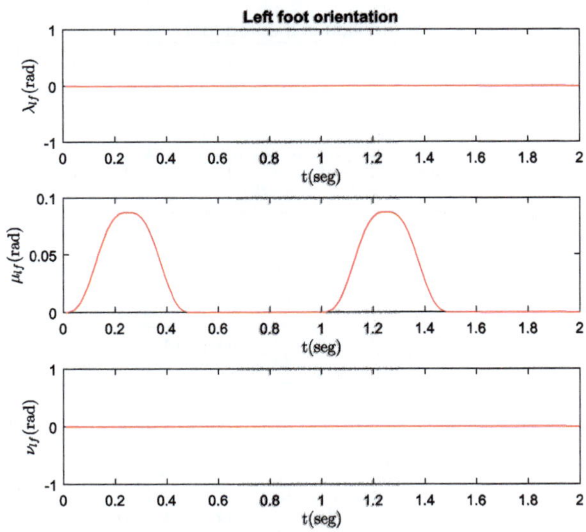

Fig. 7. Orientation left foot

proposed in [18]. At this step of the procedure, both variables θ_7^* and θ_7 are computed; the former by considering the robot as one of 12 DOF, and the second by considering the robot as one of 11 DOF. To minimize the function (6), we use the *fminunc* function in Matlab© software (Figs. 6 and 7).

5 Case Study

We apply in this Section the proposed approach for a general walk of the Nao robot in which the desired motions of the pelvis and the free foot are specified by using the pattern previously proposed in [20]. Only the rotational motion of the pelvis is synthesized as result of our approach. The functions defining the applied walking pattern are presented in Appendix A, and the equations of the inverse kinematic model are given in the Appendix B. The period of the specified walking is 0.5 seg for one step. The walking consists of 4 steps; the first step corresponds to the beginning phase and the last step is for the ending phase. The steps two and three are in the cruiser phase with constant velocity. The other parameters of the walking pattern are given in Table 2. In Figs. 4 and 7 are observed the curves of the desired motion of the pelvis and the free foot corresponding to the specified walking pattern, and a simulation for the desired motion of the pelvis and feet obtained in Matlab© is appreciated in Fig. 8. On the other hand, as result of application of our approach the joint trajectories of the Nao were obtained. A sample of postures of the robot during the walking obtained by simulation can be appreciated in the Fig. 9.

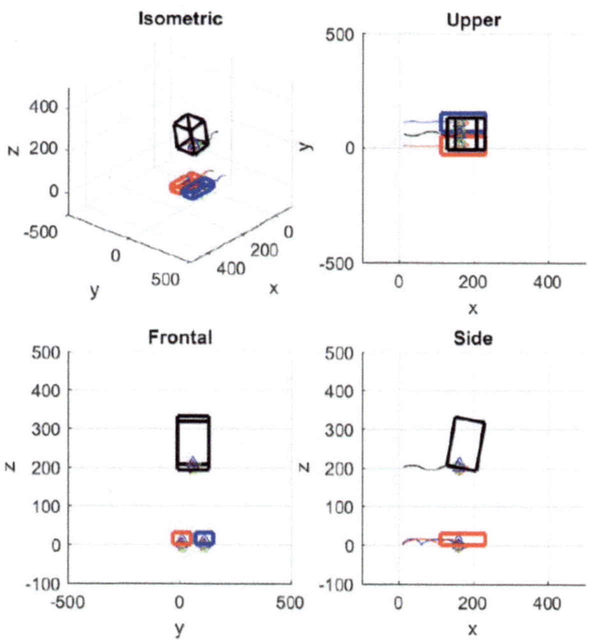

Fig. 8. Simulation of the proposed motion for the pelvis and the free foot.

Table 2. Values proposed for the parameters of the walking pattern

Parameters	Values
ns	4 steps
T_1	0.5 s
T_2	0.5 s
T_3	0.5 s
T_4	0.5 s
T	2 s
x_{pini}	10 mm
y_{pini}	60 mm
z_{pini}	200.7 mm
δx_p	150 mm
δy_p	−10 mm
δz_p	5 mm
λ_{pini}	0°
μ_{pini}	10°
ν_{pini}	0°
Parameters	Values
$\delta \lambda_p$	−5°
$\delta \mu_p$	0°
$\delta \nu_p$	0°
x_{fini}	10 mm
y_{fini}	10 mm
z_{fini}	150 mm
δx_f	150 mm
δy_f	5 mm
δz_f	15 mm
λ_{fini}	0°
μ_{fini}	0°
ν_{fini}	0°
$\delta \lambda_f$	0°
$\delta \mu_f$	−5°
$\delta \nu_f$	0°

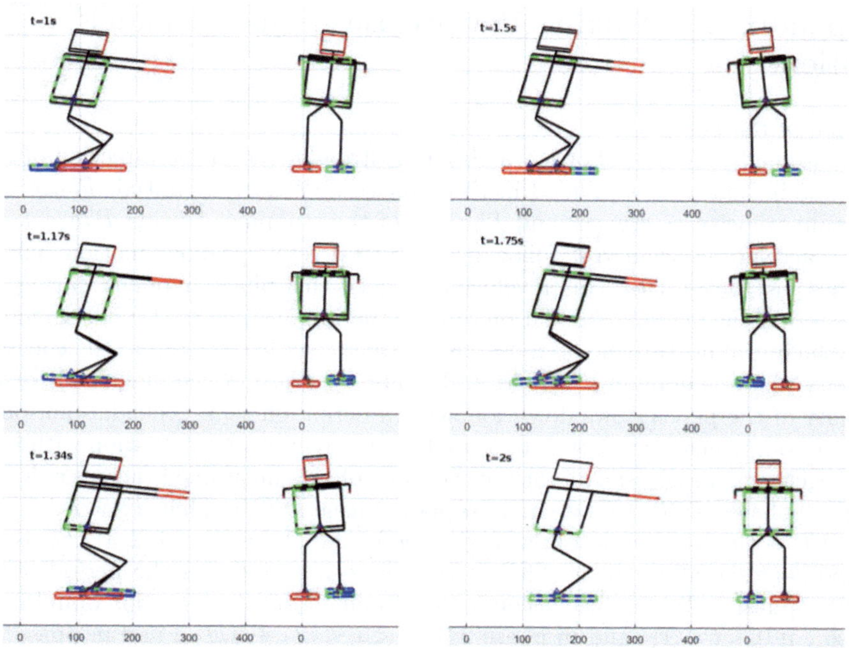

Fig. 9. Walking simulation

6 Conclusion

A study was developed in this paper for walking planning of the Nao robot. A method is proposed in order to find a suitable gait for the humanoid. The obtained results show that continuous motion of the robot and stable contact of the free foot with the floor at each step are achieved by using our approach. This one allows to solve the inverse kinematic problem of the legs with 11 DOF of the Nao humanoid for a suitable walking pattern of the robot. The motion of the free foot is fully specified while only the path of one point of the pelvis is stipulated. In this way, the translational motion of the pelvis would allow improve the control of the ZMP during the single support phase, while the rotational motion of the pelvis is synthesized in such a way that the full control for poses of the free foot is gotten. To the knowledge of the authors there are not previous studies in the literature about this subject. In future works about the gait of the Nao robot, a formulation for an optimization problem will be proposed in order to obtain the parameters of the walking pattern that improve as much as possible the behavior of the ZMP during the walking.

Acknowledgements. This work was supported by the DGTN (General Direction of the National Institute of Technology) of the Ministry of Public Education of Mexico, and the CONACyT (National Council of Research and Technology) of Mexico.

Appendix A Walking Pattern Based in Cycloidal Motions

Motion of the pelvis
It is assumed that the robot is walking in a straight line in the direction of x_M, this walking pattern is divided in three phases: *Phase 1* or initial; *Phase 2* of constant velocity or cruiser; and *Phase 3* of the stopping. During *Phase 1*, the pelvis accelerates from rest following a startup semi-cycloidal motion until it reaches just below the cruiser velocity (V_{max}); this phase happens during the first step of the walking simulation, and its period is equal to T_1; it is assumed that the initial posture of the robot has the feet together. A step in the constant velocity phase is done in a T_2 period. Finally, on *Phase 3* the magnitude of the velocity of the pelvis slows down following a detention semi-cycloidal motion in a T_3 period, which corresponds to a single step, arriving at a detention with both feet together. In each step on each of the three phases mentioned there is a double support phase (*DSP*) and a single support phase (*SSP*). Thus $T_i = T_{Si} + T_{Di}$ where T_i is the total period in a single step in the phase $i(i = 1, 2, 3)$, T_{Si} is the period corresponding to the *SSP* in the phase i, and T_{Di} is the period of the *DSP* in phase i. Taking into account the aforementioned, the robot would be in phase 1 if $0 \leq t \leq T_1$; and in phase 2 if $T_1 < t \leq (T_1 + n_p T_2)$; and in phase 3 if $(T_1 + n_p T_2) < t \leq T_T$, where n_p is the number of steps on constant velocity, t is the actual time(elapsed time since de startup) and T_T its the total time of the walking simulation ($T_T = T_1 + n_p T_2 + T_3$). The proposed movement motions for the pelvis are shown in Eqs. (A.1)–(A.7)

$$x_p(t) = \begin{cases} \frac{S_x}{2}\left(\frac{t}{T_1} - \frac{1}{\pi}sin(\frac{\pi t}{T_1})\right) & ,0 \leq t \leq T_1 \\ S_x(j - \frac{1}{2} + \frac{t'}{T_2}) & ,0 \leq t' \leq T_2 \\ nS_x + \frac{S_x}{2}\left(1 + \frac{t''}{T_3} + \frac{1}{\pi}sin(\frac{\pi t''}{T_3})\right) & ,0 \leq t'' \leq T_3 \end{cases} \tag{A.1}$$

where $j = 1, ..., n_p$

$$y_p(t) = \begin{cases} y_{po} - \delta_{yp}\left(\frac{t}{T_1} - \frac{1}{2\pi}\sin(\frac{2t\pi}{T1})\right) & ,0 \leq t \leq T_1 \\ y_{co} + 2\delta_{yp}\left(\frac{t'}{T_2} - \frac{1}{4\pi}\sin(\frac{4t'\pi}{T_2})\right) & ,0 \leq t' \leq \frac{T_2}{2} \\ y_{co} + 2\delta_{yp}\left(1 - \frac{t'}{T_2} + \frac{1}{4\pi}\sin(\frac{4t'\pi}{T_2})\right) & ,\frac{T_2}{2} \leq t' \leq T_2 \\ y_{co} + \delta_{yp}\left(\frac{t''}{T_3} - \frac{1}{2\pi}\sin(\frac{2t''\pi}{T3})\right) & ,0 \leq t'' \leq T_3 \end{cases} \tag{A.2}$$

$$z_p(t) = \begin{cases} z_{po} + \epsilon_p\delta_{zp}\left(\frac{t}{T_1} - \frac{1}{2\pi}\sin(\frac{2t\pi}{T_1})\right) & ,0 \leq t \leq T_1 \\ z_{po} + 2\epsilon_p\delta_{zp}\left(\frac{t'}{T_2} - \frac{1}{2\pi}\sin(\frac{2t'\pi}{T_2})\right) & ,0 \leq t' \leq T_2 \\ z_{po} + \epsilon_p\delta_{zp}\left(\frac{t''}{T_3} - \frac{1}{2\pi}\sin(\frac{2t''\pi}{T_3})\right) & ,0 \leq t'' \leq T_3 \end{cases} \tag{A.3}$$

$$\lambda_p(t) = \begin{cases} \lambda_{po} - \epsilon_p \delta\lambda_p \left(\frac{t}{T_1} - \frac{1}{2\pi}\sin(\frac{2t\pi}{T1})\right) & ,0 \le t \le T_1 \\ \lambda_{po} + \epsilon_p \delta\lambda_p \left(1 - 2\left(\frac{t'}{T_2} - \frac{1}{2\pi}\sin(\frac{2t'\pi}{T_2})\right)\right) & ,0 \le t' \le T_2 \\ \lambda_{po} + \epsilon_p \delta\lambda_p \left(1 - \left(\frac{t''}{T_3} - \frac{1}{2\pi}\sin(\frac{2t''\pi}{T_3})\right)\right) & ,0 \le t'' \le T_3 \end{cases} \quad \text{(A.4)}$$

$$\mu_p(t) = \begin{cases} \mu_{po} + \epsilon_p \delta\mu_p \left(\frac{t}{T_1} - \frac{1}{2\pi}\sin(\frac{2t\pi}{T_1})\right) & ,0 \le t \le T_1 \\ \mu_{po} + \epsilon_p \delta\mu_p \left(1 + 2\left(\frac{t'}{T_2} - \frac{1}{4\pi}\sin(\frac{4t'\pi}{T_2})\right)\right) & ,0 \le t' \le T_2 \\ \mu_{po} - \epsilon_p \delta\mu_p \left(\frac{t''}{T_3} - \frac{1}{2\pi}\sin(\frac{2t''\pi}{T_3})\right) & ,0 \le t'' \le T_3 \end{cases} \quad \text{(A.5)}$$

$$\nu_p(t) = \begin{cases} \nu_{po} + \epsilon_p \delta\nu \left(\frac{t}{T_1} - \frac{1}{4\pi}\sin\frac{4t\pi}{T_1})\right) & ,0 \le t \le \frac{T_1}{2} \\ \nu_{po} - \epsilon_p \left(\frac{\delta\nu_p}{2}\right)\left(1 - 4\left(\frac{t}{T_1} - \frac{1}{4\pi}\sin(\frac{4t\pi}{T1})\right)\right) & ,\frac{T_1}{2} \le t \le T_1 \\ \nu_{po} - \epsilon_p(\delta\nu_p)\left(2\left(\frac{t'}{T_2} - \frac{1}{4\pi}\sin(\frac{4t'\pi}{T_2})\right)\right) & ,0 \le t' \le \frac{T_2}{2} \end{cases} \quad \text{(A.6)}$$

$$\nu_p(t) = \begin{cases} \nu_{po} + \epsilon_p \delta\nu_p \left(2\left(1 - \frac{t'}{T_2} - \frac{1}{4\pi}\sin(\frac{4t'\pi}{T_2})\right)\right) & ,\frac{T_2}{2} \le t' \le T_2 \\ \nu_{po} + \epsilon_p \delta\nu_p \left(2\left(\frac{t''}{T_2} - \frac{1}{4\pi}\sin(\frac{4t''\pi}{T_2})\right)\right) & ,0 \le t'' \le \frac{T_3}{2} \\ \nu_{po} - \epsilon_p \delta\nu_p \left(1 - \left(\frac{t''}{T_3} - \frac{1}{4\pi}\sin(\frac{4t''\pi}{T_3})\right)\right) & ,\frac{T_3}{2} \le t'' \le T_3 \end{cases} \quad \text{(A.7)}$$

Motion of the free foot
The expressions on (A.8) and (A.10) determine, within the frame $x_M - y_M - z_M$, the position of the origin O_{12} of the frame 12 on the free foot (Fig. 2). This point is ubicated on the left ankle if the right foot is on the floor, and viceversa. The orientation of the free foot is determined in relation to the frame $x_M - y_M - z_M$ and is given by the functions (A.11) and (A.13). In the Eq. (A.8) $j = 1, ..., n_p$.

$$x_f(t) = j\delta x_f + \delta x_f \left[\frac{t}{T_{Si}} - \frac{1}{2\pi}\sin(\frac{2t\pi}{T_{Si}})\right] \quad \text{(A.8)}$$

$$y_f(t) = \begin{cases} \delta y_f\left(\sin(pi x_f(t))\right) & ,0 \le t \le T_1 \\ \delta y_f\left(\sin\left(\frac{pi}{2}\left(1 + \frac{x_f(t')}{S_x}\right)\right)\right) & ,0 \le t' \le T_2 \\ \delta y_f\left(\sin\left(pi x_f(t'')\right)\right) & ,0 \le t'' \le T_3 \end{cases} \quad \text{(A.9)}$$

$$z_f(t) = \begin{cases} \delta z_f[\frac{2t}{T_{Si}} - \frac{1}{2\pi}\sin\frac{4t\pi}{T_{S_i}}] & \text{if } t \le \frac{T_{Si}}{2} \\ \delta z_f - \delta z_f[\frac{(2t - T_{Si})}{T_{Si}} - \frac{1}{2\pi}\sin(\frac{2\pi(2t - T_{Si})}{T_{Si}})] & \text{if } t > \frac{T_{Si}}{2} \end{cases} \quad \text{(A.10)}$$

$$\lambda_f(t) = \begin{cases} \lambda_{fo} + \delta\lambda_f[\frac{2t}{T_{Si}} - \frac{1}{2\pi}\sin(\frac{4t\pi}{T_{Si}})] & \text{if } t \le \frac{T_{Si}}{2} \\ \lambda_{fo} + \delta\lambda_f\{1 - [\frac{(2t - T_{Si})}{T_{Si}} - \frac{1}{2\pi}\sin(\frac{2\pi(2t - T_{Si})}{T_{Si}})]\} & \text{if } t > \frac{T_{Si}}{2} \end{cases} \quad \text{(A.11)}$$

$$\mu_f(t) = \begin{cases} \mu_{fo} + \delta\mu_f\left[\frac{2t}{T_{Si}} - \frac{1}{2\pi}\sin(\frac{4t\pi}{T_{Si}})\right] & \text{if } t \le \frac{T_{Si}}{2} \\ \mu_{fo} + \delta\mu_f\left\{1 - \left[\frac{(2t - T_{Si})}{T_{Si}} - \frac{1}{2\pi}\sin\left(\frac{2\pi(2t - T_{Si}}{T_{Si}}\right)\right]\right\} & \text{if } t > \frac{T_{Si}}{2} \end{cases} \quad \text{(A.12)}$$

$$\nu_f(t) = \begin{cases} \nu_{fo} + \delta\nu_f\left[\frac{2t}{T_{Si}} - \frac{1}{2\pi}\sin(\frac{4t\pi}{T_{Si}})\right] & \text{if } t \le \frac{T_{Si}}{2} \\ \nu_{fo} + \delta\nu_f\left\{1 - \left[\frac{(2t - T_{Si})}{T_{Si}} - \frac{1}{2\pi}\sin\left(\frac{2\pi(2t - T_{Si}}{T_{Si}}\right)\right]\right\} & \text{if } t > \frac{T_{Si}}{2} \end{cases} \quad \text{(A.13)}$$

Appendix B Inverse Kinematic Model of a 12 DOF Biped Robot

The inverse kinematic model equations for the feet of the humanoid robot are determined by a specific position and orientation frame of the pelvis with respect to the frame $x_M - y_M - z_M$, by an homogenous matrix equation expressed as:

$$ {}^{M}_{p}T^{*} = \begin{bmatrix} t_{p11} & t_{p12} & t_{p13} \\ t_{p21} & t_{p22} & t_{p23} \\ t_{p31} & t_{p32} & t_{p33} \end{bmatrix} \tag{B.1} $$

The star in ${}^{M}_{p}T^{*}$ denotes that this matrix is known. The elements of ${}^{M}_{p}T^{*}$ are calculated from the operational coordinates defined in Eqs. (A.1)–(A.7). The first three rows and columns are determined by using the Bryant angles matrix, that is expressed as:

$$ {}^{M}_{p}R = \begin{bmatrix} c\mu c\nu & -c\mu s\nu & s\mu \\ s\lambda s\mu c\nu + c\lambda s\nu & -s\lambda s\mu s\nu + c\lambda c\nu & -s\lambda c\mu \\ -c\lambda s\mu c\nu + s\lambda s\nu & c\lambda s\mu s\nu + s\lambda c\nu & c\lambda c\mu \end{bmatrix} \tag{B.2} $$

where the angles, corresponding to the pelvis, are defined by the function (A.4)–(A.7). At the same time, the three elements of the last row from ${}^{M}_{p}T^{*}$ are the Cartesian coordinates of the origen frame of the pelvis in the frame $x_M - y_M - z_M$, that are specified by the functions (A.1)–(A.3). The inverse kinematic equations for the free foot are obtained from the homogeneous matrix equation ${}^{p}_{12}T^{*}$ that defines the reference frame 12, present in the free foot, with respect from the frame of the pelvis. This matrix is the result of the product:

$$ {}^{p}_{12}T^{*} = {}^{p}_{M}T^{*} * {}^{M}_{f}T^{*} * {}^{f}_{12}T^{*} \tag{B.3} $$

where ${}^{p}_{M}T^{*}$ is the inverse matrix given by (B.1). The homogeneous matrix equation ${}^{f}_{12}T^{*}$ from the Eq. (B.3) is:

$$ {}^{f}_{12}T^{*} = \begin{bmatrix} 0 & 0 & 1 & 0 \\ 1 & 0 & 0 & 0 \\ 0 & 1 & 0 & 0 \\ 0 & 0 & 0 & 1 \end{bmatrix} \tag{B.4} $$

As show in Fig. 2, the elements of ${}^{M}_{f}T^{*}$ are calculated by the operational coordinates defined in Eqs. (A.8)–(A.13). The orientation of this matrix is determined by using the matrix (B.2), with the Bryant angles for the free foot. Finally, the known elements that results from the matrix ${}^{p}_{12}T^{*}$ are designated as follows:

$$
{}^{p}_{12}T^{*} = \begin{bmatrix} t_{f11} & t_{f12} & t_{f13} & t_{f14} \\ t_{f21} & t_{f22} & t_{f23} & t_{f24} \\ t_{f31} & t_{f32} & t_{f33} & t_{f34} \\ 0 & 0 & 0 & 1 \end{bmatrix} \tag{B.5}
$$

The inverse kinematic model is determined by using the Eqs. (B.1) and (B.5), the geometric parameters of the humanoid robot, and the walking parameters shown in Tables 1 and 2. Next we show the equations of the inverse kinematic model, where $\bar{d_7} = (d_7/2)$:

- Support foot:

$$
\begin{aligned}
s_6 &= (t_{p23}r_6 - y_p)t_{p31} - (t_{p33}r_6 - z_p)t_{p21} - (t_{p21}t_{p32} - t_{p22}t_{p31} + t_{p31})\bar{d_7} \\
c_6 &= (t_{p23}r_6 - y_p - \bar{d_7})t_{p32} + z_p - t_{p33}r_6 \\
s_1 &= (t_{p31}c_6 - t_{p32}s_6)r_5 - t_{p32}\bar{d_7} - t_{p33}r_6 + z_p \\
c_1 &= (t_{p21}c_6 - t_{p22}s_6)r_6 - t_{p22}\bar{d}7 - t_{p23}r_6 + y_p + \bar{d_7} \\
s_5 &= (t_{p21}s_1 - t_{p31}c_1)s_6 - (t_{p22}s_1 - t_{p32}c_1)c_6 \\
c_5 &= t_{p33}c_1 - t_{p23}s_1 \\
x_{14} &= (t_{p21}c_1c_6 + t_{p31}s_1c_6 - t_{p22}c_1s_6 - t_{p32}s_1s_6)r_5 - (t_{p22}c_1 + t_{p32}s_1)\bar{d_7}+ \\
&\quad (y_p - t_{p23}r_6 + \bar{d_7})c_1 + (z_p - t_{p33}r_6)s_1 \\
z_{14} &= (t_{p11}c_6 - t_{p12}s_6)r_5 - t_{p12}\bar{d_7} - t_{p13}r_6 + x_p \\
r_{14} &= \sqrt{(x_{14})^2 + (z_{14})^2} \\
r_{tax} &= \frac{x_{14}}{r_{14}}\sqrt{d_3^2 - 0.25(r_{14})^2} + 0.5z_{14} \\
r_{tay} &= \frac{x_{14}}{2} - \frac{z_{14}}{r_{14}}\sqrt{d_3^2 - 0.25(r_{14})^2} \\
r_{fax} &= z_{14} - r_{tax} \\
r_{fay} &= x_{14} - r_{tay} \\
s_{2\bar{3}4} &= -(t_{p11}s_6 + t_{p12}c_6)c_5 - t_{p13}s_5 \\
c_{2\bar{3}4} &= t_{p11}c_6 - t_{p12}s_6 \\
\theta_1 &= atan2(s_1, c_1) \\
\theta_2 &= atan2(r_{tay}, r_{tax}) - 0.5\pi \\
\theta_3 &= atan2(r_{fay}, r_{fax} - \theta_2 - 0.5\pi \\
\theta_4 &= atan2(s_{2\bar{3}4}, c_{2\bar{3}4}) - \theta_2 - \theta_3 \\
\theta_5 &= atan2(s_5, c_5) \\
\theta_6 &= atan2(s_6, c_6)
\end{aligned}
$$

- Free foot:

$$
\begin{aligned}
s_7 &= t_{f23}z_f + (\bar{d}_7 + y_f)t_{f23} \\
c_7 &= t_{f33}x_f - t_{f13}z_f \\
s_8 &= z_f \\
c_8 &= x_f s_7 - (\bar{d}_7 - y_f)c_7 \\
x_{812} &= (x_f s_7 + (y_f - \bar{d}_7)c_7)c_8 + z_f s_8 \\
z_{812} &= (\bar{d}_7 - y_f)s_7 + x_f c_7 - r_8 \\
r_{812} &= \sqrt{(x_{812})^2 + (z_{812})^2} \\
r_{flx} &= \frac{x_{812}}{r_{812}}\sqrt{d_{10}^2 - 0.25(r_{812})^2} + 0.5z_{812} \\
r_{fly} &= \frac{x_{812}}{2} - \frac{z_{z812}}{r812}\sqrt{d_{10}^2 - 0.25(r_{812})^2} \\
r_{tlx} &= z_{812} - r_{flx} \\
r_{tly} &= x_{812} - r_{fly} \\
g_{11} &= t_{g13}c_7 - t_{f23}s_7 \\
s_{12} &= t_{f31}c_8 - (t_{f11}s_7 + t_{f21}c_7)s_8 \\
c_{12} &= t_{f32}c_8 - (t_{f12}s_7 + t_{f22}c_7)s_8 \\
\theta_7 &= atan2(s_7, c_7) \\
\theta_8 &= atan2(s_8, c_8) \\
\theta_9 &= atan2(r_{fly}, rflx) - 0.5\pi \\
\theta_{10} &= atan2(r_{tly}, rtlx) - \theta_9 - 0.5\pi \\
c_{910} &= cos(\theta_9 + \theta_{10}) \\
s_{910} &= sin(\theta_9 + \theta_{10}) \\
h_{11} &= \sqrt{(s_{910})^2 + (c_{910})^2 - (g_{11})^2} \\
\theta_{11} &= atan2((-s_{910}g_{11} - c_{910}h_{11}), (c_{910}g_{11} - s_{910}h_{11})) \\
\theta_{12} &= atan2(s_{12}, c_{12})
\end{aligned}
$$

References

1. Kato, I.: Development of WABOT 1. Biomechanism, **2**, 173–214 (1973)
2. Kornblum, J.: Meet Hondas ASIMO, a helpful Mr. Roboto. USA-Today (2000)
3. Honda unveils all-new ASIMO with significant advancements. http://hondanews.com/releases/cd29e4c7-c889-4378-a86a-3534ed615459. Accessed 30 Sep 2016
4. Kaneko, K., Kanehiro, F., Morisawa, M., Miura, K., Nakaoka, S., Kajita, S.: Cybernetic human HRP-4C. In: 9th IEEE-RAS International Conference on Humanoid Robots, pp. 7–14. IEEE (2009)
5. Diftler, M., Mehling, J., Abdallah, M., Radford, N., Bridgwater, L., Sanders, A., Askew, R., Linn, D., Yamokoski, J., Permenter, F., et al.: Robonaut 2-the first humanoid robot in space. In: IEEE International Conference on Robotics and Automation (ICRA), pp. 2178–2183. IEEE (2011)

6. Aldebaran robotics announçes NAO Next Gen humanoid robot. https://www.engadget.com/2011/12/10/aldebaran-robotics-announces-nao-next-gen-humanoid-robot-video/. Accessed 30 Sep 2016
7. Chestnutt, J., Lau, M., Cheung, G., Kuffner, J., Hodgins, J., Kanade, T.: Footstep planning for the Honda ASIMO humanoid. In: Proceedings of the IEEE International Conference on Robotics and Automation, pp. 629–634. IEEE (2005)
8. Jiwen, Z., Li, L., Ken, C.: Footstep planning for rapid path following in humanoid robots. Int. J. Humanoid Rob. **13**(4), 1650013 (2016)
9. Sakamoto, H., Katayose, H., Miyazaki, K., Nakatsu, R.: Extended-knee walk for humanoid robot with parallel link legs. Int. J. Humanoid Rob. **6**(04), 565–584 (2009)
10. Castano, J., Li, Z., Zhou, C., Tsagarakis, N., Caldwell, D.: Dynamic and reactive walking for humanoid robots based on foot placement control. Int. J. Humanoid Rob. **13**(02), 1550041 (2016)
11. Omran, S., Sakka, S., Aoustin, Y.: Effects of the COM vertical oscillation on joint torques during 3d walking of humanoids robots. Int. J. Humanoid Rob. **13**(4), 1650019 (2016)
12. Yang, L., Liu, Z., Zhang, Y.: Online walking control system for biped robot with optimized learning mechanism: an experimental study. Nonlinear Dyn. **86**(3), 2035–2047 (2016)
13. Strom, J., Slavov, G., Chown, E.: Omnidirectional walking using ZMP and preview control for the NAO humanoid robot. In: Baltes, J., Lagoudakis, M.G., Naruse, T., Ghidary, S.S. (eds.) RoboCup 2009. LNCS (LNAI), vol. 5949, pp. 378–389. Springer, Heidelberg (2010). doi:10.1007/978-3-642-11876-0_33
14. Dong, E., Wang, D., Chen, C., Tong, J.: Realization of biped robot gait planning based on NAO robot development platform. In: IEEE International Conference on Mechatronics and Automation (ICMA), pp. 1073–1077. IEEE (2016)
15. Liu, J., Urbann, O.: Bipedal walking with dynamic balance that involves three-dimensional upper body motion. Rob. Auton. Syst. **77**, 39–54 (2016)
16. Hashemi, E., Khajepour, A.: Kinematic and three-dimensional dynamic modeling of a biped robot. Proc. Inst. Mech. Eng. Part K J. Multi-body Dyn. 1464419316645243 (2016)
17. Khalil, W., Kleinfinger, J.: A new geometric notation for open and closed-loop robots. In: IEEE International Conference on Robotics and Automation, Proceedings, vol. 3, pp. 1174–1179. IEEE (1986)
18. Pamanes, J.A.: Modelado cinemtico del robot humanoide bioloid. Technical report, Facultad de la ingeniera mecnica y elctrica de la Universidad Autnoma de Coahuila (2009)
19. Fierro, J., Pamanes, J.A., Santibanez, V., Ruiz, G., Ollervides, J.: Condiciones para una marcha elemental del robot NAO. AMRob J. Rob. Theor. Appl. **4**(1), 13–18 (2014)
20. Arias, L., Olvera, L., Pamanes, J.A., Nunez, J.: Patrón de marcha 3d de tipo cicloidal para humanoides y suaplicación al robot bioloid. Revista Iberoamericana de Ingeniería Mecánica **18**, 03–22 (2014)

2 DOF Mechanism for a Variable Geometry Hybrid Wheel

Hector A. Moreno[1(✉)], Isela G. Carrera[1], J. Alfonso Pamanes[2], and Emilio Camporredondo[1]

[1] Universidad Autónoma de Coahuila, Barranquilla S/N, Monclova, Coahuila, Mexico
{h_moreno,iselacarrera,emiliocamporredondoa}@uadec.edu.mx
[2] Instituto Tecnológico de la Laguna, Blvd. Revolución y Clzda. Cuauhtémoc S/N, 27000 Torreón, Coahuila, Mexico
japamanesg@correo.itlalaguna.edu.mx

Abstract. The design of a 2 degree of freedom (DOF) mechanism for a variable geometry hybrid wheel is presented in this paper. The mechanism allows transforming a circular wheel in a hybrid one with multiple limbs. It is actuated by both rotational and a linear motors mounted on the vehicle chassis. The former rotates the wheel and the linear motor makes that the legs be extended or contracted. The highlights of the mechanism are discussed in the paper.

1 Introduction

In the last decades, some researchers have implemented the so-called hybrid wheels in mobile robots. These wheels consist of a symmetrical body with multiple limbs rotated by an actuator mounted on the chassis of the vehicle. Experiments have shown the remarkable performance of these wheels when traversing rough terrain, crossing slopes, or climbing stairs. More recently, some interesting designs of hybrid wheels that can change their shape have been proposed. Such kind of wheels can adapt better to the terrain conditions. These wheels are referred as Variable Geometry Hybrid Wheels (VGHW).

VGHWs combine the advantages of circular wheels and hybrid wheels. These wheels can be used in mobile platforms for service and field robotics. In the case of a service mobile robot, a hybrid wheel could allow the robot the access to all parts of a house or building due to its ability to climb stairs. Furthermore, the damage on the floor's surface is minimal or nonexistent when using a circular wheel, differently to the case of a hybrid wheel since there is an impact at each step.

On the other hand, in the case of a field mobile robot, it is known that circular wheels are the most efficient way to transit on plain terrains at high velocities. Alternatively, a hybrid wheel would be more suitable in traversing rough terrain, sand, sludge, and snow. Additionally, by using the limbs of the hybrid wheel as paddles, an amphibious vehicle can be constructed.

Two configurations of a VGHW actuated by the proposed mechanism in this paper are shown in Fig. 1.

In the following paragraphs some previously proposed VGHWs are described. Then, our mechanism is presented and its interesting features are discussed.

I. Chang et al. (eds.), *Advances in Automation and Robotics Research in Latin America*, Lecture Notes in Networks and Systems 13, DOI 10.1007/978-3-319-54377-2_3

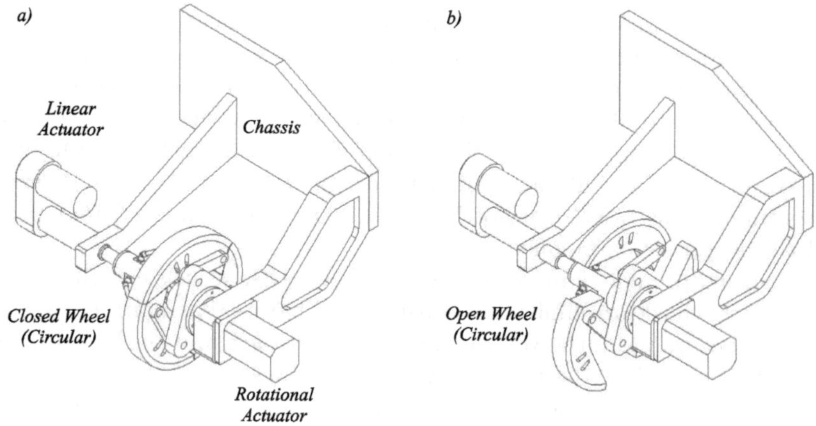

Fig. 1. Variable geometry hybrid wheel in two configurations: (a) wheel open, with the limbs flexed (b) wheel closed with the wheels extended

2 Variable Geometry Hybrid Wheels

In [1] Burt et al. present a device with two hybrids wheels that can change their geometry. The legs of the wheels consist of a slider-rod-crank mechanism. The slider provides the input motion and is implemented by a motor that rotates a shaft with external thread that is in contact with a cylindrical element (the slider) with internal thread. The slider is connected to the rod by a revolute joint in the same way the rod is connected to the crank. The crank is the element that is in contact with the ground and corresponds to a leg of the hybrid wheel. Interesting features have been documented for this design, nevertheless an inadequate operation of the hybrid wheel could occurs when its legs turn on an axis orthogonal to the wheel's axis. Indeed, collision of the wheel with obstacles would obstruct the extension/flexion of the limbs. As consequence of the same kind of motion of the leg an unmoderated wear could occurs on both elements the wheel and the floor. Such a behavior of the device become a major drawback for applications in household or factory environments.

In other work Sgherri [2] presents a wheel in which a central tapered shaft expels a number of links with rollers that linearly move through different conduits in the rim of the wheel. The tapered shaft has an external thread at the end and it is assembled to the rim by means of an internal thread. When the wheel meet an obstacle the applied torque to the drive shaft increases, then a relative rotation of the tapered shaft occurs with regard to the rim against the action of a linear spring. As consequence the limb on the wheel is expelled. The major drawback of this device is that if the vehicle meet an obstacle when moves on a slippery surface, the wheel will spin without opening the mechanism since the generated torque could not be enough to produce the extension of the links.

A vehicle with hybrid wheels that can transform from a circular shape wheels to C shape hybrid wheels was presented by Lin et al. [3]. Such a transformation is accomplished by rotating the two halves of the wheel on an axis perpendicular to the axis of rotation of the wheel. A supplementary actuator is incorporated in the wheel's rim in order to achieve the change. This mechanism has been applied in a mobile robot to climb stairs. However, at least six wheels are required in order to have three support points for the mobile robot at each step. Otherwise, the chassis of the robot will have contact with the stairs or the floor. Moreover, the structure of this wheel can not be replicated for a number of limbs (of the hybrid wheel) greater than one.

In [4] Susuki presents a model of a vehicle toy whose wheels can change their form. The wheel has three limbs that consist of a 4 bar mechanism. The extension of the limbs occurs when the vehicle has met an obstacle and the wheels cannot rotate. In this situation, the torque generated in the shaft of the wheel exceeds a certain value and overcomes the force of a set of linear springs and this makes one of the links of the mechanism rotate. The main limitation of such a device is that the opening of the wheel is not automatic and controllable. On the other hand, likewise to wheels in [2], if the vehicle is on a slippery surface the wheel will spin without opening the mechanism.

A wheel with three deployable limbs is presented in [5]. Through a pair of gears a triangular element is rotated, and this make the wheels extend. Each limb consist of a 4 bar mechanism with revolute joints. A set of links in form of "V" connect all the mechanisms of the limbs. The triangular element transmit motion to the limbs through a bar with all the revolute joints. The motion of the triangular element is generated by an electric motor placed inside the rim of the wheel. This fact difficult the implementation of the system due to the way the electric energy must be transmitted to such a motor. On the other hand, by mounting the actuator on the wheel the mass and volume of the wheel is increased.

In [6] the design of a wheel for an amphibious vehicle is presented. The wheel consists of six limbs with shape of arc. When the wheel is closed the limbs form a circular wheel. The mechanism of the wheel contains conical and planetary gears, and it is actuated by a hydraulic system. Since the mechanism contains a planetary gearbox, the implementation requires the fabrication of many pieces with good precision. This makes the mechanism complex and expensive.

Other design is considered in [7], which presents an anti-slip mechanism for wheels. The mechanisms consist of a series of telescopic cylinders that extends to provide higher traction with the ground. The extension of the cylinders is performed by the injection of hydraulic oil. A drawback of this mechanism is that it cannot be directly actuated by an electric motor. On the other hand the telescopic cylinders must have a robust design to prevent liquid leakages.

In [8] a wheel is presented with multiple deployable limbs based in the slider-rod-crank mechanism. The crank is the element that is in contact with the ground and has the form of a paddle. The purpose of this mechanism is to provide higher traction for a rice seeding machine. A handicap of this device is that the motion of the crank is performed in a plane that is not orthogonal to the rotation axis. Some problems could occurs if the ground is stiff or sticky. Indeed, the forces generated by dragging the paddles against the ground could be important.

Other work presents a wheel that can change from circular to hybrid [9]. The limbs are extended through a series of servomotors mounted in the wheel. Each servomotor actuates two limbs by means of a four bar mechanism. The limitation of this mechanism is that several servomotors are needed to achieve the transformation into a hybrid wheel. Moreover, the addition of the servomotors to the rim makes the wheel more voluminous and heavy. The main limitation of that invention is the complexity of the wiring (for power transmission and control) of the servomotors.

3 The Heise Wheel

The proposed mechanism for a VGHW is described in this section, we refer to this device as the Heise wheel. The following features of the mechanism overcomes the drawbacks of the devices considered in the previous paragraphs:

1. The kinematic chain allows that the extension of the limbs be independently controlled by a servo motor. Thus, the extension of the limbs can be set at different levels accordingly to different kind of terrain to be traversed.
2. The motion of links that have contact with the ground is achieved on a plane orthogonal to the rotation axis of the wheel. Consequently, the excessive forces are avoided in mechanism and actuators. Such forces arise in the previously proposed wheels because of dragging of limbs against the ground during their extension.
3. The actuators are placed on the chassis of the vehicle. Thus, the inertia of actuators is not included in the wheels and consequently the size and inertia of the wheels are reduced. Besides, different kinds of actuators and power transmission mechanisms can be applied to drive the wheel.
4. The number of limbs can be greater or equal to 1. Indeed, the proposed kinematic chain of a limb can be replicated as much as necessary.

Figure 2 presents the proposed mechanism of two degrees of freedom useful for implementing hybrid wheels. This mechanism is able to transform the wheel from the original circular form to a hybrid wheel composed of multiple limbs. One degree of freedom corresponds to a prismatic joint, which open or close the legs of the mechanism when actuated by a motor (electric, pneumatic or hydraulic). The other degree of freedom corresponds to a revolute joint, which rotates the hybrid wheel when actuated by a motor (electric, pneumatic or hydraulic). As shown in Fig. 2, the mechanism consists of a mobile shaft, a triangular rim, and three deployable limbs. The kinematic chain of one limb consist of binary links and the following sequence of joints: revolute (R_1), universal (U), revolute (R_2), revolute (R_3). The mobile shaft is connected to the triangular rim trough a prismatic joint (P) that allows the shaft move axially on the rim. On the other hand, the shaft is also connected to the rod by the joint R_1. The opposite extreme of the rod is connected by the joint U to the link in contact with the ground (the traction link). Note that the joint U could be replaced by a spherical (S) joint without losing the functionality of the mechanism. The link in contact

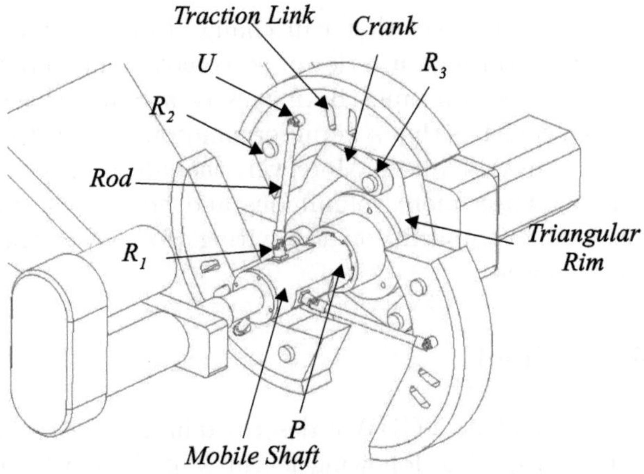

Fig. 2. Heise wheel components.

with the ground is connected by the joint R_2 to the crank. Finally, the crank is connected to the rim by means of the joint R_3. Clearly the rim is connected to the chassis by a revolute joint.

In order to make the link in contact with the ground move on a plane orthogonal to the rotation axis of the wheel, the axis of joints R_2, and R_3 should be parallel to the rotation axis of the wheel. The axis of joint R_1 must be perpendicular to the rotation axis of the wheel. Figure 3 shows the configurations of the mechanism when the wheel is open and closed. Other designs of the wheel can be obtained by adequate changes in the kinematic chain of the limbs. The designs are obtained by kinematic chains of the limbs composed of binary links with the following sequence of joints:

- *R-U-P-R*
- *R-U-R-P*
- *R-U-G*

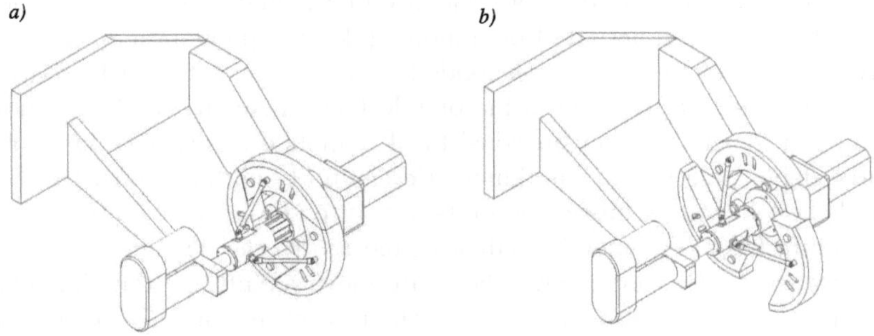

Fig. 3. Configurations of the mechanism when the wheel is closed (a) and open (b).

where U for universal joint, P for prismatic joint, R for revolute joint and G stands pin in slot joint. The first joint of the sequence is connected to mobile shaft and the last joint to the triangular rim.

4 Conclusions

Variable geometry hybrid wheels combine the advantages of circular wheels and hybrid wheels. Experiments have shown the advantages of these wheels to traverse rough terrain, crossing slopes, or climbing stairs and therefore these wheels can be used in mobile platforms for service and field robotics. This paper introduced the Heise wheel, a 2 DOF mechanism useful for implementing a variable geometry hybrid. This mechanism is able of transforming a circular wheel in a hybrid wheel with multiple limbs. The kinematic chain of the mechanism allows the extension of the limbs be independently controlled by a servo motor. On the other hand, the mechanism allows the motion of the links that have contact with the ground be performed in a plane orthogonal to the rotation axis of the wheel. This feature allows avoiding the drawback of excessive forces generated in the mechanism because of the dragging of the limbs against the ground during their extension. The mechanism allows the actuators of the wheel to be placed on the chassis of the vehicle and not inside the rim of wheels. The mechanism allows the hybrid wheel to have a number of limbs equal to 1 or any other number, by replicating the kinematic chain of the limbs. A patent application for the mechanism presented in this paper has been filed in [10].

Acknowledgments. The authors would like to thank to PRODEP Mexico and Universidad Autonoma de Coahuila for all support to this project.

References

1. Burt, I.T., Papanikolopoulos, N.P.: Adjustable diameter wheel assembly, and methods and vehicles using same. US Patent 6,860,346, 1 March 2005
2. Sgherri, G.R., La Spina, G.: Variable-diameter wheel. US 20120104834 A1, 3 May 2012
3. Lin, P.-C., Shen, S.-Y.: Mobile platform. US 8307923 B2, 13 November 2012
4. Suzuki, M.: Toy model with transforming tire mechanism. US 20110021112 A1, 27 January 2011
5. Folding combined obstacle detouring wheel. CN 102350917 A, 15 February 2012
6. 杨重明. Openable vehicle wheel and amphibious automobile employing same. CN 203157579 U, 28 August 2013
7. 胡玉法. Automotive cat-claw-shaped telescopic antiskid wheel. CN 202716669 U, 6 February 2013
8. 姬江涛, 李玉柱, 杜新武, 贺智涛, 郑治华, 高吭. Traveling mechanism of wheel and jaw combined type robot CN 202357792 U, 1 August 2012
9. 何俊虎, 刘荣, 王珂, 赵霄洋. Variable-diameter wheel leg combined wheel structure. CN 101844483 A, 29 September 2010
10. Moreno, H.A., Carrera, I.G.: Mecanismo de 2 grados de libertad para implementer una rueda hibrida de geometria variable. Patent application: MX/a/2016/013175

Experiences on the Design of a Needle Insertion Surgery Robot: Kinematic Analysis

Lisandro J. Puglisi[1(✉)], Roque Saltaren Pazmiño[1], Germán Rey Portolés[2], Cecilia E. Garcia Cena[1], and Rafael Aracil Santonja[1]

[1] Centro de Automática Y Robótica, UPM-CSIC, Jose Gutierrez de Abascal 2, 28006 Madrid, Spain
lisandro.puglisi@alumnos.upm.es
[2] Hospital RUBER Internacional, Unidad de Radiofísica, Madrid, Madrid, Spain

Abstract. In this work is presented the evolution of the concept of a needle insertion surgery robot. Five different prototypes were proposed, keeping the core concept of decoupling kinematics, and passing from a passive device to an autonomous robot. The kinematics of the mechanisms are presented. The different configurations for the inverse kinematics are introduced. The workspace of the final prototype is analyzed.

1 Introduction

Traditional free-hand needle placement has significant limitations. Physicians and surgeons often rely only upon kinesthetic feedback from the tool that they correlate with their own mental 3-D perception of anatomic structures, and in some cases rigid needles lead to excessive injurious pressure on tissues. Even when the entry point is correctly defined the tissue is deformed while the needle penetrates and the target might be missed. This leads to whether the target could be achieved with just one attempt. Each correction requires another imaging session and further reinsertion, increasing procedure's risk, discomfort, and radiation exposure [1].

Robotics technology can enhance the effectiveness of this clinical procedure by coupling complex information such as medical images [4,5]. Therefore, the design of a robot that could assist to the specialist during needle insertion procedure is the main objective of this work. More precisely, it is presented the evolution of the concept towards a robotic device presenting the kinematic modeling and highlighting the advantages and disadvantages of the prototypes.

2 Prototypes

2.1 Design Evolution

At the beginning of this project, the positioning robot was intended to be a passive device guided manually and would automatically block when the desire configuration was achieved. A classical 5R serial arm with breaks at every joint

I. Chang et al. (eds.), *Advances in Automation and Robotics Research in Latin America*, Lecture Notes in Networks and Systems 13, DOI 10.1007/978-3-319-54377-2_4

was chosen for the first prototype (see Fig. 1(a)) with a guiding tube that constrained the needle towards the insertion path placed at the end effector (EF). Although this kinematic chain provided an adequate workspace, the torque that the brakes had to support was very high due to the weight of its own structure. This problem was solved by replacing the first revolute joint by a sliding bridge, and a SCARA like structure was proposed for the second prototype (see Fig. 1(b)). Therefore, the torques at the second and third joint where only produced by the external forces applied at the EF. However, this prototype was discarded because the bridge limited the surgeon's workspace for manipulation. In order to reduce this limiting factor, the bridge was moved to a side (see Fig. 1(c)). Nevertheless, this structure didn't present enough stiffness and rigidity in order to warranty the desired precision at the EF.

The design was reformulated and the concept of an actuated device took place. The next generation of prototypes recovered the kinematic structure of the first prototype and it also kept the sliding base (see Fig. 1(d)). Additionally, in order to enhance the precision of the device it was implemented an spheric parallel mechanism 3$\underline{P}$SS-1S (see Fig. 1(e)) for the orientation of the EF.

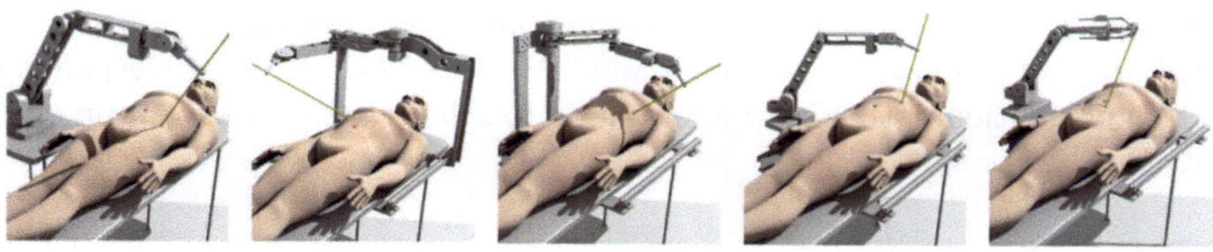

(a) (b) (c) (d) (e)

Fig. 1. Prototype concept evolution. (a): 5R. (b): Sliding bridge. (c): Sliding lateral bridge. (d): Sliding 5R. (e): Hybrid mechanism.

2.2 Kinematic Analysis

The design of the prototype evolved from a classic 5R serial mechanism into a hybrid mechanism, keeping the decoupling kinematic concept and a 2R serial planar topology in the kinematic chain.

2.2.1 Kinematic Decoupling

Considering the core structure of the prototypes, and a fixed frame O_{xyz} attached at the base of the mechanism, and a moving frame P_{uvw} at the EF as shown in Fig. 2, the relative orientation of the moving frame to the fixed frame is given by the rotation matrix ${}^{O}R_{P} = [\hat{u}, \hat{v}, \hat{w}]$, where $\hat{u}, \hat{v}, \hat{w}$ are the direction cosines of the orientation of EF. The center of the wrist P_w is located at the intersection of the axes of last two joints, and can be defined by $\overrightarrow{OP_w} = \overrightarrow{OP} - \overrightarrow{P_w P}$, which it can be expressed as:

$$P_w = P - L_{ef}\,\hat{w} \tag{1}$$

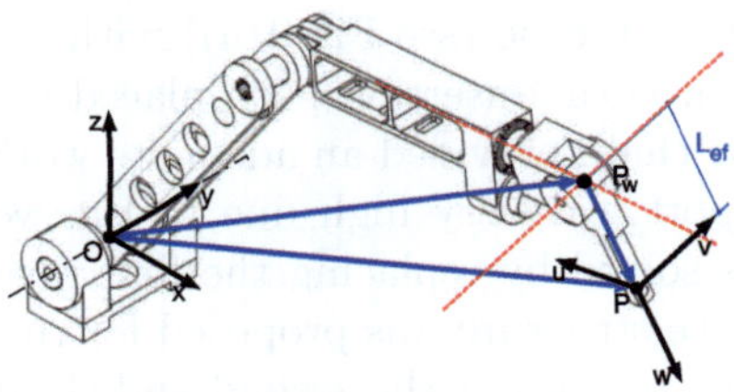

Fig. 2. Kinematic decoupling. The red dashed lines denotes the axis of the last two joints.

Given P_w and ${}^{O}R_P$, there are two problems to be solved:

1. Inverse Position Problem. Given P_w, the state of the first joints must be found.
2. Inverse Orientation Problem. Given the desire orientation of the EF (${}^{0}R_n$) and the state of the first joints, it must be found the state of the last joints that satisfies:
$$ {}^{i}R_n = \left({}^{0}R_i\right)^{-1} {}^{0}R_n \tag{2}$$
where ${}^{i}R_n = {}^{i}R_{i+1} \cdot \cdots \cdot {}^{n-1}R_n$ is the rotation matrix resulting from the contribution of the last *n-i+1* joints, and ${}^{0}R_i ={}^{0} R_1 \, {}^{1}R_2 \cdots {}^{i-1}R_i$ corresponds to the contribution of the first i joints (used to achieve the desire position of the wrist).

2.2.2 General Formulation for the Direct Kinematics

The direct kinematic problem is solved using the Successive Screw Displacement Method. This methodology is based on the definition of a reference configuration of the mechanism and the association of a finite screw to each joint [8]. The position of any point attached to the mechanism can be obtained as the premultiplication of the individual contribution of each joint as follows:

$$P_f = \left(\prod_{i=1}^{n} M_{\$_i}\right) P_o^T, \tag{3}$$

where $M_{\$_i}$ is the screw transformation matrix associated to the *ith* joint, n is the number of joints, P_o is the position of a point P at the initial configuration, and P_f is the resulting location of P after the displacements of the joints. Additionally, the orientation of the EF is given by the first three rows, and first three columns of the overall screw matrix transformation $M_{\$} = \prod_{i=1}^{n} M_{\$_i}$.

2.3 Core 2R Serial Planar: Inverse Kinematics

Taking a closer look at the 2R serial planar mechanism shown in Fig. 3, it can be seen that the position of the wrist P_w, is given by Eqs. (4) and (5), where $\theta_{12} = \theta_1 + \theta_2$.

$$P_{w_x} = d_1 \sin\theta_1 + d_2 \sin\theta_{12} \tag{4}$$

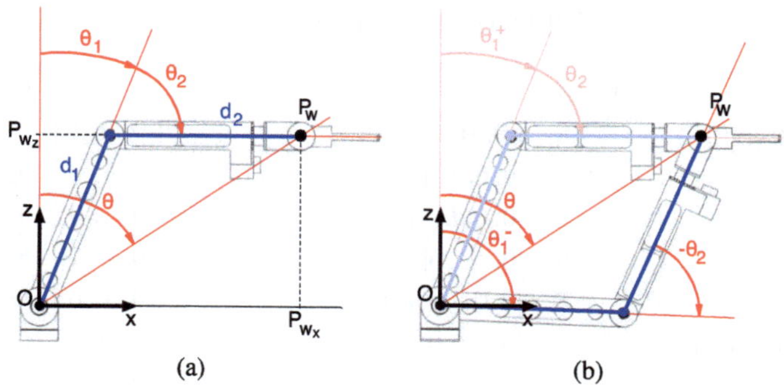

Fig. 3. 2R wire scheme. (a): Elbow up configuration. (b): Elbow down Configuration.

$$P_{w_z} = d_1 \cos\theta_1 + d_2 \cos\theta_{12} \tag{5}$$

Equating Eqs. (4) and (5) it is possible to find a relation of θ_2 independent of θ_1:

$$\theta_2 = cos^{-1}k, \tag{6}$$

where,

$$k = \frac{{P_{w_x}}^2 + {P_{w_z}}^2 - {d_1}^2 - {d_2}^2}{2\,d_1\,d_2} \tag{7}$$

Given the characteristic of the cosine function, Eq. (6) yields three distinctive possible solutions:

1. $|k| < 1$: two different real roots, leading to two possible configurations: elbow up configuration if ${\theta_2}^+ = \cos^{-1}k$, and elbow down configuration if ${\theta_2}^- = -\cos^{-1}k$ (see Fig. 3(b)).
2. $|k| = 1$: a double root, corresponding to a fully stretched or folded configuration.
3. $|k| > 1$: no real roots, then the configuration is not possible.

Taking into account Eq. (6), and expanding Eq. (4) it can be found that:

$$\theta_1 = Atan2(\sin\theta_1, \cos\theta_1) \tag{8}$$

where $Atan2$ is the four-quadrant inverse tangent and:

$$\sin\theta_1 = \frac{(d_1 + d_2\,\cos\theta_2)\,P_{w_x} - (d_2\,\sin\theta_2)\,P_{w_z}}{\Delta} \tag{9}$$

$$\cos\theta_1 = \frac{(d_2\,\sin\theta_2)\,P_{w_x} + (d_1 + d_2\,\cos\theta_2)\,P_{w_z}}{\Delta} \tag{10}$$

being $\Delta = {d_1}^2 + {d_2}^2 + 2\,d_1\,d_2\,\cos\theta_2$.

2.4 Prototype I: 5R

The first prototype possess five links plus the ground, articulated by five revolute joints, i.e. a 5R serial kinematic chain (see Fig. 4(a)).

Table 1. Screw's parameters.

	Joint i	$\$_i$	s_{0i}	s_i	θ_i	d_i
Prototype I	1	$\$_1$	[0, 0, 0]	[0, 0, 1]	θ_1	0
	2	$\$_2$	$[0, 0, L_1]$	[0, 1, 0]	θ_2	0
	3	$\$_3$	$[0, 0, L_1 + L_2]$	[0, 1, 0]	θ_3	0
	4	$\$_4$	$[0, L_3, 0]$	[0, 0, 1]	θ_4	0
	5	$\$_5$	$[0, 0, L_1 + L_2 + L_4]$	[0, 1, 0]	θ_5	0
Prototype II	1	$\$_1$	[0, 0, 0]	[1, 0, 0]	0	d_1
	2	$\$_2$	$[L_1, L_{y0}, 0]$	[0, 0, 1]	θ_2	0
	3	$\$_3$	$[L_1 + L_2, L_{y0}, 0]$	[0, 0, 1]	θ_3	0
	4	$\$_4$	$[0, L_{y0}, L_{z0} - L_3]$	[1, 0, 0]	θ_4	0
	5	$\$_5$	$[L_1 + L_2 + L_4, 0, L_{z0} - L_3]$	[0, 1, 0]	θ_5	0
Prototype V	1	$\$_1$	$[L_{x0}, -L_{y0}, 0]$	[0, 0, 1]	θ_1	0
	2	$\$_2$	$[0, -L_{y0}, L_1]$	[1, 0, 0]	θ_2	0
	3	$\$_3$	$[0, -L_{y0}, L_1 + L_2]$	[1,0, 0]	θ_3	0

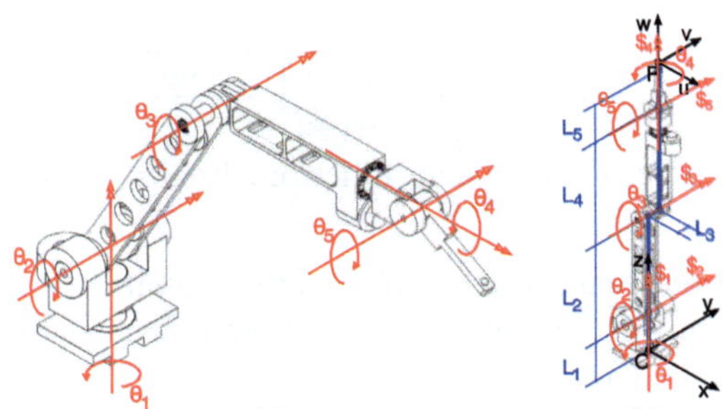

(a) (b)

Fig. 4. Prototype I: 5R. (a): 5R schematic description. (b): Initial configuration of the 5R, and screws definition.

2.4.1 Direct Kinematic

Considering the initial configuration of the mechanism as the one presented in Fig. 4(b) and the reference frame O_{xyz} attached to the base of the mechanism, then, the screw's parameters associated to the joints can be found (see Table 1) and the initial position of the EF is given by $P_{ef0} = [0, L_3, L_1 + L_2 + L_4 + L_5, 1]$ where L_i is the longitude of the *i-th* link.

2.4.2 Inverse Kinematics

Considering the mechanism at any given configuration as shown in Fig. 5(a), it can be seen that the core 2R kinematic chain will be always constrained to a plane Ω, where the direction of its normal will depend of the state of the first joint. The projection of the position of the wrist $P_w = [P_{w_x}, P_{w_y}, P_{w_z}]$ on the

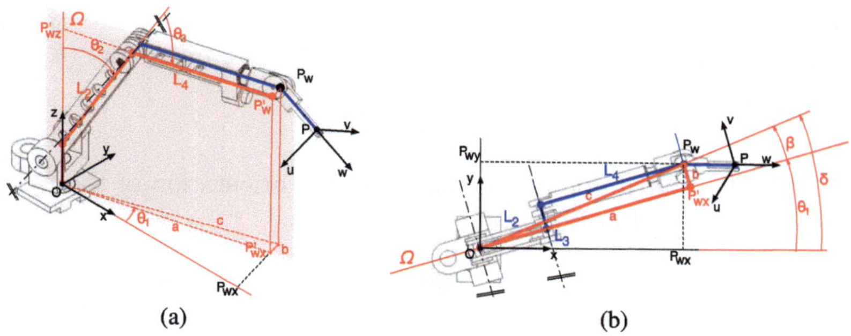

Fig. 5. Kinematic analysis of the 5R prototype. (a): general configuration. (b): top view.

plane Ω is given by $P'_w = [P'_{w_x}, P'_{w_z}]$, and the projection of points P_w and P'_w on the plane $z = 0$, defines the triangle $\overset{\triangle}{abc}$ (see Fig. 5(b)). The first joint can be found as $\theta_1 = \delta - \beta$, and it can be expressed in terms of the elements of the triangle $\overset{\triangle}{abc}$ as Eq. (11):

$$\theta_1 = Atan2(P_{w_y}, P_{w_x}) - \sin^{-1}\left(\frac{b}{c}\right), \tag{11}$$

where $c = \sqrt{{P_{w_x}}^2 + {P_{w_z}}^2}$ and $b = L_3$.

The state of the second and third joints can be found by considering the 2R serial kinematic chain conformed by link L_2 and the projection of link L_4 on the plane Ω and customizing the equations shown in Sect. 2.3. Hence, the state of the third joint is given by Eq. (12).

$$\theta_3 = \cos^{-1} k, \tag{12}$$

where,

$$k = \frac{{P'_{w_x}}^2 + {P'_{w_z}}^2 - {L_2}^2 - {L_4}^2}{2\, L_2\, L_4}. \tag{13}$$

The state of the second joint is given by Eq. (14).

$$\theta_2 = Atan2(\sin\theta_2, \cos\theta_2), \tag{14}$$

where,

$$\sin\theta_2 = \frac{(L_2 + L_4\, \cos\theta_3)\, P'_{w_x} - (L_4\, \sin\theta_3)\, P'_{w_z}}{\Delta} \tag{15}$$

$$\cos\theta_2 = \frac{(d_2 .\sin\theta_3)\, P'_{w_x} + (L_2 + L_4\, \cos\theta_3)\, P'_{w_z}}{\Delta} \tag{16}$$

and $\Delta = {L_2}^2 + {L_4}^2 + 2\, L_2\, L_4\, \cos\theta_3$.

Given the state of the first three joints (θ_1, θ_2 and θ_3), and the desired orientation at the EF (${}^{O}R_5$), then the state of the last two joints (θ_4 and θ_5), must satisfy Eq. (17).

$$ {}^{3}R_5 = \left({}^{O}R_3\right)^{-1} {}^{O}R_5 \tag{17} $$

where ${}^{3}R_5$ is the contribution of θ_4 and θ_5 to the orientation of the EF, whose elements are presented in Eq. (18).

$$ {}^{3}R_5 = \begin{bmatrix} r_{1,1} & r_{1,2} & r_{1,3} \\ r_{2,1} & r_{2,2} & r_{2,3} \\ r_{3,1} & r_{3,2} & r_{3,3} \end{bmatrix} = \begin{bmatrix} \cos\theta_4 \cos\theta_5 & -\sin\theta_4 & \cos\theta_4 \sin\theta_5 \\ \sin\theta_4 \cos\theta_5 & \cos\theta_4 & \sin\theta_4 \sin\theta_5 \\ -\sin\theta_5 & 0 & \cos\theta_5 \end{bmatrix} \tag{18} $$

Then, the state of the forth and fifth joints are given by Eqs. (19)–(20).

$$ \theta_5 = Atan2(-r_{3,1}, r_{3,3}) \tag{19} $$

$$ \theta_4 = Atan2(-r_{2,1}, r_{2,2}) \tag{20} $$

2.4.3 Inverse Position Multiple Solutions

The core 2R serial mechanism admits two possible solutions, the elbow up and elbow down configuration. Adding a third revolution joint whose axis is perpendicular to the axes of the 2R introduces a new pair of elbow up-down configurations that depend on the state of this additional joint (see Fig. 6). The four configurations are defined by the angular relations summed up in Table 2.

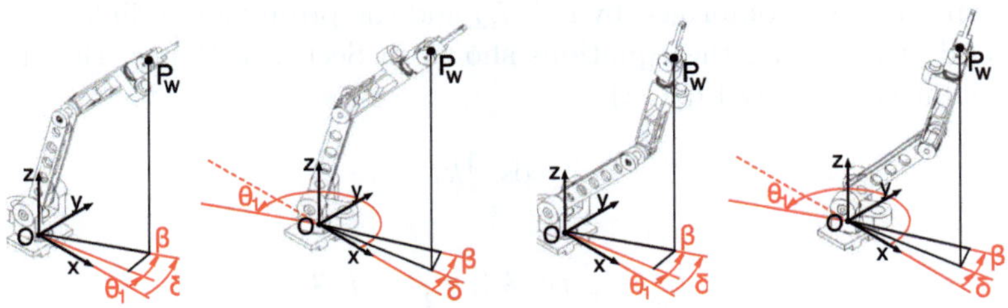

Fig. 6. 3R Configurations. (a): Elbow Up-Left. (b): Elbow Up-Right. (c): Elbow Down-Left. (d): Elbow Down-Right.

2.5 Prototype II: Sliding Bridge

The first revolution joint was replaced by a prismatic joint, and the base of the serial arm was placed over a bridge like structure as shown in Fig. 7(a), turning it into a 1P-4R kinematic chain, with the position of the wrist constrained to a plane parallel to the OXY plane, and allows 3DoF of orientation of the EF.

Table 2. Possible configurations of the 3R mechanism

θ_1	θ_2	θ_3	Configuration
$\delta - \beta$	θ_2^+	θ_3^+	Elbow Left-Up
	θ_2^-	θ_3^-	Elbow Left-Down
$\delta + \beta + \pi$	$-\theta_2^+$	$-\theta_3^+$	Elbow Right-Up
	$-\theta_2^-$	$-\theta_3^-$	Elbow Right-Down

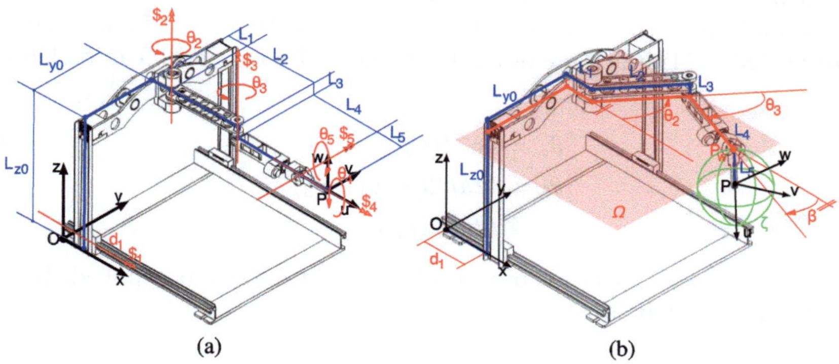

Fig. 7. II. (a): kinematic description. (b): general configuration.

2.5.1 Direct Kinematics

Considering the posture illustrated in Fig. 7(a), the position of the EF is given by $P_{ef_0} = [L_1 + L_2 + L_4 + L_5, L_{y0}, L_{z0} - L_3, 1]$, and the screw parameters are found and summed up in Table 1.

2.5.2 Inverse Kinematics

The first three joints defines a planar serial mechanism, constraining the movement of the wrist over the plane Ω (see Fig. 7(b)), defined by:

$$\Omega : \quad z = z_\Omega = L_{z_0} - L_3. \tag{21}$$

The sphere ς centered at the EF with radio L_5 must intersect the plane Ω, i.e. the position of the EF ${}^{O}P_{ef} = [P_x, P_y, P_z]$ must satisfy:

$$L5 < |P_z - z_\Omega| . \tag{22}$$

Considering that the position of EF is described by ${}^{O}P_{ef} = [P_x, P_y, P_z]$, and its orientation by the rotation matrix:

$${}^{O}R_{ef} = \begin{bmatrix} r_{1,1} & r_{1,2} & r_{1,3} \\ r_{2,1} & r_{2,2} & r_{2,3} \\ r_{3,1} & r_{3,2} & r_{3,3} \end{bmatrix} . \tag{23}$$

Then, from Fig. 7(b), the orientation of the EF can be expressed as the successive rotation matrix given by Eq. (24):

$$\begin{aligned} {}^{O}R_{ef} &= R_w(\beta)\, R_u(\theta_4)\, R_v(\theta_5) \\ &= \begin{bmatrix} C\beta C\theta_5 - S\beta S\theta_4 S\theta_5 & S\beta C\theta_4 & C\beta S\theta_5 + S\beta S\theta_4 C\theta_5 \\ S\beta C\theta_5 - C\beta S\theta_4 S\theta_5 & C\beta C\theta_4 & S\beta S\theta_5 + C\beta S\theta_4 C\theta_5 \\ -C\theta_4 S\theta_5 & S\theta_4 & C\theta_4 C\theta_5 \end{bmatrix}, \end{aligned} \tag{24}$$

where S and C are a simplified notation for sin and cos, and $\beta = \theta_2 + \theta_3$ is the contribution of the second and third joints to the orientation of the EF. From Eqs. (23)–(24), the state of the forth and fifth joint and β can be found as follows:

$$\theta_4 = \sin^{-1}(r_{3,2}) \tag{25}$$

$$\theta_5 = Atan2(-r_{3,1}, r_{3,3}) \tag{26}$$

$$\beta = Atan2(r_{1,2}, r_{2,2}). \tag{27}$$

However, if $\sin\theta_4 = \pm 1$ the expression for θ_5 and β can't be applied. Instead, Eqs. (28)–(29) must be used.

$$\theta_5 = Atan2\left(\frac{r_{23} - r_{11}}{2}, \frac{r_{13} - r_{21}}{2}\right) \tag{28}$$

$$\beta = Atan2\left(\frac{r_{21} + r_{13}}{2}, \frac{r_{11} + r_{33}}{2}\right) \tag{29}$$

Given the position and orientation of the EF in terms of the cosine direction matrix ${}^{O}R_{ef} = [\hat{u}, \hat{v}, \hat{w}]$ the position of the wrist is given by Eq. (30).

$${}^{O}P_w = {}^{O}P_{ef} - L_5\hat{u} \tag{30}$$

Analyzing the projection of the kinematic chain over the plane $z = 0$ (see Fig. 7(b)), the x and y coordinates of the position of the wrist are given by:

$$P_{w_x} = d_1 + L_1 + L_2\cos\theta_2 + L_4\cos\beta \tag{31}$$

$$P_{w_y} = L_{yo} + L_2\sin\theta_2 + L_4\sin\beta \tag{32}$$

Rearranging these last two expressions and remembering that $\beta = \theta_2 + \theta_3$, the state of the three first joints can be found according Eqs. (33)–(35).

$$\theta_2 = \sin^{-1}\left(\frac{P_{w_y} - L_{yo} - L_4\sin\beta}{L_2}\right) \tag{33}$$

$$\theta_3 = \beta - \theta_2 \tag{34}$$

$$d_1 = P_{w_x} - L_1 - L_2\cos\theta_2 - L_4\cos\beta \tag{35}$$

2.6 Prototype V: Sliding Hybrid

The last two joints of Prototype I provided the orientation of the EF, in Prototype V they were replaced by a 3<u>P</u>SS-1S spherical parallel mechanism (SPM,

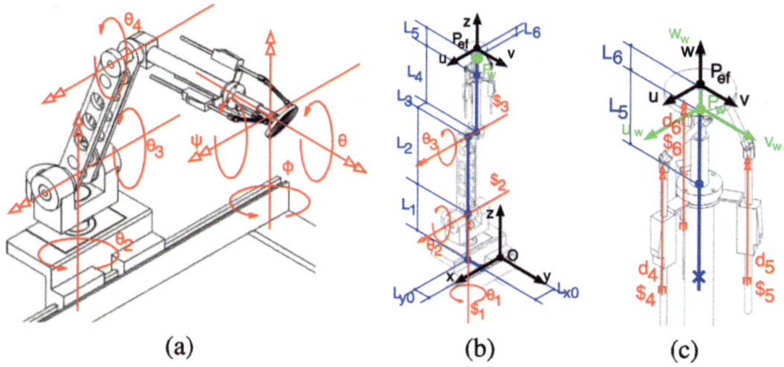

Fig. 8. Prototype V. (a): kinematic description. (b): initial configuration. (c): 3PSS-1S.

see Fig. 8(c)). This SPM posses a moving platform connected to a base platform by four legs. Three of them posses a Prismatic-Spheric- Spheric kinematic chain (where the prismatic joint is actuated), and a fixed leg that connects the moving platform with the base by a spheric joint. This low weight structure provides 3DoF for orientation, high accuracy and stiffness. The design of the SPM is fully described in [6] and its kinematic analysis is detailed in [7].

2.6.1 Direct Kinematic

Considering the initial configuration presented in Fig. 8(b), the position of the EF ${}^{O}P_{ef}$ can be found as follows:

$$ {}^{O}P_{ef} = M_{\$}\, {}^{O}P_{w_0}^{T} + M_{d_{456}}\, {}^{w}L_{ef_0}^{T}, \tag{36}$$

where $M_{\$} = \Pi_{i=1}^{3} M_{\$_i}$ is the overall screw displacement matrix associated to the movement of the first three joints whose screw's parameter are summed up in Table 1. ${}^{O}P_{w_0} = [L_{x_0} - L_3, -L_{y_0}, L_1 + L_2 + L_4 + L_5, 1]$ is the position of the wrist at the initial configuration expressed in homogenous coordinate referenced at O_{xyz}. $M_{d_{456}}$ is the contribution of the SPM to the orientation of the EF according to the state of the prismatic actuators d_4, d_5 and d_6 (see [7] for details). ${}^{w}L_{ef_0} = [0, 0, L_6, 1]$ is the position of the EF at its initial configuration expressed at the reference frame attached at the wrist P_{uvw}. The orientation of the EF is given by:

$$ {}^{O}R = R_{\$} R_{d_{456}} \tag{37}$$

where $R_{\$}$ is the contribution to the orientation of the EF of the first three joints and $R_{d_{456}}$ is the contribution of the SPM to the orientation of the EF.

2.6.2 Inverse Kinematic

Considering the mechanism as shown in Fig. 8(b), the position of the EF given by ${}^{O}P_{ef}$ and its orientation expressed in terms of the direct cosine matrix ${}^{O}R_{uvw} = [\hat{u}, \hat{v}, \hat{w}]$, the position of the wrist of the mechanism is given by:

$$ {}^{O}Pw = {}^{O}P_{ef} - L_6 \hat{w} \tag{38}$$

As it can be seen in Fig. 8(b) and (c), the kinematic chain associated to the positioning of the wrist is the 3R kinematic chain of the prototype 5R, and its inverse kinematic solutions is detailed in Sect. 2.4. In order to achieve the desire orientation at the EF, the parallel mechanism must provide the orientation given by Eq. (39). The inverse kinematic solution of the parallel mechanism is detailed in [7].

$$R_{d_{456}} = R_{\$}^{-1}\,{}^{O}R_{uvw} \tag{39}$$

2.6.3 Workspace

The maximum workspace of the mechanism is defined by all those positions that the EF can reach at least with one orientation. In this representation the orientation capabilities of the mechanism are lost. Therefore, an orientation density map is also analyzed. This map considers all the possible orientations that the EF can reach at a given position.

The maximum workspace of the mechanism is presented in Fig. 9(a) and it is generated after testing more than 450.000 positions of the EF, considering the elbow left-up configuration only. As it can be seen the workspace has an spheric type morphology, and the maximum reachable configurations is given by the maximum length of the mechanism. The orientation density map is generated after the evaluation of 257 different orientations for each given position of the EF and it is presented for different slices of the maximum workspace in Fig. 9.

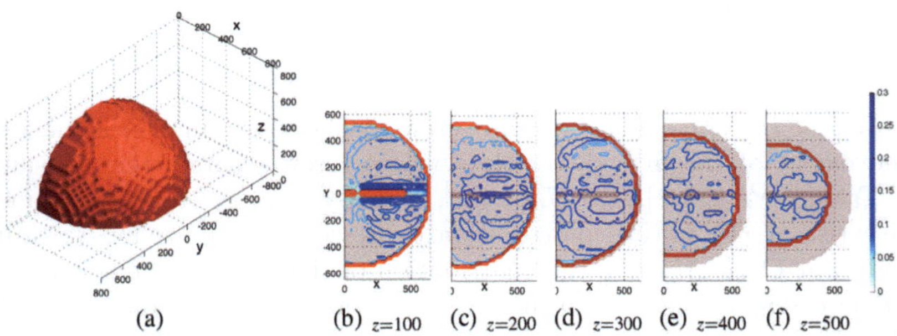

Fig. 9. Workspace. (a): Maximum workspace. (b)–(f): Orientation Density Maps. The red contour defines the boundaries of the workspace slice, and the blue scaled contours defines the probability of finding a solution for the given position of the EF.

3 Conclusion and Discussions

In this work it is summed up the evolution of the concept for a needle insertion surgery robot intended to assist and guide the needle during the insertion procedure. The concept passed through five different prototypes. The final prototype consists of an hybrid mechanism: a 3R serial kinematic chain for positioning its wrist, and a 3PSS-1S spheric parallel mechanism for the orientation of the EF.

This mechanism provides of 6DoF movement plus and additional sliding based that allows a first gross setup. The mobility of the mechanism is redundant for the task, since it is only required 5DoF. However, this redundancy allows to deal with the reduced workspace of the spherical wrist avoiding its singularities while providing multiples configurations of the mechanism for the same path of insertion.

The density map indicates that the mechanism can not reach all the orientations for all the position of the wrist. This limitation is primarily caused by the limited orientation capabilities of the spheric wrist (see [7] for further details). But it also indicates that the mechanism has better orientation capabilities at the center of its workspace. This criteria can be used when the location of the base must be chosen.

Acknowledgements. The authors would like to thank the financial support of Spanish Government CICYT Project Ref. DPI2014-57220-C2-1-P, DPI2013-49527-EXP Universidad Politécnica de Madrid Project Ref. AL14-PID-15 and also to Comunidad de Madrid who supports the project ROBOCITY2030-III P2013/MIT-2748.

References

1. Fichtinger, G., Deguet, A., Masamune, K., Balogh, E., Fischer, G.S., Mathieu, H., Taylor, R.H., Zinreich, S.J., Fayad, L.M.: Image overlay guidance for needle insertion in CT scanner. IEEE Trans. Biomed. Eng. **52**(8), 1415–1424 (2005)
2. Fichtinger, G., Kazanzides, P., Okamura, A.M., Hager, G.D., Whitcomb, L.L., Taylor, R.H.: Surgical and interventional robotics Part II: surgical CAD-CAM systems. IEEE Robot. Autom. Mag. **15**(3), 94–102 (2008)
3. Glozman, D., Shoham, M.: Image-guided robotic flexible needle steering. IEEE Trans. Rob. **23**(3), 459–467 (2007)
4. Guthart, G.S., Salisbury, J.K.: The Intuitive telesurgery system: overview and application. In: Proceedings ICRA, Millennium Conference, IEEE International Conference on Robotics and Automation, vol. 1, pp. 618–621 (2000)
5. Kazanzides, P., Fichtinger, G., Hager, G.D., Okamura, A.M., Whitcomb, L.L., Taylor, R.H.: Surgical and interventional robotics - core concepts, technology, and design. IEEE Robot. Autom. Mag. **15**(2), 122–130 (2008)
6. Puglisi, L.J., Saltaren, R.J., Moreno, H.A., Crdenas, P.F., Garcia, C., Aracil, R.: Dimensional synthesis of a spherical parallel manipulator based on the evaluation of global performance indexes. Robot. Auton. Syst. **60**(8), 1037–1045 (2012)
7. Puglisi, L.J., Saltaren, R.J., Portols, G.R., Moreno, H., Crdenas, P.F., Garcia, C.: Design and kinematic analysis of 3PSS-1S wrist for needle insertion guidance. Robot. Auton. Syst. **61**(5), 417–427 (2013)
8. Tsai, L.W.: Robot Analysis: The Mechanics of Serial and Parallel Manipulators. Wiley, New York (1999)

Dynamic Analysis of the ROAD Robot During the Sit to Stand Task

Isela G. Carrera(✉), Hector A. Moreno, Jose F. Flores, Jesus A. Esquivel, and Mario A. Barrera

Universidad Autonoma de Coahuila,
Barranquilla S/N, Monclova, Coahuila, Mexico
iselacarrera@uadec.edu.mx

Abstract. There is a large part of the population that needs technical assistance to accomplish with activities of daily living, among them are the elderly and those with a physical disability. We found that a technical assistance that is commonly needed by those users is to help to stand up or sit down. This work presents the dynamic analysis of a robotic device that helps with this task. The analysis presented here can be useful for simulation, design, path planning and control of the mechanism.

1 Introduction

In this work the mathematical model of the ROAD robot is presented. The purpose of the ROAD robot is to assist the user in the standing up, sitting down and locomotion tasks. Different concepts for this propose has been explored and are presented in [1]. Other different mechanisms have been proposed to assist a person to stand up from a sit position, some of the most significant are the standing up wheel chairs, for example the TEC Robotic Mobilization Device from Matia Robotics [4] that is a type of bipedestation that helps to transfer a person from one place to another, other similar work is QOLO [5] that has a bending forward innovative mechanisms and follows the user motion intention. All these devices help a person to stand up but are intended for users that have full lower limbs disability. The purpose of the ROAD robot concept is to assist a person who still has some strength in his legs and can be rehabilitated by therapy or to reduce the loss of ability due to aging.

This robot is intended to interact with the user in the environment of a hospital or home. The identification of users needs and the engineering requirements for this concept are described in a previous publication [2]. The robot is suspended from the ceiling through tracks (See Fig. 1). The robots mechanism is composed of three links connected by three joints. They are: a traction system on the rails, a rotational joint with vertical axis, and the actuated prismatic joint that moves up and down the arm support. The rail traction system is located on the ceiling and supports the entire structure.

This work presents the dynamic analysis of the ROAD mechanism. The analysis presented here is useful for simulation, design, path planning and control of the robot. For analysis of the robot, the Lagrange method is applied

I. Chang et al. (eds.), *Advances in Automation and Robotics Research in Latin America*,
Lecture Notes in Networks and Systems 13, DOI 10.1007/978-3-319-54377-2_5

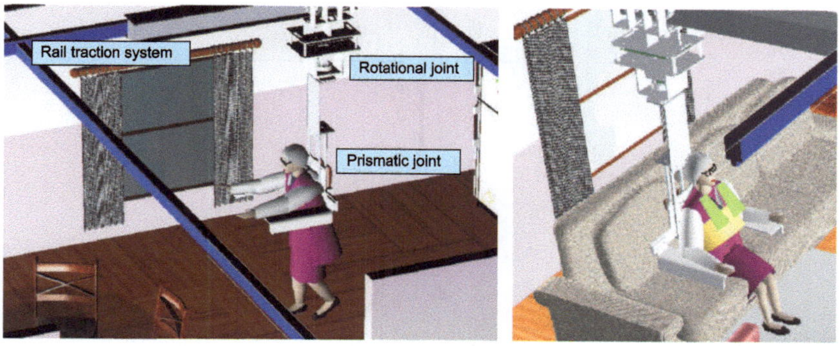

Fig. 1. ROAD components

because it allows to formulate the equations of motion according to the set of joint coordinates. Additionally, the modeling of a user during the standing up task is performed using the Newton-Euler equations of motion in the plane. The analysis presented in this work provides a dynamic model which allows to calculate the forces required at the robot actuators to assist the user to perform the task of standing up.

2 Kinematics Analysis

In this section the kinematic model of ROAD is described. For the analysis consider the robots kinematic scheme shown in Fig. 2. The fixed reference frame is located in the ceiling. For our purposes, the plane $x - y$ is parallel to the sagittal plane of the user. The joint variable q_1 corresponds to the prismatic joint that provides the horizontal movement of the system. The joint variable q_2 corresponds to the rotational joint whose axis is parallel to the axis y of the reference frame. Finally, the joint variable q_3 corresponds to the joint that generates vertical movement.

2.1 Direct Kinematics

The direct kinematics of position consist in determining the position of the user's the attachment point $\mathbf{r}_t = \left[r_{tx}\; r_{ty}\; r_{tz}\right]^T$, given the vector of joint coordinates, $\mathbf{q} = \left[q_1\; q_2\; q_3\right]^T$. From the Fig. 2, the vector $\mathbf{r}_t$ can be obtained as follows:

$$\mathbf{r}_t = \begin{bmatrix} q_1 + d_2\cos(q_2) \\ -(q_3 + d_1) \\ d_2\sin(q_2) \end{bmatrix} \tag{1}$$

The parameters d_1 and d_2 depend of the geometric characteristics of the system.

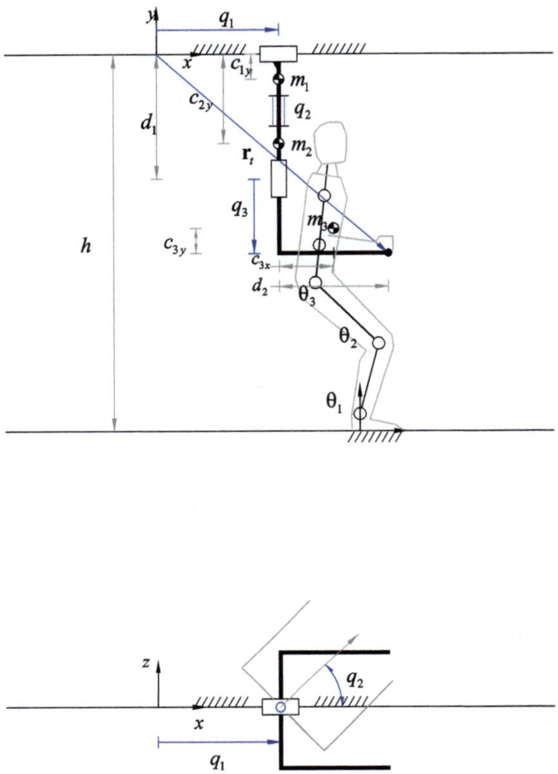

Fig. 2. Robot ROAD scheme

The direct differential analysis consists in determining the user's the attachment point velocity, given the actuator velocities. Deriving (2), we can get the differential model of the system as follows:

$$\dot{\mathbf{r}}_t = \mathbf{J}\dot{\mathbf{q}} \tag{2}$$

where $\dot{\mathbf{r}}_t$ and $\dot{\mathbf{q}}$ denote time derivatives of $\mathbf{r}_t$ and $\mathbf{q}$, respectively. The Jacobian matrix is:

$$\mathbf{J} = \begin{bmatrix} 1 & -d_2 \sin(q_2) & 0 \\ 0 & 0 & -1 \\ 0 & d_2 \cos(q_2) & 0 \end{bmatrix} \tag{3}$$

Deriving (2), the acceleration of the user's the attachment point, $\ddot{\mathbf{r}}_t$, can be calculated given the velocity and joint accelerations, $\dot{\mathbf{q}}$ y $\ddot{\mathbf{q}}$. The direct analysis of acceleration is:

$$\ddot{\mathbf{r}}_t = \mathbf{J}\ddot{\mathbf{q}} + \dot{\mathbf{J}}\dot{\mathbf{q}} \tag{4}$$

where the matrix $\dot{\mathbf{J}}$ is the time derivative of (3) and has the following form:

$$\dot{\mathbf{J}} = \begin{bmatrix} 0 & -d_2\cos(q_2)\dot{q_2} & 0 \\ 0 & 0 & 0 \\ 0 & -d_2\sin(q_2)\dot{q_2} & 0 \end{bmatrix} \tag{5}$$

2.2 Inverse Kinematics

The inverse kinematics of position consist in determining the joint variables, q_1, q_2 and q_3, given the position of the user's the attachment point $\mathbf{r}_t$. The solution is obtained straight forward:

$$q_1 = r_{tx} - d_2\sqrt{1 - \left(\frac{r_{tz}}{d_2}\right)^2} \tag{6}$$

$$q_2 = \arcsin\left(\frac{r_{tz}}{d_2}\right) \tag{7}$$

and

$$q_3 = d_1 - r_{ty} \tag{8}$$

In order to obtain the velocities required for a desired velocity in the user's the attachment point, we solve the Eq. (2) for $\dot{\mathbf{q}}$:

$$\dot{\mathbf{q}} = \mathbf{J}^{-1}\dot{\mathbf{r}}_t \tag{9}$$

In this case the inverse of the Jacobian matrix is:

$$\mathbf{J}^{-1} = \begin{bmatrix} 1 & 0 & -\tan(q_2) \\ 0 & 0 & 1/d_2\cos(q_2) \\ 0 & -1 & 0 \end{bmatrix} \tag{10}$$

The inverse kinematics analysis of acceleration is obtained in the following way:

$$\ddot{\mathbf{q}} = \mathbf{J}^{-1}(\ddot{\mathbf{r}}_t - \dot{\mathbf{J}}\dot{\mathbf{q}}) \tag{11}$$

3 Dynamic Analysis

This section presents the dynamic model of ROAD. For analysis, the method of Lagrange is applied because it allows to formulate the equations of motion using a set of generalized coordinates (i.e. the joint variables). Lagrange equations are given as follows [3]:

$$\frac{d}{dt}\left[\frac{\partial L}{\partial \dot{q}_i}\right] - \frac{\partial L}{\partial q_i} = \tau_i \tag{12}$$

For $i = 1, 2, 3$, the Lagrangian function is defined as the difference between kinetic and potential energy:

$$L = K - U \tag{13}$$

From the scheme in Fig. 2, the kinetic energy for each link can be written:

$$K_1 = \frac{1}{2} m_1 \dot{q}_1^2 \tag{14}$$

$$K_2 = \frac{1}{2} m_2 \dot{q}_1^2 + \frac{1}{2} I_2 \dot{q}_2^2 \tag{15}$$

$$K_3 = \frac{1}{2} m_3 \dot{q}_1^2 - m_3 \dot{q}_1 \dot{q}_2 c_{3x} \sin(q_2) + \frac{1}{2} m_3 \dot{q}_2^2 c_{3x}^2 + \frac{1}{2} I_3 \dot{q}_2^2 + \frac{1}{2} m_3 \dot{q}_3^2 \tag{16}$$

And the potential energy:

$$U_1 = -m_1 c_{1y} g \tag{17}$$

$$U_2 = -m_2 c_{2y} g \tag{18}$$

$$U_3 = -m_1 (d_1 + q_3 - c_{3y}) g \tag{19}$$

where m_1, m_2, m_3 are the masses of the three links; I_2 and I_3 are the moments of inertia of the links 2 and 3 on the axis y; c_{1y}, c_{2y}, c_{3y} and c_{3x} are the coordinates of the center of mass of each link; and g is the gravity.

With the above expressions the Lagrangian function of the system is obtained as follows:

$$\begin{aligned} L =& \frac{1}{2} m_1 \dot{q}_1^2 + \frac{1}{2} m_2 \dot{q}_1^2 + \frac{1}{2} I_2 \dot{q}_2^2 + \frac{1}{2} m_3 \dot{q}_1^2 \\ & - m_3 \dot{q}_1 \dot{q}_2 c_{3x} \sin(q_2) + \frac{1}{2} m_3 \dot{q}_2^2 c_{3x}^2 + \frac{1}{2} I_3 \dot{q}_2^2 + \frac{1}{2} m_3 \dot{q}_3^2 \\ & + m_1 c_{1y} g + m_2 c_{2y} g + m_3 (d_1 + q_3 - c_{3y}) g \end{aligned} \tag{20}$$

For the first joint variable we have:

$$\frac{\partial L}{\partial \dot{q}_1} = m_1 \dot{q}_1 + m_2 \dot{q}_1 + m_3 \dot{q}_1 - m_3 \dot{q}_2 c_{3x} \sin(q_2) \tag{21}$$

$$\frac{d}{dt} \left[\frac{\partial L}{\partial \dot{q}_1} \right] = m_1 \ddot{q}_1 + m_2 \ddot{q}_1 + m_3 \ddot{q}_1 - m_3 \ddot{q}_2 c_{3x} \sin(q_2) - m_3 \dot{q}_2^2 c_{3x} \cos(q_2) \tag{22}$$

$$\frac{\partial L}{\partial q_1} = 0 \tag{23}$$

For the second joint variable:

$$\frac{\partial L}{\partial \dot{q}_2} = I_2 \dot{q}_2 - m_3 \dot{q}_1 c_{3x} \sin(q_2) + m_3 \dot{q}_2 c_{3x}^2 + I_3 \dot{q}_2 \tag{24}$$

$$\frac{d}{dt} \left[\frac{\partial L}{\partial \dot{q}_2} \right] = (I_2 + m_3 + I_3) \ddot{q}_2 - m_3 \ddot{q}_1 c_{3x} \sin(q_2) - m_3 \dot{q}_1 c_{3x} \cos(q_2) \dot{q}_2 \tag{25}$$

$$\frac{\partial L}{\partial q_2} = -m_3 \dot{q}_1 \dot{q}_2 c_{3x} \cos(q_2) \tag{26}$$

Finally, for the third joint variable:

$$\frac{\partial L}{\partial \dot{q}_3} = m_3 \dot{q}_3 \tag{27}$$

$$\frac{d}{dt}\left[\frac{\partial L}{\partial \dot{q}_3}\right] = m_3 \ddot{q}_3 \tag{28}$$

$$\frac{\partial L}{\partial q_3} = m_3 g \tag{29}$$

From the Eq. (12), the ROAD equations of motion are obtained as follows:

$$\tau_1 = (m_1 + m_2 + m_3)\ddot{q}_1 - m_3 \ddot{q}_2 c_{3x} \sin(q_2) - m_3 \dot{q}_2^2 c_{3x} \cos(q_2) \tag{30}$$

$$\tau_2 = (I_2 + m_3 c_{3x} + I_3)\ddot{q}_2 - m_3 \ddot{q}_1 c_{3x} \sin(q_2) \tag{31}$$

$$\tau_3 = m_3 \ddot{q}_3 - m_3 g \tag{32}$$

The dynamic model of the mechanism in the joint space can be expressed in the following way:

$$\boldsymbol{\tau} = \mathbf{M}\ddot{\mathbf{q}} + \mathbf{C}\dot{\mathbf{q}} + \mathbf{g}_q \tag{33}$$

The elements of each matrix are obtained arranging the Eqs. (30)–(31) in matrix form. $\mathbf{M}$ is the inertia matrix, $\mathbf{C}$ is the centrifugal forces and coriolis matrix, and $\mathbf{g}$ is the gravitational force vector. For the ROAD robot, the inertial matrix has the next structure:

$$\mathbf{M} = \begin{bmatrix} (m_1 + m_2 + m_3) & m_3 c_{3x} \sin(q_2) & 0 \\ m_3 c_{3x} \sin(q_2) & (I_2 + m_3 c_{3x} + I_3) & 0 \\ 0 & 0 & m_3 \end{bmatrix} \tag{34}$$

the centrifugal force and Coriolis matrix is:

$$\mathbf{C} = \begin{bmatrix} 0 & m_3 \dot{q}_2 c_{3x} \cos(q_2) & 0 \\ 0 & 0 & 0 \\ 0 & 0 & 0 \end{bmatrix} \tag{35}$$

and the gravitational forces vector:

$$\mathbf{g}_q = \begin{bmatrix} 0 \\ 0 \\ m_3 g \end{bmatrix} \tag{36}$$

4 Modeling a Person During Standing Up Task

This section presents the mathematical modeling of a person during the standing up task. In this case we use the Newton-Euler equations of motion in the plane for the analysis. From the analysis presented here the forces required in the end effector of the robot for performing the task of standing up are obtained.

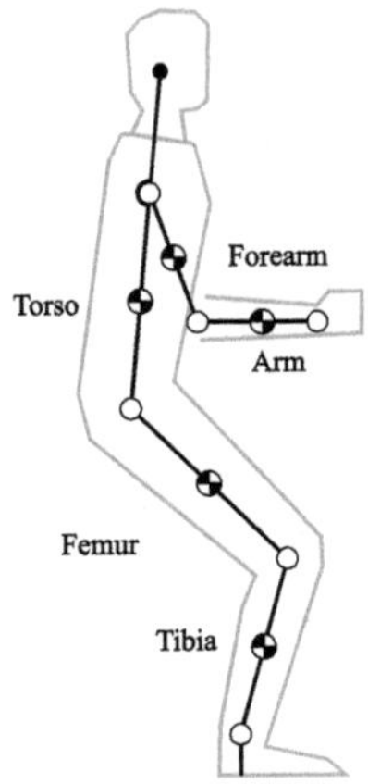

Fig. 3. User scheme

The strategy for solving the problem consist in considering the user's body as kinematic chain, i.e. a set of links connected by rotational joints. For the motion in the plane we consider as links the tibia, femur, torso, forearm, and arm. The head is considered attached to torso. Only rotational joints axes are perpendicular to the plane $x - y$ are considered, see Fig. 3.

Considering the symmetry of the mechanism, the analysis is performed by separating the legs (left and right) and the arms (left and right) from the trunk, Fig. 4 shows this idea. The analysis considers that there are torques in the joints generated by the user. These torques are τ_a, τ_b, τ_c, τ_d, and τ_e, for the ankle, the knee, the hip, the shoulder and the elbow, respectively. The torques can be calculated from a biomechanics analysis of the standing up task [2]. Subsequently it is necessary to determine the percentage of force that the patient has and can generate during the movement. Finally, the necessary force is determined at the anchor point of the user that the robot must provide.

Figure 5 presents a diagram of a user leg. The length of the tibia and femur are denoted by l_t and l_f, respectively. The distance of the centers of mass of each body of adjacent joints are denoted by c_t and c_f. On the other hand m_t and I_t represents the mass and the moment of inertia of tibia. The same meaning m_f and I_f for femur.

In Fig. 6 the free-body diagram of tibia is shown. The sum of moments at point A is given by the following equation:

$$\sum M_A = I_t\alpha_t + \mathbf{r}_{ct} \times m_t\mathbf{a}_t \tag{37}$$

$$\tau_a - \tau_b - \mathbf{r}_{lt} \times \mathbf{f}_b + \mathbf{r}_{ct} \times m_t\mathbf{g} = I_t\alpha_t + \mathbf{r}_{ct} \times m_t\mathbf{a}_t \tag{38}$$

where

$$\mathbf{r}_{ct} = \begin{bmatrix} c_t\cos(\theta_1) \\ c_t\sin(\theta_1) \end{bmatrix}, \mathbf{r}_{lt} = \begin{bmatrix} l_t\cos(\theta_1) \\ l_t\sin(\theta_1) \end{bmatrix}, \mathbf{a}_t = \begin{bmatrix} a_{tx} \\ a_{ty} \end{bmatrix}, \mathbf{g} = \begin{bmatrix} 0 \\ -g \end{bmatrix} \text{and} \mathbf{f}_b = \begin{bmatrix} f_{bx} \\ f_{by} \end{bmatrix}$$

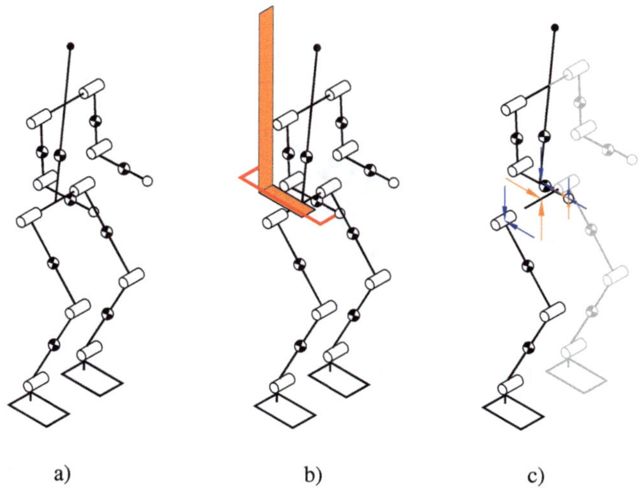

Fig. 4. User kinematic scheme and the forces that interact with the robot, (a) the body considered as kinematic chain, (b) points of contact between the robot and user, (c) separation of the legs and internal and external forces

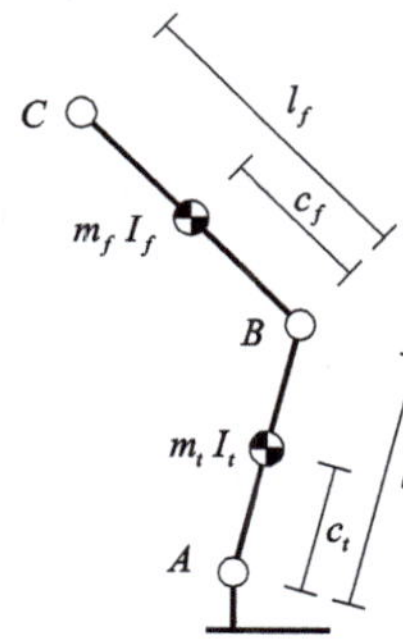

Fig. 5. Leg Diagram

α_t is the angular acceleration, $\mathbf{a}_t$ is the aceleration of tibia center of mass, and $\mathbf{f}_b$ is the lineal force in point B, corresponding to the knee joint. As it was stablished before g is the gravity aceleration. On the other hand τ_a and τ_b are the torques generated by the user in joints ankle and knee, respectively.

We can expand the Eq. (38) as follows:

$$\begin{aligned}\tau_a - \tau_b - f_{by} l_t \cos(\theta_1) + f_{bx} l_t \sin(\theta_1) - m_t g c_t \cos(\theta_1) = \\ I_t \alpha_t + m_t a_{tx} c_t \cos(\theta_1) - m_t a_{ty} c_t \sin(\theta_1)\end{aligned} \tag{39}$$

Solving for the terms that contains elements of $\mathbf{f}_b$ we have:

$$-f_{by} l_t \cos(\theta_1) + f_{by} l_t \sin(\theta_1) = I_t \alpha_t + m_t c_t [(a_{tx} + g)\cos(\theta_1) - a_{ty} \sin(\theta_1)] - \tau_a + \tau_b \tag{40}$$

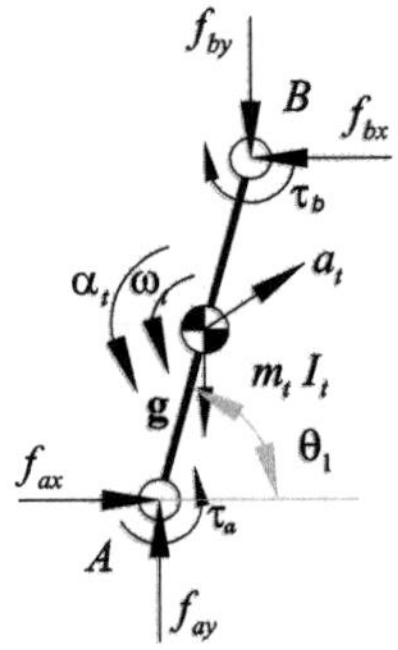

Fig. 6. Free body diagram of the tibia

For analysis of the femur consider the free-body diagram shown in Fig. 7. The sum of moments at point C is:

$$\sum M_C = I_f \alpha_f + \mathbf{r}_{cf} \times m_f \mathbf{a}_f$$

$$\tau_b - \tau_c - \mathbf{r}_{lf} \times \mathbf{f}_b + \mathbf{r}_{cf} \times m_f \mathbf{g} = I_f \alpha_f + \mathbf{r}_{cf} \times m_f \mathbf{a}_f \tag{41}$$

where

$$\mathbf{r}_{cf} = \begin{bmatrix} c_f \cos(\gamma) \\ c_f \sin(\gamma) \end{bmatrix}, \mathbf{r}_{lf} = \begin{bmatrix} l_f \cos(\gamma) \\ l_f \sin(\gamma) \end{bmatrix} \text{ and } \mathbf{a}_f = \begin{bmatrix} a_{fx} \\ a_{fy} \end{bmatrix}$$

α_f is the angular acceleration and $\mathbf{a}_f$ this is the femur acceleration of center of mass.

The Eq. (41) can be expanded as follows:

$$\begin{aligned} \tau_b - \tau_c - f_{by} l_f \cos(\gamma) + f_{bx} l_f \sin(\gamma) - m_f g c_f \cos(\gamma) = \\ I_f \alpha_f + m_t a_{fy} c_f \cos(\gamma) - m_f a_{fx} c_f \sin(\gamma) \end{aligned} \tag{42}$$

By solving for the terms that contain the elements of $\mathbf{f}_b$ we have:

$$\begin{aligned} -f_{by} l_f \cos(\gamma) + f_{bx} l_f \sin(\gamma) = \\ I_f \alpha_f + m_f c_f [(a_{fx} + g)\cos(\gamma) - a_{fy} \sin(\gamma)] - \tau_b + \tau_c \end{aligned} \tag{43}$$

Taking the Eqs. (40) and (43) a system of equations can be obtained as follows:

$$\mathbf{L}\mathbf{f}_b = \mathbf{T}_b \tag{44}$$

where:

$$\mathbf{L} = \begin{bmatrix} l_t \sin(\theta_1) & -l_t \cos(\theta_1) \\ l_f \sin(\gamma) & -l_f \cos(\gamma) \end{bmatrix} \tag{45}$$

and

$$\mathbf{T}_b = \begin{bmatrix} I_t \alpha_t + m_t c_t [(a_{tx} + g)\cos(\theta_1) - a_{ty} \sin(\theta_1)] \tau_a + \tau_b \\ I_f \alpha_f + m_f c_f [(a_{fx} + g)\cos(\gamma) - a_{fy} \sin(\gamma)] - \tau_b + \tau_c \end{bmatrix} \tag{46}$$

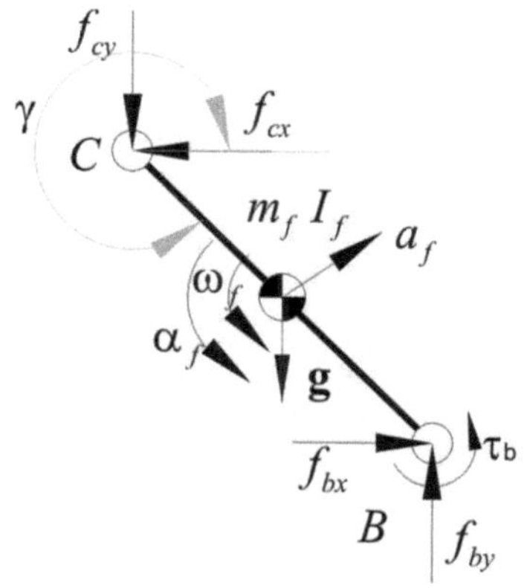

Fig. 7. Free body diagram of the femur

The vector $\mathbf{f}_b$ is obtained by simply by inverting the $\mathbf{L}$ matrix in Eq. (44):

$$\mathbf{f}_b = \mathbf{L}^{-1}\mathbf{T}_b$$

A symbolic expression of the inverse of: $\mathbf{L}$ is the following:

$$\mathbf{L}^{-1} = \frac{1}{\sin(\theta_1 - \gamma)} \begin{bmatrix} \cos(\gamma)/l_t & -\cos(\theta_1)/l_f \\ \sin(\gamma)/l_t & -\sin(\theta_1)/l_f \end{bmatrix} \tag{47}$$

Once $\mathbf{f}_b$ is obtained, the force required at point C to carry the user's legs can be computed in the following way:

$$\mathbf{f}_c = \mathbf{f}_b + m_f\mathbf{g} - m_f\mathbf{a}_f \tag{48}$$

In the same manner the force at the ankle can be obtained.

For the analysis of the torso and arms of the user, consider the diagram in Fig. 8. In this figure l_o, l_a and l_b denotes the length of the torso, forearm and arm, respectively. On the other hand, c_o, c_a and c_b denotes the position of mass center of each link respectively to an adjacent joint. The inertia parameters of each link are represented through m_o, I_o, m_a, I_a, m_b e I_b. For analysis we consider only mass and inertia of the trunk corresponding to an arm.

The free body diagrams of the three links which compose the system are presented in Fig. 9.

To solve the problem, the sum of moments is performed in the links of the trunk and forearm, at points C and E respectively. For the trunk, it is (see Fig. 9a):

$$\sum M_C = I_o\alpha_o/2 + \mathbf{r}_{co} \times m_o\mathbf{a}_o/2$$

Expanding the previous equation:

$$\tau_c - \tau_d - \mathbf{r}_{lo} \times \mathbf{f}_d + \mathbf{r}_{co} \times m_o\mathbf{g}/2 = I_o\alpha_o/2 + \mathbf{r}_{co} \times m_o\mathbf{a}_o/2 \tag{49}$$

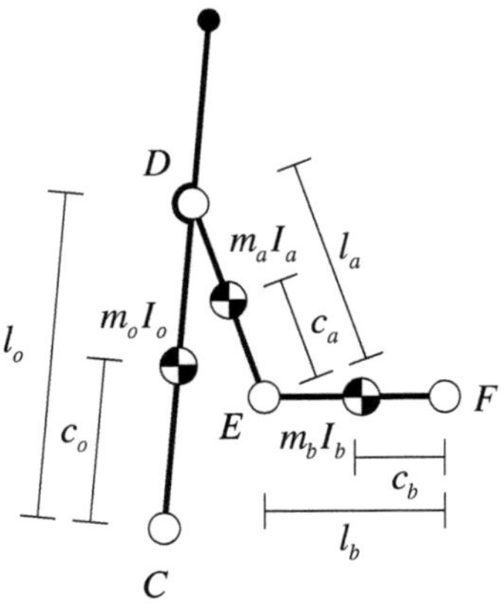

Fig. 8. Diagram of the trunk and arm

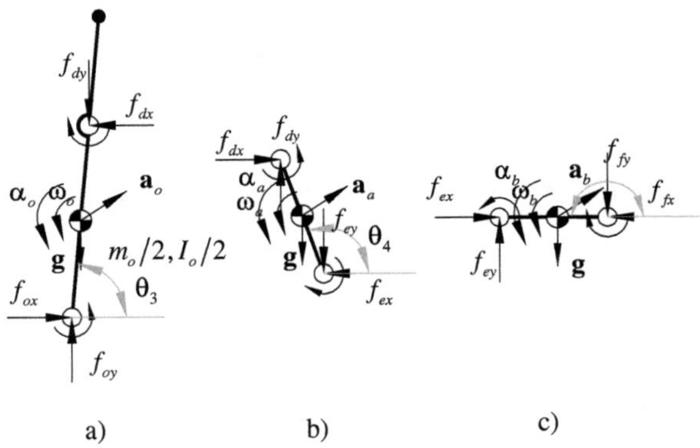

Fig. 9. Free body diagrams of torso and arm links

where

$$\mathbf{r}_{co} = \begin{bmatrix} c_o\cos(\theta_3) \\ c_o\sin(\theta_3) \end{bmatrix}, \mathbf{r}_{lo} = \begin{bmatrix} l_o\cos(\theta_3) \\ l_o\sin(\theta_3) \end{bmatrix}, \mathbf{a}_o = \begin{bmatrix} a_{ox} \\ a_{oy} \end{bmatrix} \text{ and } \mathbf{f}_d = \begin{bmatrix} f_{dx} \\ f_{dy} \end{bmatrix}$$

α_o is the angular acceleration, $\mathbf{a}_o$ is torso center of mass acceleration, $\mathbf{f}_d$ is the linear force at the point D.
Expanding the last expression:

$$\begin{aligned} -f_{dy}l_o\cos(\theta_3) + f_{dx}l_o\sin(\theta_3) = I_o\alpha_o/2 \\ +m_oc_o[(a_{ox}+g)\cos(\theta_3) - a_{oy}\sin(\theta_3)]/2 - \tau_c + \tau_d \end{aligned} \tag{50}$$

For the forearm the sum of moments is as follows (see Fig. 9b):

$$\sum M_E = I_a\alpha_a + \mathbf{r}_{ca} \times m_a\mathbf{a}_a$$

expanding

$$\tau_d - \tau_e - \mathbf{r}_{la} \times \mathbf{f}_d + \mathbf{r}_{ca} \times m_a \mathbf{g} = I_a \alpha_a + \mathbf{r}_{ca} \times m_a \mathbf{a}_a \quad (51)$$

where

$$\mathbf{r}_{ca} = \begin{bmatrix} c_a \cos(\theta_4) \\ c_a \sin(\theta_4) \end{bmatrix}, \mathbf{r}_{la} = \begin{bmatrix} l_a \cos(\theta_4) \\ l_a \sin(\theta_4) \end{bmatrix} \text{ and } \mathbf{a}_a = \begin{bmatrix} a_{ox} \\ a_{oy} \end{bmatrix}$$

α_a is the angular acceleration $\mathbf{a}_a$ is the forearm center of mass acceleration. Expanding the last expression:

$$\begin{aligned} -f_{dy} l_a \cos(\theta_4) + f_{dy} l_a \sin(\theta_4) = I_a \alpha_a + m_a c_a [(a_{ax} + g)\cos(\theta_4) - \\ a_{ay} \sin(\theta_4)] - \tau_d + \tau_e \end{aligned} \quad (52)$$

From Eqs. (50) and (52) the following matrix equation can be obtained:

$$\mathbf{L}_2 \mathbf{f}_d = \mathbf{T}_d \quad (53)$$

where

$$\mathbf{L}_2 = \begin{bmatrix} l_o \sin(\theta_3) & -l_o \cos(\theta_3) \\ l_a \sin(\theta_4) & -l_a \cos(\theta_4) \end{bmatrix} \quad (54)$$

$$\mathbf{T}_d = \begin{bmatrix} I_o \alpha_o / 2 + m_o c_o [(a_{ox} + g)\cos(\theta_3) - a_{oy} \sin(\theta_3)]/2 - \tau_c + \tau_d \\ I_a \alpha_a + m_a c_a [(a_{ax} + g)\cos(\theta_4) - a_{ay} \sin(\theta_4)] - \tau_d + \tau_e \end{bmatrix} \quad (55)$$

By solving for $\mathbf{f}_d$ it can be found all the resulting forces:

$$\mathbf{f}_o = 2\mathbf{f}_d + m_o \mathbf{g} - m_o \mathbf{a}_o \quad (56)$$

$$\mathbf{f}_e = \mathbf{f}_d + m_a \mathbf{g} - m_a \mathbf{a}_a \quad (57)$$

$$\mathbf{f}_f = \mathbf{f}_e + m_b \mathbf{g} - m_b \mathbf{a}_b \quad (58)$$

where $\mathbf{a}_b$ is the arm acceleration.

Finally the total force that the mechanism should apply over the user to perform the desired movement it is as follows:

$$\mathbf{f}_H = 2\mathbf{f}_c + \mathbf{f}_o + 2\mathbf{f}_f \quad (59)$$

A particular case is the one which the upper body user does not move, it means it remains fixed due the type of harness used. In this case we have:

$$\mathbf{f}_o + 2\mathbf{f}_f = (m_o + 2m_a + 2m_b)(\mathbf{g} + \mathbf{a}_o) \quad (60)$$

The dynamic model of the robot interacting with the user for the standing up motion can be simplified and can be written as follows:

$$\boldsymbol{\tau}^* = \mathbf{M}^* \ddot{\mathbf{q}}^* + \mathbf{g}^* + \mathbf{J}^{*T} \mathbf{f}_H \quad (61)$$

where

$$\boldsymbol{\tau}^* = \begin{bmatrix} \tau_1 \\ \tau_3 \end{bmatrix} y \; \ddot{\mathbf{q}}^* = \begin{bmatrix} q_1 \\ q_3 \end{bmatrix}$$

In this case, the inertia matrix has the next structure:

$$\mathbf{M}^* = \begin{bmatrix} (m_1 + m_2 + m_3) & 0 \\ 0 & m_3 \end{bmatrix} \tag{62}$$

and the gravitational forces vector:

$$\mathbf{g}^* = \begin{bmatrix} 0 \\ m_3 g \end{bmatrix} \tag{63}$$

The Jacobian Matrix $\mathbf{J}^*$ is given by:

$$\mathbf{J}^* = \begin{bmatrix} 1 & 0 \\ 0 & -1 \end{bmatrix} \tag{64}$$

5 Conclusions

This work presented the dynamic analysis of the ROAD robot during the sit to stand task. For the analysis of the mechanism the Lagrange method is applied because it allows to formulate the equations of motion according to the set of joint coordinates. The modeling of a user during the standing up task is performed using the Newton-Euler equations of motion in the plane. The analysis presented in this work provides a mathematical model which allows to calculate the forces required at the robot actuators to assist the user during the standing up.

The analysis was performed considering the symmetry of the users body and by separating the legs (left and right) and the arms (left and right) from the trunk. The analysis considers that there are torques in the joints generated by the user. These torques are obtained from a biomechanics analysis of the standing up task. The analysis determines the required forces the robot must provide to complete the task.

The analysis presented here can be used for simulation, design, path planning and control of the robot. Future works are the experimental validation of the model through a prototype testbed and its use in model based control techniques.

Acknowledgements. The authors would like to thank to PRODEP Mexico and Universidad Autonoma de Coahuila for all support to this project.

References

1. Carrera, I., Moreno, H., Saltaren, R., Aracil, R.: Soluciones de la Robotica a la Asistenciay Rehabilitacin de Personas Discapacitadas en Espaa, DTR4all Congress, Madrid, Spain, June 2011
2. Carrera, I., Moreno, H.A., Saltarn, R., et al.: ROAD: domestic assistant and rehabilitation robot. Med. Biol. Eng. Comput. **49**, 1201 (2011). doi:10.1007/s11517-011-0805-4
3. Tsai, L.W., Analysis, R.: The Mechanics of Serial and Parallel Manipulators. Wiley, New York, ISBN:978-0-471-32593-2, 22 Feb 1999
4. Robotic Movilization Device. http://www.matiarobotics.com/about.html
5. QOLO: The James Dyson Award. http://www.jamesdysonaward.org/es/projects/qolo/

A Modular Robotic System for Assessment and Exercise of Human Movement

José Baca[1(✉)], Mohan Sai Ambati[1], Prithviraj Dasgupta[1], and Mukul Mukherjee[2]

[1] Computer Science Department, University of Nebraska at Omaha, Omaha, USA
{jbacagarcia,bwoosley,pdasgupta}@unomaha.edu
[2] Biomechanics Research Building, University of Nebraska at Omaha, Omaha, USA
mmukherjee@unomaha.edu

Abstract. This project targets the problem of developing a wearable modular robotic system (This work was supported by the Peter Kiewit Institute and NASA Nebraska Space Grant.), for assessing human movement and providing different types of exercises for the user. The system attempts to provide not only a variety of exercises (concentric, eccentric, assisted and resisted), but also to assess the change in variability of the movement as the subject shows functional improvement. The system will not only be useful for patients with sensorimotor problems such as stroke, Parkinson's, cerebral palsy, but also for special populations such as astronauts who spend long periods of time in space and experience muscle atrophy. In this work, a first prototype of a modular robot is presented along with preliminary test results from basic active and passive wrist exercises that show the feasibility of this type of systems for assessment and exercise of human movement.

1 Introduction

On average, every 40 s, someone in the United States has a stroke which makes stroke the leading cause of long-term disabilities in the United States affecting 795,000 people annually [8]. The direct and indirect cost of stroke is about $38.6 billion [8]. Goal-directed Rehabilitation interventions can significantly improve the restoration of motor function. Even after years of research, stroke rehabilitation still falls short when trying to develop treatments that are quick, effective, and inexpensive. In addition, post-stroke patients face a huge economic burden. Moreover, traditional rehabilitation paradigms require both equipment and assistance, and are labor intensive for therapists. Immobility of the patients also demands the presence of caregivers, which greatly increase the cost [19].

Therefore, a therapy that could be provided in a home-based environment has the advantage of being more motivating, less strenuous for the patient and the caregiver, and could significantly reduce cost in comparison to hospital-based therapy [1,20]. If the therapy can be assisted with a robotic system, the duration, repeatability, number of training sessions and techniques could be easily modified according to the therapist. Moreover, If the same robotic system can be used

I. Chang et al. (eds.), *Advances in Automation and Robotics Research in Latin America*, Lecture Notes in Networks and Systems 13, DOI 10.1007/978-3-319-54377-2_6

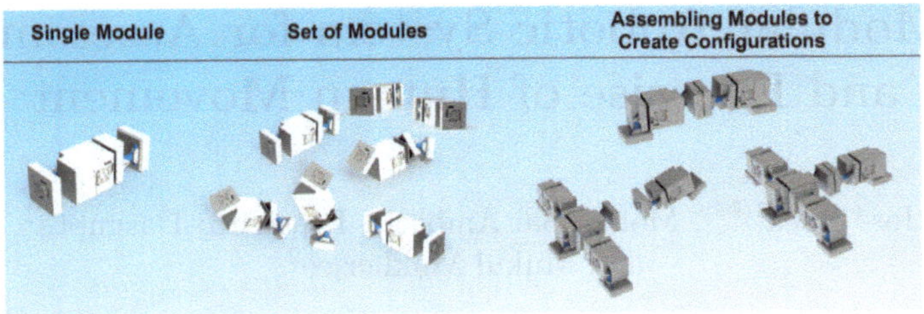

(a)

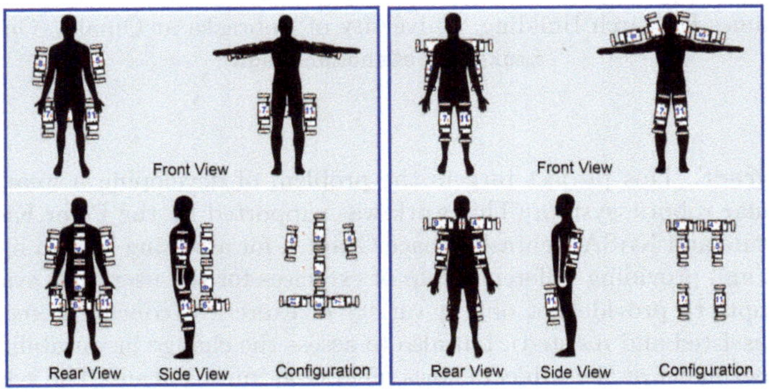

(b)

Fig. 1. (a) Set of ModRED II modules. (b) By assembling modules, it is possible to create configurations that can be worn by the user and exercise different muscles and joints.

to treat different joints in the human body (e.g., wrist, elbow, shoulder, knee or ankle), then its versatility will directly impact the overall cost of the treatments and it will increase the patient comfort by having the opportunity to use the system at home.

This work presents the use of a wearable modular robotic system, for assessing human movement and providing different types of exercises for the user. The aim is to be able to provide not only a variety of exercises (concentric, eccentric, assisted and resisted), but also to assess the change in variability of the movement as the subject shows functional improvement. The system will not only be useful for patients with sensorimotor problems such as stroke, parkinsons, cerebral palsy, but also for special populations such as astronauts who spend long periods of time in space and experience muscle atrophy. A modular robotic system is a system composed of a set of modules that can be connected to each other to create different configurations [30]. We propose to use a modular robotic system to build configurations that can be worn by the user to exercise diverse group of muscles, as show in Fig. 1.

2 Related Work

Robot-assisted therapy has proven to be able to recover patients motor functions [3,7,13,16–18,22,25,28,33]. For example, in the case of a patients arm, the robotic system is able to position the patients arm and hand at the desired points in space with different motion patterns, considering relevant degrees of freedom and their restrictions. In this field, there are two types of robotic systems that are being used to address the challenge, i.e., a robotic platform that is attached to a surface (table, wall, floor, etc.) and has an end-effector tool that is connected to the patients hand, forearm, etc. [10,12,15], and an exoskeleton-type robot that resembles the anatomy of the human body and is attached to the patients body [23]. Exoskeleton robots are preferred compared to end-effector type robots, particularly for automating neurorehabilitation [3,17,18]. In the exoskeleton robots, there is a close alignment of the anatomical axes of the human joints with the corresponding mechanical axes of the exoskeleton. This is an advantage over end-effector type of robots. All human joint angles and torques of interest can be directly measured and individually controlled [18]. These robotic systems target one joint at a time, and for each joint (wrist, elbow, shoulder, knee or ankle) there are different robotic designs. Both types of robots have a fixed configuration (kinematic parameters cannot be changed) and do not target multiple joints at the same time. Some of these systems assist only the upper body joints with passive or/and active exercises, and some others assist only the lower body joints, as shown in Table 1.

Table 1. Robot-assisted systems: End-Effector robot type - Exoskeleton robot type

	Joint type	Passive exercises	Active exercises
Upper body	Wrist	[4,9][b] [5,26][a] [21][b] [12,27,32][a]	[9][b] [5,12,26,27,32][a]
	Fore arm	[3,4,9,21][b] [12][a]	[3,9,21][b] [12][a]
	Elbow	[3,4,9,18,21][b] [12][a]	[3,9,18,21][b] [12][a]
	Shoulder	[4,18,21][b] [12][a]	[18,21][b] [12][a]
Lower body	Knee	[32][b] [14][a]	[32][b] [14][a]
	Ankle	[33][a] [6,24][b] [31][a]	[33][a] [6,24,29][b] [31][a]
	Hip		[29][b]

[a]End-Effector robot type
[b]Exoskeleton robot type

3 ModRED II System

The proposed solution is based on the development of a wearable modular robotic system that can be used to provide muscular strength training. The target application relates the use of the ModRED II (Modular Robot for Exploration and Discovery) platform to create special configurations that can be used to exercise the muscles and joints [2,11].

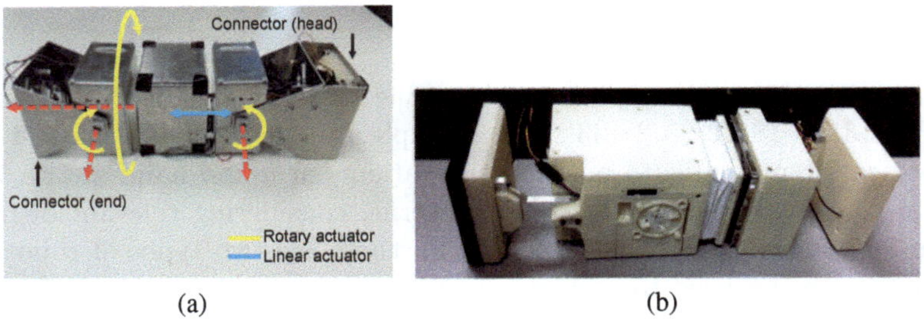

Fig. 2. (a) A ModRED module has 3 rotational and 1 prismatic DOF. (b) ModRED II module.

ModRED II is the second generation of a modular robot classified as a homogeneous system with modules with four degrees of freedom (DOF) - three rotational and one prismatic, as shown in Fig. 2(a). Each module is capable of producing pitch, yaw, roll and one extension DOF. Each DOF is independently actuated and each module has a dimension of 14.85 in × 4.53 in × 4.53 in. Each module contains its own processing, sensing, and communication boards, and it is mainly powered by Li-ion batteries. The length of the module can be increased by 1.57 in due to the linear actuator in the modules design. Each module contains 4 docking mechanisms that provide the possibility of docking modules together and forming chain and lattice type configurations. This feature allows us to have flexibility when designing new robots or structures.

4 Wearable Modular Robotic System

In order to monitor and track muscle activity, a variety of sensors have been added to a new type of module that will work as a link between the patient/user and ModRED II, as shown in Fig. 3(a). This module will collect long-duration position data, sense muscle activity, and close the loop between the sensed data and the DOF (motors) of the configuration. This module contains one BeagleBone Black that takes analog input data from a force sensor resistor (FSR), an electromyography (EMG) device and a flex sensor. It transmits data to the other modules comprising the configuration via XBee with the purpose to control the motors inside each of the modules, as show in Fig. 3(b).

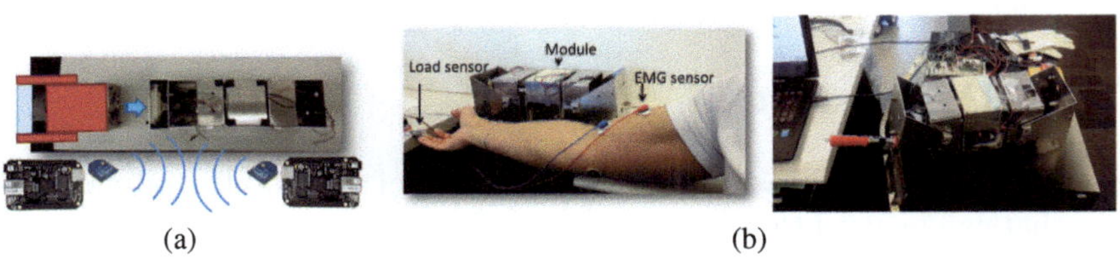

Fig. 3. (a) Communication between modules is done via XBee. (b) Prototype of new module with sensors on-board.

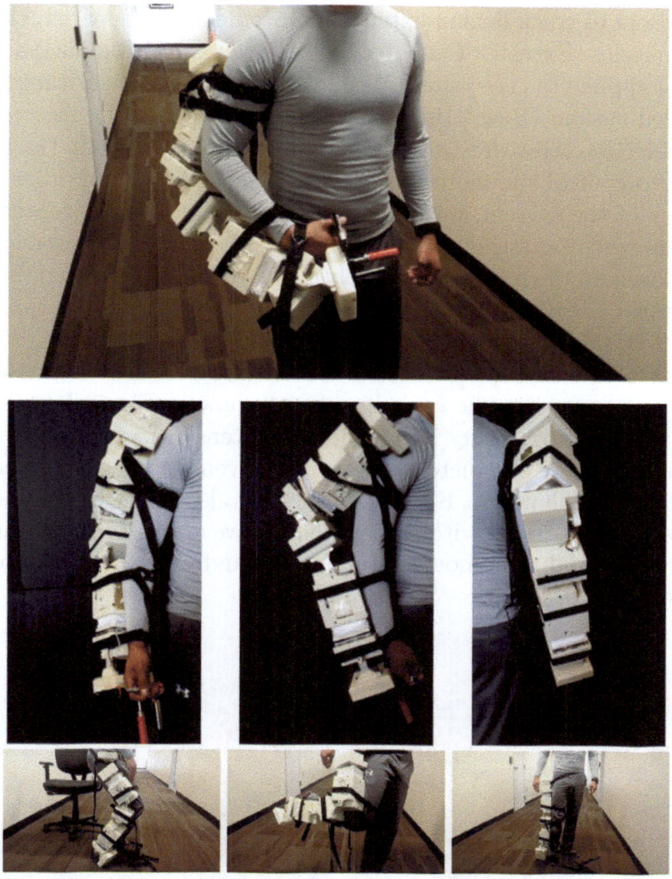

Fig. 4. Two ModRED II modules (in a chain-type configuration) have been attached to the upper and lower limbs of the human body with the purpose of analyzing their workspace and performance when working with different muscles and joints.

We propose to use ModRED II to build configurations that can be worn by the user to exercise diverse group of muscles and joints, as shown in Fig. 4.

5 Experiments

The objective of creating new configurations is to create a system capable of stimulating the muscles by the effect of pushing or pulling against resistance. The possibility of varying the resistance of the system by increasing/decreasing the force and the position of each actuated DOF can bring more benefits than just working with elastic resistance bands.

For this experiment, we have built a prototype that targets the wrist joint and it can be used to perform basic active-passive exercises. We have integrated different sensors to track flexion-extension movements of the wrist, the electrical activity of muscles (Extensor Carpi Radialis, Flexor Carpi Radialis) and the force that is being applied to the system. We have developed an algorithm that runs in the embedded system (BeagleBone Black) of ModRED that takes analog input data from the force sensor resistor (FSR), the electromyography (EMG) and the Flex sensor to control the motors of the attached modules. In order to validate the efficiency of the system, we measure the effort/movement generated by the muscle/joints when performing specific exercises with the proposed system. The analysis is based on the measurements obtained via FSR (newtons), EMG (volts) and Flex sensor (degrees). For comparison purposes we execute same exercises using the off-the-shelf system, Biodex system 4 Pro.

Four sets of flexion and extension movements of wrist in active and passive modes were noted in the experiments, as show in Fig. 5. In passive mode, speed and type of movement (i.e., flexion or extension) is specified to the system, experiments were conducted by varying movement speed (10 degree/sec to 2.5 degree/sec), as shown in Fig. 6. In active mode, speed of the movement is specified to the system and with constant speed we are varying the force applied by the individual and the amount of effort that individual using to perform the exercise, as shown in Fig. 7.

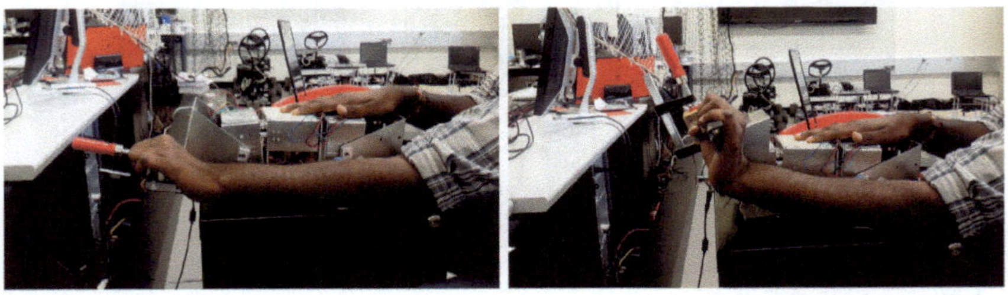

(a)

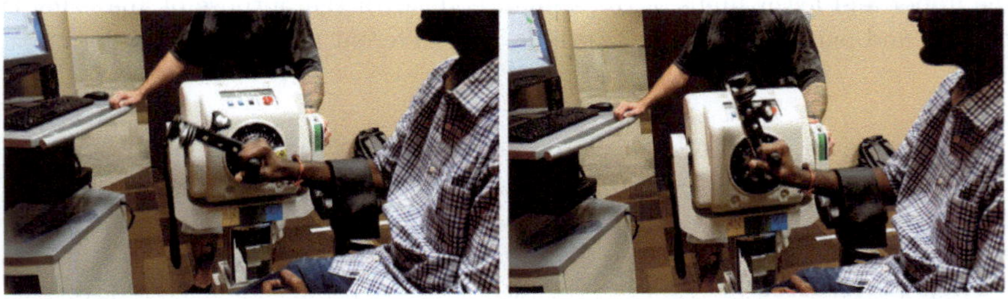

(b)

Fig. 5. (a) Basic exercise for wrist rehabilitation. Using the developed prototype, active and passive exercises were executed and data from the sensors was collected. (b) The force and joint angle measured in each mode is compared to same exercise performed by Biodex system 4 Pro.

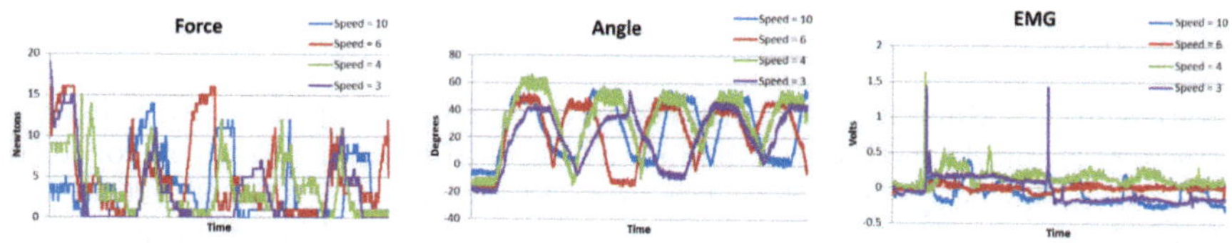

Fig. 6. Passive mode: User does not use the force to move the system. User is assisted by the module. In passive mode speed and direction are the parameters given to the program.

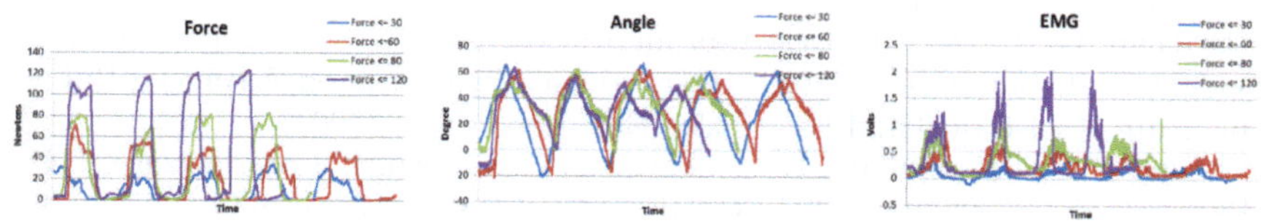

Fig. 7. Active mode: user applies force against the resistance generated by the module's DOF. In active mode speed is the only parameter given to the program.

6 Conclusions

Our preliminary results suggest that it is feasible to use modular robots for human body exercises. Experimental results with the ModRED II platform show the reliability of the device when performing active and passive wrist exercises. The capability of a modular system to be used simultaneously in different muscle groups and joints brings clear advantages in terms of versatility and cost. This system has the potential to be used in a home-based environment which has the advantage of being more motivating, less strenuous for the patient and the caregiver, and could significantly reduce cost in comparison to hospital-based therapy. If the therapy could be assisted with a robotic system, the duration, repeatability, and number of training sessions could be incremented according to the therapist. The system reduces the overall cost of the treatment and increases the patient comfort by having the system at home. We would like to highlight that there are several challenges that need to be addressed to successfully use modular robots for exercising such as the uncertainty to target a specific muscle group with a robot configuration, the adequate muscle stimulation and defining a limited number of modules available within the set. As future work, we plan to increase the intelligence of the modular robotic system to provide active-passive exercise routines that can self-adapt in real-time to the changing patient's ability with respect to the goals. The proposed project can be further extended for various other applications, such as wearable suit for astronauts facing long-duration space missions.

References

1. Anderson, C., Mhurchu, C.N., Rubenach, S., Clark, M., Spencer, C., Winsor, A.: Home or hospital for stroke rehabilitation? Results of a randomized controlled trial: II: cost minimization analysis at 6 months. Stroke **31**(5), 1032–1037 (2000)
2. Baca, J., Hossain, S., Dasgupta, P., Nelson, C., Dutta, A.: Modred: hardware design and reconfiguration planning for a high dexterity modular self-reconfigurable robot for extra-terrestrial exploration. Robot. Auton. Syst. **62**(7), 1002–1015 (2013)
3. Burgar, C.G., Lum, P.S., Shor, P.C., Van der Loos, H.F.M.: Development of robots for rehabilitation therapy: the palo alto va/stanford experience. J. Rehabil. Res. Dev. **37**(6), 663–674 (2000)
4. Carignan, C., Liszka, M.: Design of an arm exoskeleton with scapula motion for shoulder rehabilitation. In: 12th International Conference on Advanced Robotics. Institute of Electrical and Electronics Engineers. IEEE (2005)
5. Fischer, H.C., Triandafilou, K.M., Thielbar, K.O., Ochoa, J.M., Lazzaro, E.D.C., Pacholski, K.A., Kamper, D.G.: Use of a portable assistive glove to facilitate rehabilitation in stroke survivors with severe hand impairment. IEEE Trans. Neural Syst. Rehabil. Eng. **24**(3), 344–351 (2016)
6. Goodman, R.N., Rietschel, J., Barton, J.E., Krebs, H.I., Macko, R.F., Forrester, L.W., Roy, A.: Clinical application of a modular ankle robot for stroke rehabilitation. NeuroRehabilitation **33**, 85–97 (2013)
7. Gao, F., Ren, Y., Roth, E.J., Harvey, R., Zhang, L.-Q.: Effects of repeated ankle stretching on calf muscle-tendon, ankle biomechanical properties in stroke survivors. Clin. Biomech. **26**(5), 516–522 (2011)
8. Go, A.S., Mozaffarian, D., Roger, V.L., Benjamin, E.J., Berry, J.D., Blaha, M.J., Dai, S., Ford, E.S., Fox, C.S., Franco, S., Fullerton, H.J., Gillespie, C., Hailpern, S.M., Heit, J.A., Howard, V.J., Huffman, M.D., Judd, S.E., Kissela, B.M., Kittner, S.J., Lackland, D.T., Lichtman, J.H., Lisabeth, L.D., Mackey, R.H., Magid, D.J., Marcus, G.M., Marelli, A., Matchar, D.B., McGuire, D.K., Mohler, E.R., Moy, C.S., Mussolino, M.E., Neumar, R.W., Nichol, G., Pandey, D.K., Paynter, N.P., Reeves, M.J., Sorlie, P.D., Stein, J., Towfighi, A., Turan, T.N., Virani, S.S., Wong, N.D., Woo, D., Turner, M.B.: Heart disease, stroke statistics-2014 update: a report from the american heart association. Circulation **129**(3), e28–e292 (2013)
9. Gupta, A., O'Malley, M.K.: Robotic exoskeletons for upper extremity rehabilitation. In: Rehabilitation Robotics. InTech (2007)
10. Hesse, S., Werner, C., Pohl, M., Rueckriem, S., Mehrholz, J., Lingnau, M.L.: Computerized arm training improves the motor control of the severely affected arm after stroke: a single-blinded randomized trial in two centers. Stroke **36**(9), 1960–1966 (2005)
11. Hossain, S.G.M., Nelson, C.A., Dasgupta, P.: Hardware design and testing of modred a modular self-reconfigurable robot system. In: Proceedings of ASME/IEEE International Conference on Reconfigurable Mechanisms and Robots (2012)
12. Krebs, H.I., Hogan, N., Aisen, M.L., Volpe, B.T.: Robot-aided neurorehabilitation. IEEE Trans. Rehabil. Eng. **6**(1), 75–87 (1998)
13. Krebs, H.I., Volpe, B.T., Williams, D., Celestino, J., Charles, S.K., Lynch, D., Hogan, N.: Robot-aided neurorehabilitation: a robot for wrist rehabilitation. IEEE Trans. Neural Syst. Rehabil. Eng. **15**(3), 327–335 (2007)
14. Lee, S., Kim, C.: Development of a robot-aided neuromuscular rehabilitation method using perturbing forces. In: IEEE International Symposium on Robotics, Institute of Electrical and Electronics Engineers. IEEE (2013)

15. Loureiro, R., Amirabdollahian, F., Topping, M., Driessen, B., William, H.: Upper limb robot mediated stroke therapy-gentle/s approach. Auton. Robots **15**(1), 35–51 (2003)
16. Lynch, D., Ferraro, M., Krol, J., Trudell, C.M., Christos, P., Volpe, B.T.: Continuous passive motion improves shoulder joint integrity following stroke. Clin. Rehabil. **19**(6), 594–599 (2005)
17. Nef, T., Guidali, M., Riener, R.: Armin iii - arm therapy exoskeleton with an ergonomic shoulder actuation. Appl. Bionics Biomech. **6**(2), 127–142 (2009)
18. Nef, T., Mihelj, M., Riener, R.: Armin: a robot for patient-cooperative arm therapy. Med. Biol. Eng. Comput. **45**(9), 887–900 (2007)
19. O'Dell, M.W., Lin, C.-C.D., Harrison, V.: Stroke rehabilitation: strategies to enhance motor recovery. Annu. Rev. Med. **60**(1), 55–68 (2009)
20. Patel, S., Park, H., Bonato, P., Chan, L., Rodgers, M.: A review of wearable sensors and systems with application in rehabilitation. J. NeuroEngineering Rehabil. **9**(1), 21 (2012)
21. Ren, Y., Kang, S.H., Park, H.-S., Wu, Y.-N., Zhang, L.-Q.: Developing a multi-joint upper limb exoskeleton robot for diagnosis, therapy, and outcome evaluation in neurorehabilitation. IEEE Trans. Neural Syst. Rehabil. Eng. **21**(3), 490–499 (2013)
22. Riener, R., Nef, T., Colombo, G.: Robot-aided neurorehabilitation of the upper extremities. Med. Biol. Eng. Comput. **43**(1), 2–10 (2005)
23. Sanchez, R.J., Liu, J., Rao, S., Shah, P., Smith, R., Rahman, T., Cramer, S.C., Bobrow, J.E., Reinkensmeyer, D.J.: Automating arm movement training following severe stroke: functional exercises with quantitative feedback in a gravity-reduced environment. IEEE Trans. Neural Syst. Rehabil. Eng. **14**(3), 378–389 (2006)
24. Seel, T., Raisch, J., Schauer, T.: IMU-based joint angle measurement for gait analysis. Sensors **14**(4), 6891–6909 (2014)
25. Selles, R.W., Li, X., Lin, F., Chung, S.G., Roth, E.J., Zhang, L.-Q.: Feedback-controlled and programmed stretching of the ankle plantarflexors and dorsiflexors in stroke: effects of a 4-week intervention program. Arch. Phys. Med. Rehabil. **86**(12), 2330–2336 (2005)
26. Stein, J., Bishop, L., Gillen, G., Helbok, R.: A pilot study of robotic-assisted exercise for hand weakness after stroke. In: International Conference on Rehabilitation Robotics. Institute of Electrical and Electronics Engineers (IEEE), June 2011
27. Wang, C., Lu, Z., Wang, Y., Li, M., Duan, L., Shen, Y., Wei, J., Shi, Q., Zecca, M., Li, W., Wu, Z.: Development of a rehabilitation robot for hand and wrist rehabilitation training. In: 2015 IEEE International Conference on Information and Automation. Institute of Electrical and Electronics Engineers. IEEE, August 2015
28. Wu, Y.-N., Hwang, M., Ren, Y., Gaebler-Spira, D., Zhang, L.-Q.: Combined passive stretching and active movement rehabilitation of lower-limb impairments in children with cerebral palsy using a portable robot. Neurorehabilitation Neural Repair **25**(4), 378–385 (2011)
29. Yang, W., Yang, C.J., Wei, Q.X.: Design of an anthropomorphic lower extremity exoskeleton with compatible joints. In: IEEE International Conference on Robotics and Biomimetics. Institute of Electrical and Electronics Engineers. IEEE, December 2014
30. Yim, M., Shen, W.-M., Salemi, B., Rus, D., Moll, M., Lipson, H., Klavins, E., Chirikjian, G.S.: Modular self-reconfigurable robot systems [grand challenges of robotics]. IEEE Robot. Autom. Mag. **14**(1), 43–52 (2007)

31. Yurkewich, A., Atashzar, S.F., Ayad, A., Patel, R.V.: A six-degree-of-freedom robotic system for lower extremity rehabilitation. In: IEEE International Conference on Rehabilitation Robotics. Institute of Electrical and Electronics Engineers. IEEE, August 2015
32. Zhang, F., Fu, Y., Wang, T., Zhang, Q., Wang, S., Guo, B.: Research on sensing and measuring system for a hand rehabilitation robot. In: 2013 IEEE International Conference on Robotics and Biomimetics. Institute of Electrical and Electronics Engineers. IEEE, December 2013
33. Zhang, L.-Q., Chung, S.G., Bai, Z., Xu, D., van Rey, E.M.T., Rogers, M.W., Johnson, M.E., Roth, E.J.: Intelligent stretching of ankle joints with contracture, spasticity. IEEE Trans. Neural Syst. Rehabil. Eng. **10**(3), 149–157 (2002)

Analysis of the Influence of External Actuators on the Glenohumeral Joint Movements

Marie André Destarac[1,2(✉)], Cecilia E. García Cena[1,2], Adrián Mérida Martínez[1], Luis J. Monge Chamorro[1,2], and Roque Saltarén Pazmiño[1]

[1] Centre for Automation and Robotics UPM-CSIC, C/ José Gutiérrez Abascal 2, 28006 Madrid, Spain
manded@gmail.com, {cecilia.garcia,roquejacinto.saltaren}@upm.es, a.merida91@gmail.com, luisj.mchamorro@gmail.com

[2] Aura Innovative Robotics, C/Batalla de Belchite 5, 2ºB, 28045 Madrid, Spain

Abstract. This work presents the simulation and analysis of the influence of MX-64 and RX-64 servomotors from Robotis® in human shoulder movements related to glenohumeral joint for the case of a subject with upper brachial plexus injury. The model of each motor was introduced in a 3D musculoskeletal model of the upper limb in order to compare their response and contribution in three different movements. The length change in muscles, range of movement and muscle force was obtained and compare for a healthy subject and a patient with upper brachial plexus injury. The results demonstrate the feasibility of using these servomotors in an exoskeleton for the rehabilitation process of the injury, although the RX-64 has characteristics that make it more suitable for a rehabilitation exoskeleton.

1 Introduction

Musculoskeletal injuries are one of the leading reasons of disability, the most common cause of severe long-term pain and they affect more than 1.7 billion people worldwide [1]. Disorders of the upper extremities specifically limit the independence of affected subjects and they can cause by excessive stretching, overexertion, strains, accidents, sports injuries and others [2].

The group of nerves that control movement and sensation in the upper limb is the brachial plexus. In children, the principal cause of injury in this nerve structure occurs by stretching or tearing during vaginal birth [3]. In adults is an increasingly common clinical problem produced by occupational and traffic accidents, falls and others causes [4]. Specific treatment should be performed, followed by a rehabilitation therapy.

In recent years, several examples of the use of robotic exoskeletons can be found in the rehabilitation of upper limb injuries [5, 6]. Nevertheless, there are a few examples of them developed specifically for the rehabilitation of patients with brachial plexus injury [7, 8]. An exoskeleton can be used with the intent to make the rehabilitation process more intense and controlled.

I. Chang et al. (eds.), *Advances in Automation and Robotics Research in Latin America*, Lecture Notes in Networks and Systems 13, DOI 10.1007/978-3-319-54377-2_7

In a previous works, we presented a three-dimensional musculoskeletal model of the upper limb [9] in order to study the unilateral upper brachial plexus injury. We also analyzed the influence of two different motors in the elbow flexion-extension movement [10] in order to verify the feasibility of using these servomotors in an exoskeleton for the rehabilitation process of the upper brachial plexus injury [11].

In the literature on arm rehabilitation [12, 13] the movements related to the glenohumeral joint are very important, because they are vital for important activities such as feeding or personal cleanliness.

The aim of this work is to extend the analysis of the dynamic response of those two servomotors and study their influence in the biomechanical movements related to the glenohumeral joint, which is seriously affected by the injury. This study aims to select the optimum servomotor for an exoskeleton taking into account the injury parameters and the musculoskeletal model of the patient.

2 Materials and Methods

Simulink® software was used to modeling and simulate the response of the motors chosen for this study. The rehabilitation therapies for upper brachial plexus injury was simulated with a musculoskeletal model that can be executed in Simulink®.

2.1 Musculoskeletal Model

The model was developed using the 2.2 version of *MusculoSkeletal Modeling Software* (MSMS), a free software created by the University of Southern California [14], which offer different tools to perform animations and simulations of the biomechanical behavior of musculoskeletal models.

Previous work of our research group are based on the development of a musculoskeletal model of the upper limb which have 12 muscles represented by 17 segments [9]. The muscle architecture parameters were configured for a European adult male and the

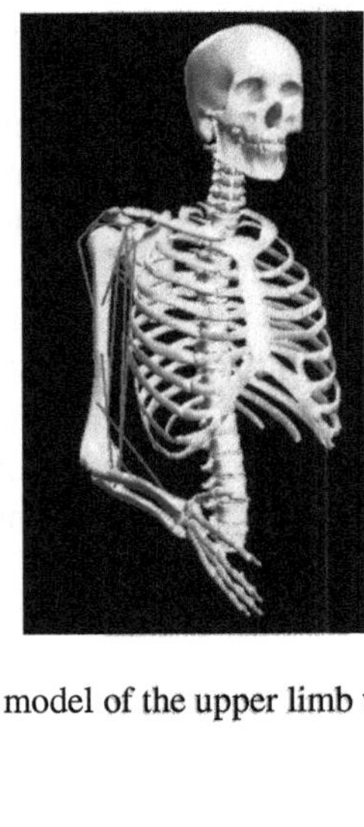

Fig. 1. Musculoskeletal model of the upper limb with 12 muscles and 7 dof.

active joints (degrees-of-freedom) are four: 1 for elbow flexion-extension and 3 for glenohumeral joint (shoulder external/internal rotation, flexion-extension and abduction-adduction).

Each joint range of motion has been restricted according to the values provided by Kapandji [15]. The movement of the scapula and clavicle has been disabled to maintain the stability of the model [16]. In Fig. 1 is shown the 3D musculoskeletal model.

2.2 Servomotors Models

The servomotors selected for this study are the Dynamixel MX-64 and RX-64 of Robotis®. The MX-64 is a position, speed, acceleration, and torque controllable servomotor, which integrates a gain configurable PID controller as an internal control [17].

To modeling the physical components of the MX-64, we distinguish DC motor, a gear and a PID controller. The DC motor is modeled following the classic patterns of a permanent magnet motor, due to the transformation of the input voltage in torque, and also is modeling the electromotive force. We use Simulink® to modeling and simulate the response of both motors [10].

The RX-64 integrate an unknown control method, but the manufacturer gives a diagram with the torque response regarding to the error in angular position, as is shown in Fig. 2 [16].

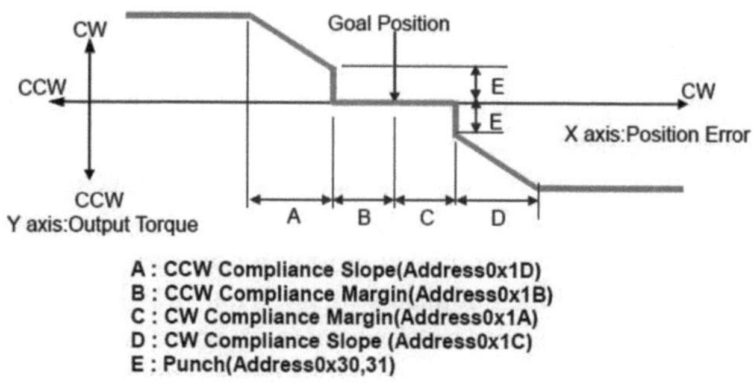

Fig. 2. Compliance diagram of RX-64 servomotor from Robotis [18].

The response offered by this servomotor is incompatible with a PID controller, because the torque response not use the same mathematical function in all the operational phases. Therefore, we can model the dynamic response in Simulink using switches to change the dynamic response of the motor according to the operational phase.

2.3 Procedure for Simulations

MSMS has a simulation tool to convert the musculoskeletal model in a Simulink® blocks diagram. One of the blocks has as input parameters the activation signals of each muscle of the model and will produce (as output) a determined force to move the segment to which they are attached.

After converting the model to its representation in Simulink®, we proceeded to activate the muscles involved in each of the 3 dof of the glenohumeral joint, as is shown in Table 1, to study their behavior and its effect on the dynamics of the musculoskeletal model.

Table 1. Activated muscles for movement simulations

Movement	Agonists muscles	Antagonists muscles
Flexion-extension	Deltoid clavicular	Deltoid scapular
External-internal rotation	Deltoid scapular, Infraspinatus, Teres minor	Deltoid clavicular
Abduction-adduction	Deltoid (all segments), Supraspinatus	Infraspinatus, Teres minor, Teres major

A pulse train was used to activate them, simulating the electrical stimulation technique or FES (Functional Electrical Stimulation) [19]. For all the muscles a pulse train with amplitude of 1 and a period of 2 s, pulse width of 50% and active delay of 1 s was selected. It is important to note that MSMS restricts the range of the excitation signal from 0 to 1, and this range represents the percentage of muscle activation. This means that muscles are active at 100%, and would be the case of a healthy subject.

This means that the agonist muscles shown in Table 1 are activated when the simulations start, while antagonists are disabled. Then, the antagonists are active while agonists are deactivated, and so on until the simulation finalize.

In [20, 21] several experiments were carried out with newly innervated muscles following brachial plexus reconstruction. These subjects showed 20% contraction capacity, compared to a healthy muscle. Rehabilitation aims to gradually increase this percentage. In order to simulate this case, the activation signals were reducing its amplitude to 20%, i.e. 0.2.

To this point, we will have data related to joint torque of a healthy subject and the case of a patient with upper brachial plexus injury. This allows to do comparisons for both cases. The next step is incorporate each motor diagram to actuate in some joints (shoulder external-internal rotation, flexion-extension and abduction-adduction) of the musculoskeletal model for the case of a patient with upper brachial plexus injury, as can

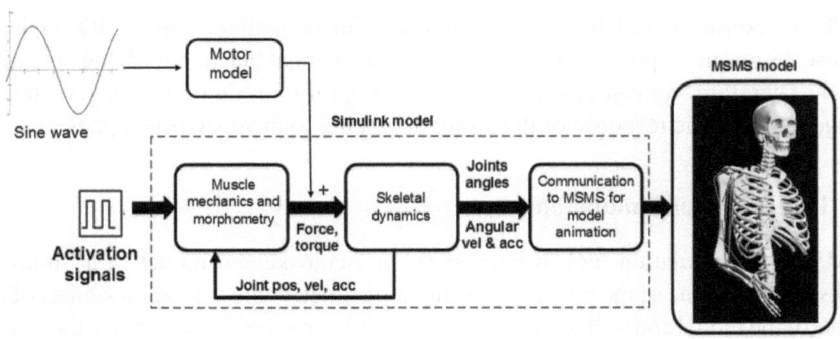

Fig. 3. Musculoskeletal model simulation pathway with the influence of a servomotor response.

be seen in Fig. 3. The output torque from the servomotor model adds to the resultant torque obtained from the muscles activation.

The signal introduced in the motor model have a sinusoidal form with a similar amplitude and frequency of the signal produced by each of the three movements studied in this work.

3 Results

The simulation of shoulder abduction-adduction, flexion-extension and external-internal rotation was made for the case of a healthy subject and a patient with upper brachial plexus injury. Figure 4 shows the range of motion reached in each case for the abduction-adduction of the shoulder and is very obvious the difference between them. For the other two studied movements, the difference is also quite noticeable.

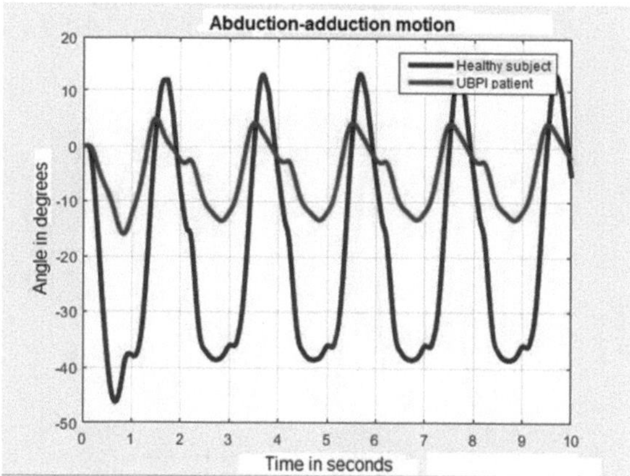

Fig. 4. Simulation of shoulder abduction-adduction for a healthy subject and a patient with upper brachial plexus injury (UBPI).

The arm abduction-adduction movement has a range between 0° and –80°, where 0° represents the neutral position (the arm is down along one side of the body) and –180° represents total arm abduction. However, the musculoskeletal model used did not include all the muscles of the upper limb and the movement of the scapula and clavicle had been disabled. For that reason, for the case of a healthy subject was only possible to simulate the first stage of movement of abduction/adduction which has a range between 0° and –60°, as is shown in Fig. 5.

The next step was to introduce the MX-64 and RX-64 model into de upper brachial plexus injury's musculoskeletal Simulink® blocks diagram. With the objective of analyze their influence in the biomechanical dynamics, we compared the results of their integration into the model, as is shown in Fig. 6.

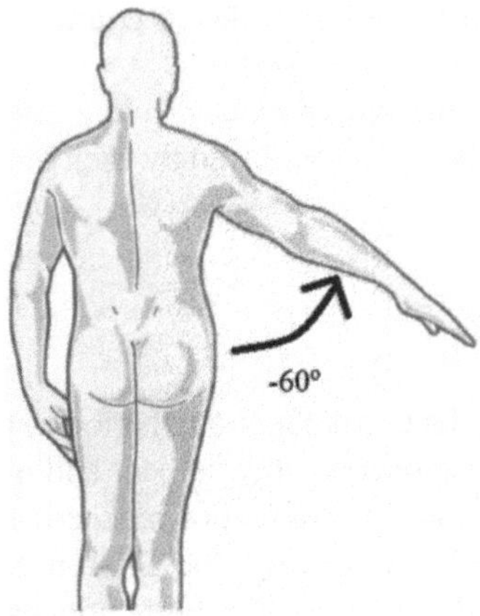

Fig. 5. The first stage of the shoulder abduction, which has a range between 0° to –60° [15].

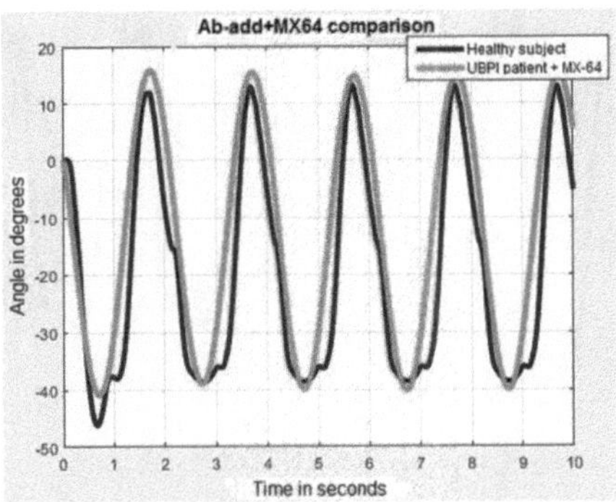

a)

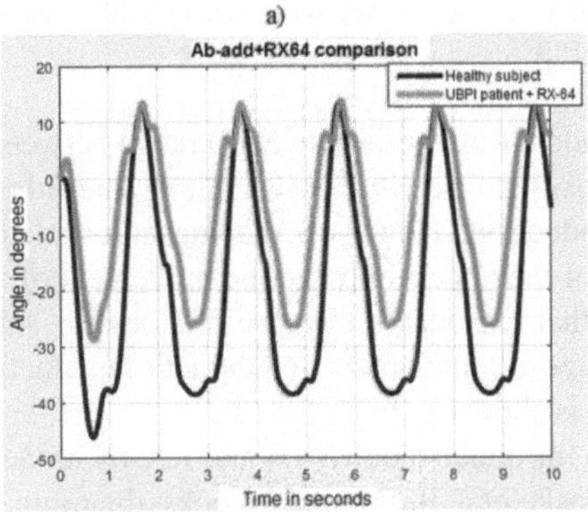

b)

Fig. 6. Comparison of shoulder abduction-adduction for a healthy subject and an upper brachial plexus injury case (UBPI) with the (a) MX-64 and (b) RX-64 influence.

The arm flexion-extension movement has a range between –50 to 180°, where –50° represents the fully extension and 180° represents total arm flexion, as can be seen in Fig. 7. In this work is possible to simulate only the first stage of the movement, which has an approximate range of 0° to 50° for the case of a healthy subject, because the musculoskeletal model does not include all the muscles that perform the movement.

Figure 8 shows the graphs obtained when the MX-64 (Fig. 8a) and RX-64 (Fig. 8b) were introduced to simulate the flexion-extension of the shoulder made for the case of a patient with upper brachial plexus injury. We also simulate the same movement for the case of a healthy subject and in both cases, the movement starts with the arm in the neutral position at 0°.

The arm external-internal rotation has a range between –80 to 90°, where –80° represents the fully external rotation and 90° represents total internal rotation, as can be seen in Fig. 9. To measure the amplitude of this movement, the elbow should be flexed at 90° so the forearm is then in the sagittal plane (Fig. 9a).

The block diagram of each motor was introduced into the musculoskeletal model of the upper brachial plexus injury and the resulting graph was compared with the simulation of a healthy subject, as is shown in Fig. 10.

Other results were obtained in order to do a better comparison of the influence of each actuator in every movement, such the length change in the muscles and the force response.

4 Discussion

Through simulations it was possible to compare the influence of the MX-64 and RX-64 motors on the movements of the glenohumeral joint of a patient with upper brachial plexus injury. In Fig. 4 is obvious the difference in range of motion that can reach a

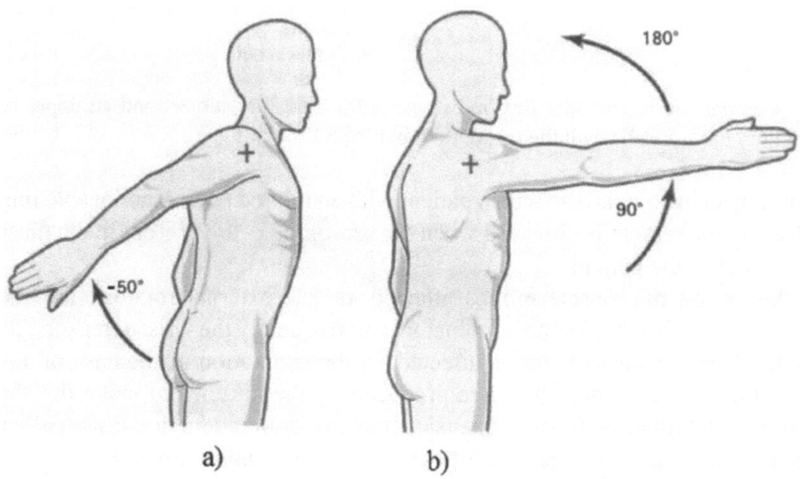

Fig. 7. The shoulder movement of (a) extension and (b) flexion are performed in the sagittal plane [15].

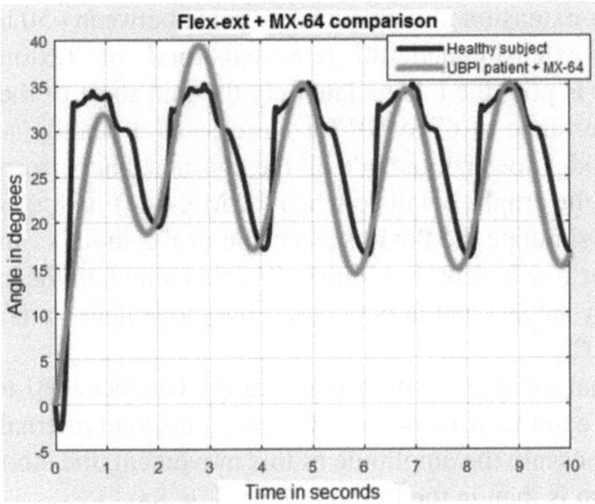

a)

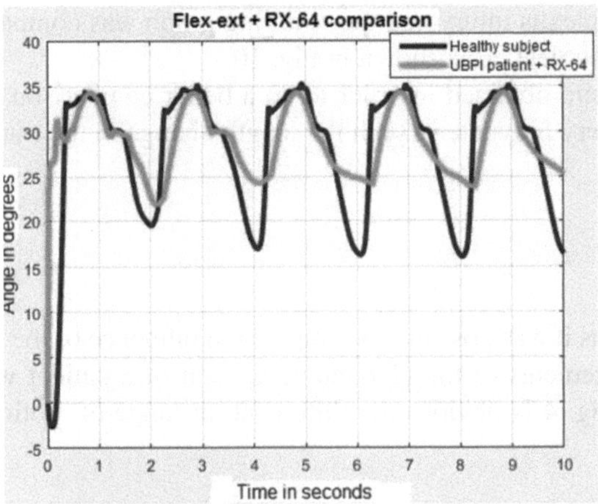

b)

Fig. 8. Comparison of shoulder flexion-extension for a healthy subject and an upper brachial plexus injury case (UBPI) with the (a) MX-64 and (b) RX-64 influence.

healthy subject in comparison with a patient with an injured arm. A noticeable improvement for any movement is observed when the servomotors are present in the musculoskeletal model of the patient.

In this lesion the muscle groups affected are the external rotators, the shoulder abductors, elbow flexors, forearm supinators and frequently the wrist extensors. In other words, the superior musculature is affected. In the simulation of the case of a patient with brachial plexus injury, the force produced by the muscles to make the shoulder external/internal rotation, flexion-extension and abduction-adduction is not sufficient to completely move the arm, which justifies the use of a rehabilitation device.

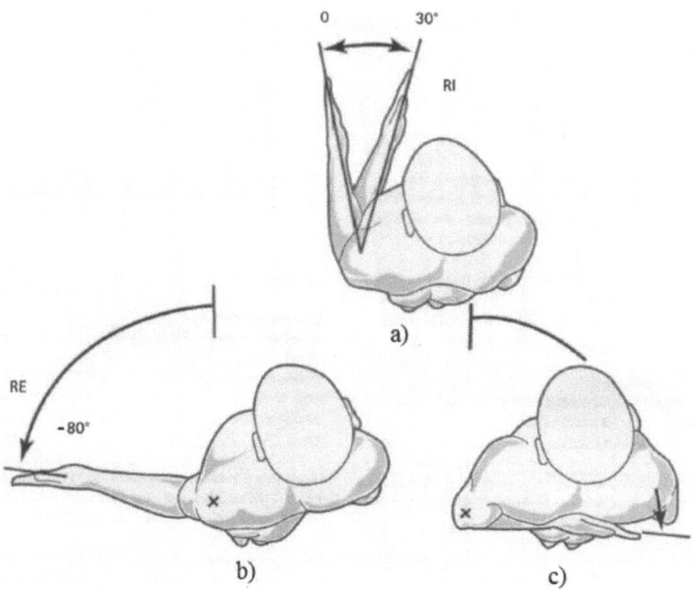

Fig. 9. The shoulder internal rotation (RI) starts at (a) the neutral position in which the elbow should be flexed at 90° so the forearm is then in the sagittal plane. The (b) external rotation (RE) has a maximal range of –80°. In (c) is shown the second stage of the RI, with a total range of a 110° [15].

For healthy muscles, the model shows that they can change enough the length to abduct, lift and rotate the arm in complete range of motions. In the simulation of the case of a patient with upper brachial plexus injury, the length change is lower and the kinematic and dynamic of the movement is seriously affected.

In Figs. 6, 8 and 10 it can be seen the differences in the response obtained with each of the servomotors. The MX-64 offers a quick and proactive response and a high value of compliance is achieved. With this motor is possible to obtain a satisfactory reproduction of the desired movement and an effective correction of path deviations.

The RX-64 model, however, provides an error margin to reach the desire trajectory without trying to correct the error. The width of this margin is configurable, as well the slope compliance. This gives to RX-64 characteristics that place it in a position of advantage over the MX-64 for their implementation in a rehabilitation exoskeleton, because patients not always can accurately perform a movement or fully reach the desire range of motion due of the injury.

The signals showed in Figs. 6, 8 and 10 are stabilized between the second and third peak, because in all cases the first peak has influenced by the initial position of the musculoskeletal model and its inertia. Other point to discuss is why the signal with the motor influence is brought forward in comparison with healthy subject's signal. The reason is that the input signal used as reference in the motor block diagram was a sinusoidal signal manually configured.

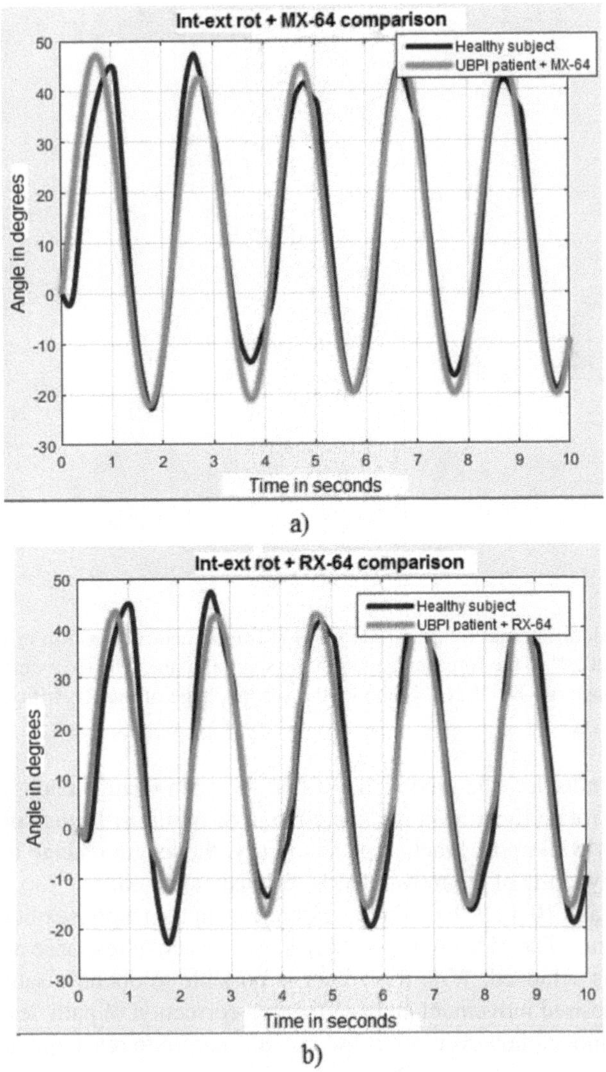

Fig. 10. Comparison of shoulder internal-external rotation for a healthy subject and an upper brachial plexus injury case (UBPI) with the (a) MX-64 and (b) RX-64 influence.

Another important aspect to mention is that in the simulations of the healthy subject cannot be reached the movement ranges established by the medical literature, because the model is incomplete. In the shoulder flexion-extension is most noticeable because are missing most of the muscles that perform that movement.

5 Conclusion and Future Work

Although the developed musculoskeletal model is not complete, since it does not include all the muscles involved in the movements of the upper limb, it is a good prototype which allows different movements, including those relevant to the rehabilitation of the brachial plexus. It also helps to reveal the dynamic and kinematic consequences produced by an affected or atrophied muscle.

The simulation of the contributions of MX-64 and RX-64 motors in biomechanical movements give important results which verified the feasibility of using the servomotors in an exoskeleton for the rehabilitation process of the upper brachial plexus injury.

The compliance characteristics of the RX-64 motor make it more suitable for a rehabilitation exoskeleton. However, as a future work is necessary to make some tests with the physical devices to verify this point. The improvement of the musculoskeletal model is also an important point, adding the missing muscles related to the movement of the shoulder. Other data can be obtained and study, as the joints moments.

Acknowledgments. This work is supported by grants from the Spanish Government CICYT Project Ref. DPI2014-57220-C2-1-P, the Spanish Ministry of Economy and Competitiveness Ref. DI-14-06967 and by the Comunidad de Madrid who supports the project ROBOCITY2030-III Ref. S2013/MIT-2748.

References

1. Murray, C., Vos, T., Naghavi, M., et al.: Global burden of diseases, injuries and risk factors study 2013. The Lancet, 22 July 2014
2. Maciejasz, P., Eschweiler, J., Gerlach-Hahn, K., et al.: A survey on robotic devices for upper limb rehabilitation. J. Neuroeng. Rehab. (2014). doi:10.1186/1743-0003-11-3
3. Sutcliffe, T.: Brachial plexus injury in the newborn. Neo Rev. Offic. J. Amer. Acad. Pediatr. **8**(6), 239–246 (2007)
4. Siqueira, M., Martins, R.: Surgical treatment of adult traumatic brachial plexus injuries. Arq. Neuropsiquiatr. **69**(3), 528–535 (2011)
5. Loureiro, R., Harwin, W., Nagai, K., et al.: Advances in upper limb stroke rehabilitation: a technology push. Med. Bio Eng. Comput. **49**(10), 1103–1118 (2011)
6. Shing, H., Quan, S.: Exoskeleton robots for upper-limb rehabilitation: State of the art and future prospects. Med. Eng. Phys. **34**(3), 261–268 (2012)
7. Slack M, Berbrayer D.: A myoelectrically controlled wrist-hand orthosis for brachial plexus injury: a case study. J. Prosth. Orth. **4**(3) (1992)
8. Ragonesi, D., Rahman, T., Sample, W., et al.: Dynamic simulation and experimental validation of an upper extremity powered orthosis. In: IEEE/ASME International Conference on Advanced Intelligent Mechatronics, pp. 1–6 (2010)
9. Destarac, M.A., García, C.E., Saltarén, R.J., et al.: Modeling and simulation of upper brachial plexus injury. IEEE Syst. J. **10**(3), 912–921 (2016)
10. Monge, L.J., Destarac, M.A., García, C.E., et al.: Modelling and simulation of servomotors for a rehabilitation exoskeleton. In: Fernández, R., Montes, H. (eds.) Open Conference on Future Trends in Robotics. Consejo Superior de Investigaciones Científicas, Madrid, Spain (2016)

11. Monge, L.J., García, C.E., Destarac, M.A., Pazmiñoet, R.S.: Simulation of rehabilitation therapies for brachial plexus injury under the influence of external actuators. In: Ibáñez, J., González-Vargas, J., Azorín, J.M., Akay, M., Ponset, J.L. (eds.) Converging Clinical and Engineering Research on Neurorehabilitation II. Biosystems & Biorobotics (BIOSYSROB), vol. 15, pp. 1043–1047. Springer International Publishing, Cham (2016)
12. Midha, R.: Nerve transfers for severe brachial plexus injury: a review. Neurosurg. Focus **16**(5), 1–2 (2004)
13. Crouch, D., Li, Z., Barnwell, J., et al.: Computer simulation of nerve transfer strategies for restoring shoulder function after adult C5 and C6 root avulsion injuries. J. Hand Surg. Am. **36**(10), 1644–1651 (2011)
14. Khachani, M., Davoodi, R., Loeb, G.: Musculo-skeletal modeling software (msms) for biomechanics and virtual rehabilitation. In: Proceedings of American Society of Biomechanics Conference (2007)
15. Kapandji, A.I.: Fisiología Articular del miembro superior, 6th edn. Editorial Médica Panamericana, Madrid (2007)
16. de Groot, J., Brand, R.: A three-dimensional regression model of the shoulder rhythm. Clin. Biomech. (Bristol, Avon) **16,** 735–743 (2001)
17. Dynamixel MX-64 Product Manual. Robotis Inc.
18. Dynamixel RX-64 Product Manual. Robotis Inc.
19. Davoodi, R.: MSMS User's Manual, version 2.2. Medical Device Development Facility, Biomedical Engineering Department, University of Southern California, pp. 102–106 (2012)
20. Bertelli, J.A., Ghizoni, M.F.: Nerve root grafting and distal nerve transfers for C5-C6 brachial plexus injuries. J. Hand Surg. Am. **35**(5), 769–775 (2010)
21. Midha, R.: Nerve transfers for severe brachial plexus injury: a review. Neurosurg. Focus **16**(5) (2004)

Design of a 3D-Printable Powered Prosthetic Hand for Transmetacarpal Amputees

Renato Mio, Luis Ccorimanya, Kevin M. Flores, Giancarlo Salazar, and Dante Elías(✉)

Laboratory of Biomechanics and Applied Robotics, Pontificia Universidad Católica del Perú, Av. Universitaria 1801, 15088 Lima, Peru
delias@pucp.pe

Abstract. A great number of new proposals for prosthetic hands made by 3D printing have been developed. These prostheses are either body powered for partial hand amputees, or myoelectric powered prostheses for transradial amputees. However, there are no current studies to develop powered 3D printed prostheses for transmetacarpal or partial hand amputees, probably because at this level of amputation there is little space to fit actuators and their associated electronics; nevertheless, it is not an impossible task. For that reason, in this work, it is proposed a design of a hand prosthesis aimed for transmetacarpal amputees and powered by DC micromotors. Additionally, a method for customizing prosthetic fingers to match a user specific anthropometry is shown. Finally, sensors and actuators selection is explained, and a basic control scheme is tested on the prototype.

1 Introduction

In Peru, the main cause for partial hand amputations are work accidents, especially in the manufacture industry [1] Affected workers usually cannot afford conventional mechanical prostheses and even less a myoelectric model. Besides, the state healthcare program does not cover for upper limb prostheses and its reach does not go beyond Lima, the capital city. In consequence, most partial hand amputees adapt to their limb loss and the majority of them has to switch jobs.

There are many open-source designs of 3D-printed body-powered prostheses aimed to help partial hand amputees, such as the Talon Hand [2], Flexy Hand [3], and the Raptor Hand (see Fig. 1). The latter is the most studied regarding its effectiveness in recovering limb functionality [4, 5]. However, studies show that it can perform only basic open/close movements and fails in any grip involving thumb rotation. On the other hand, there are also 3D-printed powered prostheses such as Exiii Hackberry [6], Galileo Bionic Hand [7] and Rehand [8] with actuators that make them more dexterous, but are aimed at transradial amputees. Partial hand amputees should also have the option to choose a prosthesis that allows to move each finger independently and is still accessible.

This work's aim is to develop a powered hand to be used as a prosthesis for transmetacarpal amputees that can be manufactured by means of 3D-printing technology,

I. Chang et al. (eds.), *Advances in Automation and Robotics Research in Latin America*, Lecture Notes in Networks and Systems 13, DOI 10.1007/978-3-319-54377-2_8

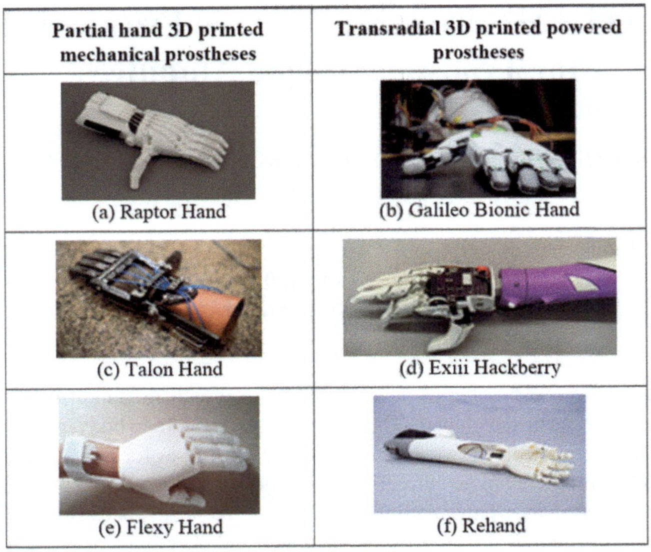

Fig. 1. 3D printed prostheses for partial hand and transradial amputees.

and that still has functionality comparable to state-of-the-art prosthetic hands, thus paving the way for affordable and functional prostheses.

2 Prosthesis Requirements

Prosthetic hand design is a difficult task given the many factors to consider and the trade-offs between them. Some attempt to design highly dexterous and biomimetic hands [9]. However, the number of actuators required make the hand heavy and large, which is not suitable for a prosthesis, at least not until actuator technologies reach the required size without compromising their force. To avoid oversizing, most hand prostheses utilize underactuated finger mechanisms such as cables and pulleys [10, 11], bar linkages [12, 13] or differential transmission [14]. By these means, a single actuator can control more than one degree-of-freedom (DoF) considering geometric constraints for the non-actuated joints.

In the case of transmetacarpal amputees, there is even less space available to fit all the components of the prosthesis. To achieve a functional design, a throughout review of the literature was needed to determine the specifications that the prototype should meet considering the spatial constraints. The requirements considered for the hand prosthesis are meant to ensure functionality and are listed in the following sections.

2.1 Mechanical Requirements

Finger mechanism (finger refers to all digits except for the thumb) for the flexion-extension movement should be underactuated to avoid using many actuators and allow adaptability to grab objects of different sizes. However, adding adaptive finger joints results in less load capacity due to friction and transmission efficiency [15]. A compromise must be done, and therefore the metacarpophalangeal (MCP) and proximal interphalangeal (PIP) joints will be adaptive while the distal interphalangeal (DIP) joint will be rigid.

For the fingers and thumb flexion-extension, transmission should be non-backdrivable to allow blocking and holding a position, thus saving energy. Additionally, it should be compact with a high reduction ratio. The mechanism that meets this requirements is the worm drive [16], which is also used by commercial and research prosthetic hands [17, 18].

The prosthesis force should be sufficient to hold vertically a bottle of approximately 500 g. This is an arbitrary value that empirically ensures the prototype's ability to grab common objects of daily life activities. Torque at the MCP joint of the fingers should be 0.1 Nm on average to be comparable to that of the human hand [15], while the torque for thumb flexion-extension has to be of 0.3 Nm so it can oppose to the forces exerted by the other fingers. Besides, finger open/close velocity at the MCP joint should be comparable to that of the human hand, which is 180 degrees per second (30 RPM).

The prosthetic hand's weight should be of 400 (ideal) to 500 (maximum), as proposed by Belter [19]. Also, hand design should be as anthropomorphic as the space available lets us and dimensions should be based on the specific user.

Fabrication should consist mainly in 3D-printed parts for ease of manufacturing. These parts should be designed for additive manufacturing and their orientation during printing must be carefully chosen to minimize the anisotropy in tensile strength due to the fused deposition modelling fabrication.

2.2 Control Requirements

Position control for each finger at every moment is necessary, and for that reason the control system should be stable at all times and capable to quickly perform and switch between different positions according to the user's needs. Also, the system should be able to handle smooth transitions between each position.

The architecture of the electronic system requires a host controller with a main control loop for the entire system that will send the desired values for the flexion-extension (and rotation for the thumb) angles of each finger to the corresponding local microcontroller. Each local controller will control of a finger independently from the others. The functions to be performed by the local microcontroller are low-level motor control and position reading of each finger.

2.3 Electronic Requirements

The dimensions of the position sensor, its associated electronics and the printed circuit board (PCB) where the electronic components are soldered, should be the minimum possible due to the limited physical space. Also, the actuators should be compact with a high volumetric power and fulfill the requirements of force and velocity aforementioned. The position sensor to be selected has to preferably measure the absolute angular position to minimize reading errors.

3 Design of Fingers and Thumb

The design process is the same for the digits II (index), III (middle), IV (ring) and V (little) due to the fact that they have the same DoF. The design of the digit I (thumb) is treated separately, although it uses the same transmission mechanisms. These designs are explained in the following sections.

3.1 Anthropometry

Fingers and thumb were designed to fit a subject with congenital amputation of hand at transmetacarpal level. Each finger phalanx length and breadth was measured from the user's healthy hand. The dimensions were verified by fitting them in a scaled image of the hand as shown in Fig. 2.

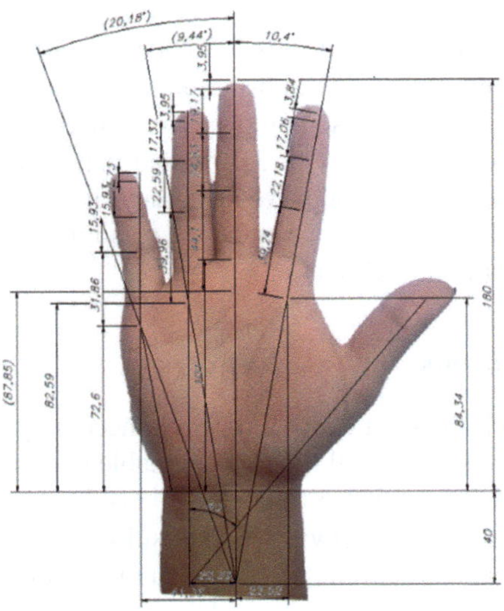

Fig. 2. Dimensions (in mm) measured of a partial hand amputee's healthy hand.

The level of customization of the design should be, nonetheless, limited to just measure the necessary dimensions to match the user's healthy hand. For this matter, dimensions and positions of the fingers were parameterized as described in the next section.

3.2 Finger Parameterization

Customizing the prosthesis fingers will require to define each of the phalanges length and the positions of the finger joints. The process to determine the lengths is as follows:

(a) Each finger proximal phalanx length (pp) is utilized to determine the medial and distal phalanges lengths by using the phalange length as percent of hand length [20].

(b) Length of medial phalanx (mp) and distal phalanx (dp) for each finger are calculated based on the corresponding proximal phalanx length using ratios between them. The ratio from pp to mp to dp are 2.3:1.3:1 for the index, middle and ring fingers; and 2:1:1 for the little finger [21].
(c) The MCP joints position on the coronal or front plane of the hand are calculated from the fingertip position for a standard hand. The positions of the MCP joints in the sagittal plane are determined by the condition that when the fingers are fully flexed all the fingertips lie in the same plane [22].
(d) The computed lengths for the proximal phalanges are then used as input to determine each segment length of a four-bar linkage mechanism, which allows for underactuated motion of the MCP and PIP joints, while the DIP joint is fixed and flexed 20° (see Fig. 3).

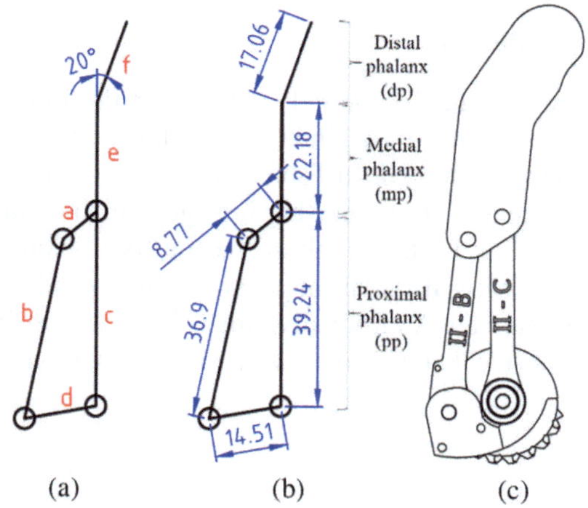

Fig. 3. (a) Four-bar linkage and phalanges dimensions for parameterization. (b) Dimensions in millimeters of the segments determined for the subject's index finger. (c) The dimensions translated to the CAD design of the finger.

The links' dimensions obtained for the parameterized fingers are summarized in Table 1 and a kinematic model was built using MATLAB ® and Robotics Toolbox with this data together with the thumb dimensions (see Fig. 4).

Table 1. Length of links for each finger in millimeters

Link/Finger	Index	Middle	Ring	Little
a	8,77	9,85	8,93	7,12
b	36,91	41,48	37,58	29,96
c = pp	39,24	44,11	39,96	31,86
d	14,51	16,31	14,78	11,78
e = mp	22,17	24,92	22,58	15,93
f = dp	17,06	19,17	17,37	15,93

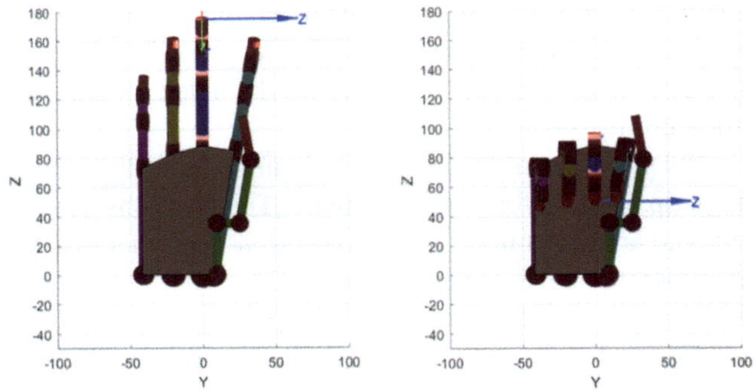

Fig. 4. Hand kinematic model based on the anthropometric data.

3.3 Fingers and Palm Design

The CAD models were designed on Autodesk Inventor Professional 2016 for quick parametric modification of each part. All links were initially designed for one finger and then modified according to the lengths in Table 1. Finger proximal phalanx in Fig. 3c is composed of the links B and C of the four-bar linkage, which include ball bearings on the MCP joint side. The link D (the crank of the four-bar linkage) is coupled to the rotation of the worm gear. Medial and distal phalanges are a rigid element with a cavity for the silicon rubber finger pad. Additionally, link C has a diametrically polarized disc magnet on the MCP joint side that will rotate with link C to measure the joint angle with a Hall Effect sensor.

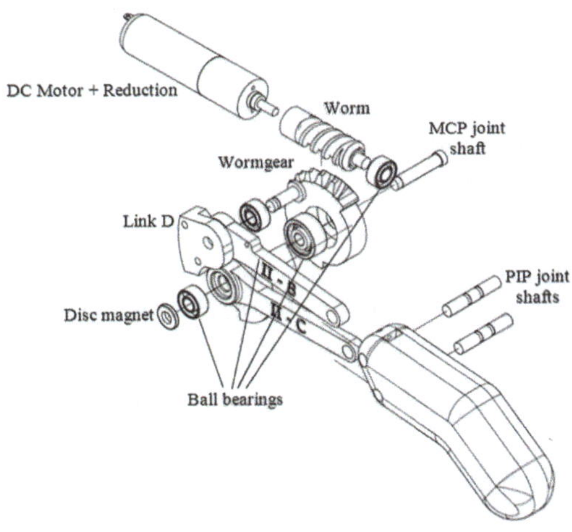

Fig. 5. Finger assembly exploded view.

The movement of each finger is powered by the DC micromotor, whose shaft is coupled with the worm drive mechanism (reduction ratio of 10:1). The micromotor has a reduction gearbox of 16:1, which combined with the worm drive gives an output velocity of 30 RPM and an output torque of 100 mNm. The finger components are shown in Fig. 5.

Thumb components are shown in Fig. 6. In this case the motor for finger flexion-extension (FE) is embedded on the thumb distal phalanx and the abduction-adduction (AA) actuator uses cylindrical gears. The FE DC motor is coupled with the worm's rotation while the worm gear is fixed on the metacarpal (see Fig. 6b). The rotation gears transmission consists of three cylindrical gear with 1:1 transmission ratios. These are located in such a manner that the motor can move the thumb without interference or collision with the user's stump.

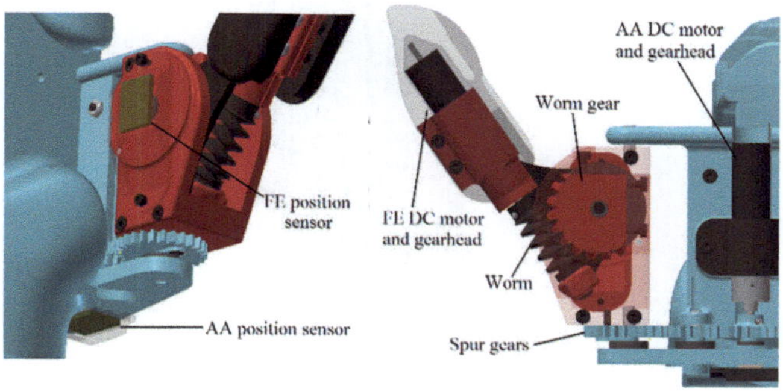

Fig. 6. (a) Thumb positions sensors' locations. (b) Thumb flexion and rotation mechanisms and actuators.

For the design of the palm, MCP joints' positions were calculated and imported to Autodesk Inventor. The palm was then designed in a way that the MCP joint shafts supports are properly located and centered. The back of the hand is designed to hold the actuators, worm and bearings properly aligned with the MCP axis. All the parts were assembled on a structure that mimics the stump and prosthetic socket, which was designed using the 3D scan of the partial hand amputee. Implemented prototype can be seen in Fig. 11.

4 Hardware and Software

The next sections describe the sensors and actuators selected, the electronic architecture and the control system developed.

4.1 Sensors and Actuators

For the absolute position sensing, a Triaxis Rotary Position Sensor MLX90316 from Melexis [23] was selected; it is a Hall Effect sensor with 12 bits of angular resolution and a PWM output from 10% to 90%, which is linearly proportional to a near magnet rotation. A diametrically magnetized Neodymium magnet 6 mm O.D. × 3 mm I.D. × 1 mm thick from First4Magnets [24] was chosen to be the rotary magnet.

As the little, ring, middle and index fingers have one active DoF (FE), and the thumb has two active DoF (FE and AA), there is one position sensor mounted on a 0.8 mm PCB per finger and two for the thumb. In the little, ring, middle and index fingers, the position sensor and the magnet are located in the palm collinear to the MCP joint, as shown in Fig. 7. In the thumb, the position sensor for the carpometacarpal (CMC) joint rotation is located below the first spur gear (the driver) while its magnet is coupled with the gear. The second position sensor together with its corresponding magnet are located in the MCP joint, as depicted in Fig. 6a.

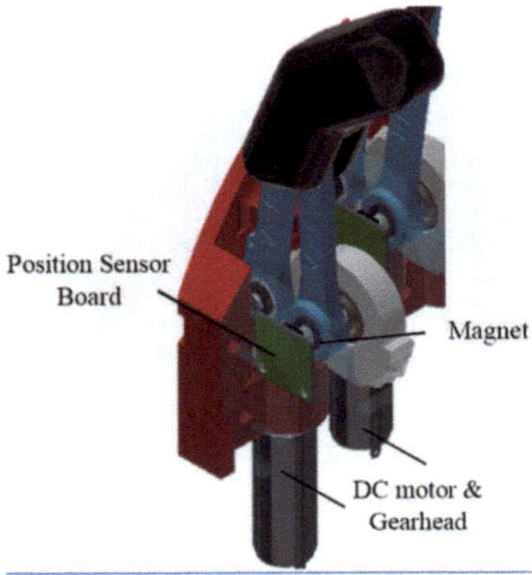

Fig. 7. Localization of the magnet and position sensor for the fingers flexion-extension and the DC micromotors.

During the fingers and thumb movement, the magnet rotates with its respective joint axis while the position sensor is fixed to the hand. In this way, the position sensor is capable of measuring the absolute angular position. In addition, during its calibration, the maximum error was 1°.

The actuators were chosen according to the requested torques for the movement of each finger. It was decided to use a DC motor but due to the fact that the torque needed was relatively high, a gearhead was added. First, to flex/extend the little, ring, middle and index fingers, the DC micromotor Series 1024 SR and the gearhead Series 10/1 16:1, both from Faulhaber [25], were selected. These motors are located in the back of the palm and are coupled with the worms of the worm drive mechanism, as shown Fig. 7. For the thumb rotation movement, the DC micromotor series 1016G and the gearhead series 10/1 256:1 were selected, while for the thumb flexion-extension of a DCX micromotor 10L with a gearhead GPX 10 16:1 from MaxonMotor [26] were selected. These micromotors were located above of the thumb CMC, and above of the thumb MCP respectively, as shown in Fig. 6b.

4.2 Architecture

The architecture of the system is based on a low-cost and high performance microcontroller unit. The block diagram proposed in Fig. 8 includes: the main circuit, finger control circuit and sensing devices mounted on the finger.

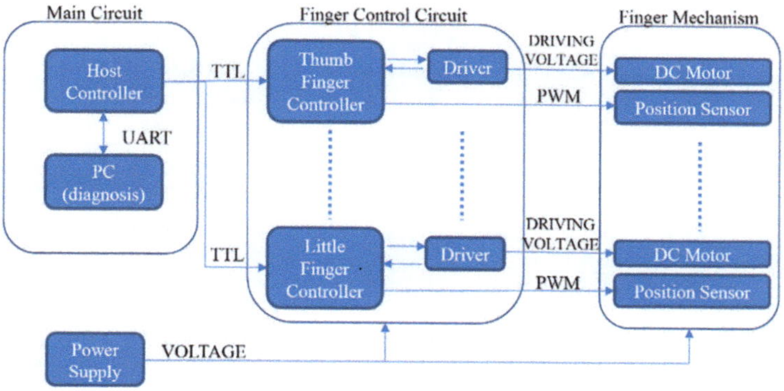

Fig. 8. Electronic system architecture of the prosthesis.

The main circuit consists in a host controller (ATmega328P) that sends control signals to the fingers control circuit and establishes communication with a PC. In the finger control circuit, each finger has a local controller unit (also ATmega328P) and a motor driver (DRV8838). This controller is responsible for the collection and distribution of the information of interest for motion control of each finger. For this task, is important that the sensing devices selected for position control are correctly mounted on the finger. The power supply has two output voltage levels: 6 V for the DC micromotors and 5 V for microcontroller units, motor drivers and position sensors.

4.3 Control and Communication System

The host controller is connected to a computer via UART communication. The computer runs a user interface developed in Visual Studio 2015, as shown in Fig. 9. This interface allows to control the desired movements for the hand, such as flexing (ON) or extending (OFF) a finger to a predetermined final position or moving the whole hand to make a specific gesture. The gestures can be rest position, power grasp, lateral grasp, index and thumb pinch and point with index finger.

The host controller identifies the desired movement that is activated from the interface. After that, the host sends a digital signal to the corresponding local controller. This digital signal triggers one of two movements of each finger: rest position or a predefined final position for a specific gesture.

The local controller, after reading the signal sent by the host controller and obtaining the desired angles of movement for the finger, runs a position control algorithm based on a Proportional-Derivative controller (PD). This control scheme is executed according

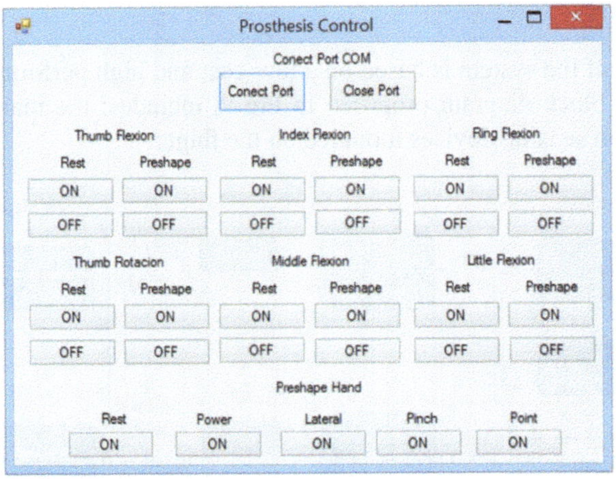

Fig. 9. Position control interface.

to Eq. 1, where K_p is the proportional constant, K_d is the derivative constant, $e(t)$ is the position error and $u(t)$ is the output of the controller.

$$u(t) = K_p e(t) + K_d \frac{de(t)}{dt} \tag{1}$$

The control scheme is shown in Fig. 10. Each local controller implements this control loop independently of each other where θ_d is the desired position, θ_m is the measured position and θ is the real position.

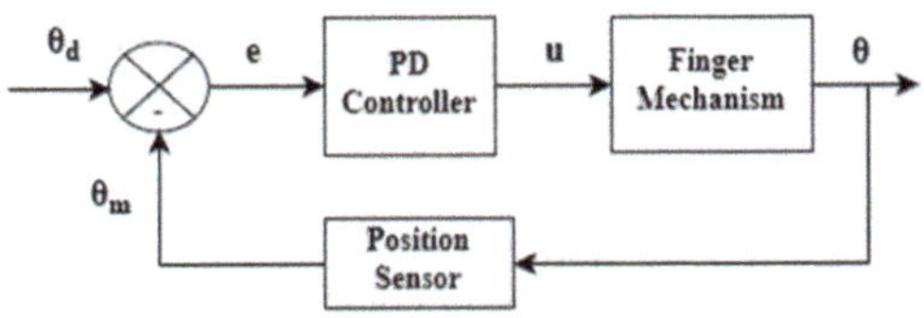

Fig. 10. Position control scheme.

5 Implementation and Results

The 3D printer chosen for fabrication was an Ultimaker 2 Extended+ and the material selected was ABS plastic (24.3% tensile elongation [27]), which is more ductile and less prone to break at impact than PLA (6% tensile elongation [28]), and therefore is more suitable for the prosthesis. The CAD models of the finger links, the palm, shafts, thumb parts and the hand base structure that mimics the stump of the subject were exported from Autodesk Inventor to STL format and then sliced in Cura 2.1.2 with high quality resolution settings (0.1 mm layer height and 100% material density). Other components include DC motors, ball bearings, screws, bolts, Seeger rings and the electronic components. The assembled prototype, including the base structure but without the electronic components that were outside the hand, weighs 364 g, which is below the ideal of 400 g.

The position control algorithm was developed for each local controller by modelling the DC motor, gearhead and mechanism (worm drive or spur gear) for each finger and then, using this model, the parameters of the PD controller were tuned to have a system response time less than 0.5 s. This algorithm has an angular position error of ±1°. For the evaluation, the angles at the MCP joint of each finger were varied changing the target position from 0° (full extension) to 85° (near to full flexion) and for the thumb it was varied from 0° (full extension) to 40° (full flexion). The accuracy of the final positions and the velocity profile were measured by video analysis using markers and Kinovea version 0.8.15, as shown in Fig. 11. The values of the positions measured are shown in Table 2, while the individual finger speeds for the five fingers are summarized in Table 3.

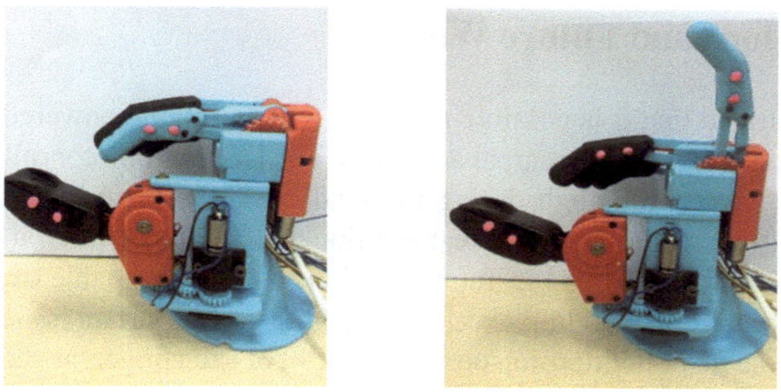

Fig. 11. Flexion and extension velocity measurement for the index finger.

Table 2. Fingers' flexion/extension position measurements

Finger	Number of trials	Average position (°)	σ(°)	Average error (°)
Index flexion	3	86.47	0.12	1.47
Index extension	3	88.8	0.85	3.8
Middle flexion	3	84.7	1.95	-0.3
Middle extension	3	84.83	1.89	-0.17
Ring flexion	3	88.72	1.08	3.72
Ring extension	3	87.52	0.78	2.52
Little flexion	3	85.13	2.12	0.13
Little extension	3	88.37	2.38	3.37
Thumb flexion	3	43.03	0.42	3.03
Thumb extension	3	43.77	1.05	3.77

Table 3. Fingers' flexion/extension speeds

Finger	Number of trials	Average speed (°/s)	σ(°)
Index flexion	3	116.62	2
Index extension	3	156.32	3.39
Middle flexion	3	146.44	2.82
Middle extension	3	110.24	4.26
Ring flexion	3	163.91	5.42
Ring extension	3	160.70	3.78
Little flexion	3	142.82	6.57
Little extension	3	156.40	3.65
Thumb flexion	3	79.22	1.42
Thumb extension	3	128.34	6.93

6 Conclusions and Future Work

It was demonstrated that it is possible to design a 3D printable powered prosthesis for the available space in transmetacarpal amputees. The DC micromotors allow for reliable control of the fingers, although during the evaluation the electric consumption was 150 mA while the prosthesis was in rest position and 2150 mA when it was executing a movement, which means that in future versions a battery capable of supplying this amount of energy without adding excessive weight is an important issue to address.

Regarding the flexion-extension position control of each finger, as shown in Table 2, the maximum position error reached was 3.8° (less than 5%) in the index finger extension. On the other hand, flexion-extension speeds of each finger, shown in Table 3, exhibit values lower than the ideal but still acceptable according to [19]. The ring finger achieves the highest speed while other fingers, like the index and middle, had their speeds reduced due to higher friction forces. The thumb had similar issues with flexion speed. These apparently random variations are mainly caused by fabrication dimensional errors due to the tolerances inherent to 3D-printing and the ABS plastic deformation during its cooling after being extruded. It is clear that this design is functional, but it could be improved by modifying dimensions and changing the design of specific parts to compensate for 3D printing dimensional errors. In the redesign process, the finger links, the palm and other visible parts will also be modified to make the hand more anthropomorphic and aesthetic.

Besides the aforementioned, future work involves upgrading the electronics to fit it with an embedded controller inside the hand. Additionally, in order to evaluate the use of the prototype by a subject, myoelectric signals acquisition and processing will be implemented together with a force-position control to make the prosthesis able to interact with objects of different stiffness.

Acknowledgments. This work was supported by the National Council of Science, Technology and Innovation (CONCYTEC), an entity of the Government of Peru, for a Basic and Applied Research Grant, with grant number 160-2015.

References

1. Camacho, H.T.: Años acumulados de vida productiva potencial perdidos en pacientes amputados por accidentes de trabajo: Instituto Nacional de Rehabilitación 2003–2007. Universidad Nacional Mayor de San Marcos, Lima (2010)
2. e-NABLE: Talon Hand 2.X - Enabling The Future (2015). http://enablingthefuture.org/upper-limb-prosthetics/talon-hand/. Accessed 26 December 2016
3. e-NABLE: The Flexy Hand and Flexy Hand 2 - Enabling The Future (2015). http://enablingthefuture.org/upper-limb-prosthetics/the-flexy-hand/. Accessed 26 December 2016
4. Borjas, R., Flores, W.: Developing a human prosthesis using a 3D printer in Honduras. In: IEEE 35th Central American and Panama Convention (CONCAPAN XXXV), Tegucigalpa (2015)
5. Dally, C., Johnson, D., Canon, M. Ritter, S., Mehta, K.: Characteristics of a 3D-Printed prosthetic hand for use in developing countries. In: IEEE 2015 Global Humanitarian Technology Conference, Seattle (2015)
6. exiii Inc.: HACKberry Open source community (2006). http://exiii-hackberry.com/. Accessed 26 December 2016
7. Fajardo, J., Lemus, A., Rohmer, E.: Galileo bionic hand: sEMG actived approaches for a multifunction upper-limb posthetic. In: IEEE 35th Central American and Panama Convention (COPACAN XXXV), Tegucigalpa (2015)
8. Yoshikawa, M., Sato, R., Higashihara, T., Ogasawara, T., Kawashima, N.: Rehand: realistic electric prosthetic hand created with a 3D printer. In: 37th Annual International Conference of the IEEE Engineering in Medicine and Biology Society (EMBC), Milan, 2015
9. Xu, Z., Todorov, E.: Design of a highly biomimetic anthropomorphic robotic hand towards artificial limb regeneration. In: IEEE International Conference on Robotics and Automation (ICRA), Stockholm (2016)
10. Dechev, N., Cleghorn, W., Naumann, S.: Multi-segmented finger design of an experimental prosthetic hand. In: Proceedings of the Sixth National Applied Mechanisms & Robotics Conference, Toronto (1999)
11. Xu, Z., Kumar, V., Matsuoka, Y., Torodov, E.: Design of an anthropomorphic robotic finger system with biomimetic artificial joints. In: 4th IEEE RAS & EMBS International Conference on Biomedical Robotics and Biomechatronics (BioRob), Roma (2012)
12. Ceccarelli, M., Rodríguez, N., Carbone, G., Lopez-Cajùn, C.: An optimal design of driving mechanism in a 1 degree of freedom (d.o.f.) anthropomorphic finger. Appl. Bionics Biomech. **2**(2), 103–110 (2004)
13. Rea, P.: On the design of underactuated finger mechanisms for robotic hands. In: Advances in Mechatronics, Shanghai, InTech 2011, pp. 131–155 (2011)
14. Liu, H., Meusel, P., Hirzinger, G., Jin, M., Liu, Y., Xie, Z.: The modular multisensory DLR-HIT-Hand: hardware and software architecture. IEEE/ASME Trans. Mechatron. **13**(4), 461–469 (2008)
15. Kargov, A., Pylatiuk, C., Martin, J., Schulz, S., Döderlein, L.: A comparison of the grip force distribution in natural hands and in prosthetic hands. Disabil. Rehabil. **26**(12), 705–711 (2004)
16. Plooij, M., Mathijssen, G., Cherelle, P., Lefeber, D., Vanderborght, B.: Review of locking devices used in robotics. IEEE Robot. Autom. Mag. **22**(1), 106–117 (2015)
17. Vincent Systems GmbH, “VINCENTevolution 2” (2016). http://vincentsystems.de/en/prosthetics/vincent-evolution-2/. Accessed 6 April 2016
18. Touch Bionics: i-limb quantum product sheet, Enero 2016. http://www.touchbionics.com/sites/default/files/files/MA01336%20rev%202%20January%202016%20i-limb%20quantum%20product%20sheet_USsize.pdf. Accessed 16 March 2016

19. Belter, J.T., Segil, J.L., Dollar, A.M., Richard, W.F.: Mechanical design and performance specifications of anthropomorphic prosthetic hands: a review. J. Rehabil. Res. Dev. **50**(5), 599–618 (2013)
20. Frievalds, A.: Biomechanics of the Upper Limbs: Mechanics, Modeling and Muskuloskeletal Injuries, p. 204. CRC Press LLC, Florida (2004)
21. Hutchinson, A.L., Hutchinson, R.L.: Fibonacci, littler, and the hand: a brief review. Hand **5**(4), 364–368 (2010)
22. Gerruti, G., Chablat, D., Gouaillier, D., Sakka, S.: Design method for an anthropomorphic hand able to gesture and grasp. In: 2015 IEEE International Conference on Robotics and Automation (ICRA), Seattle, Washington (2015)
23. Melexis: Melexis, Inpired Engineering. https://www.melexis.com/en. Accessed 25 July 2016
24. First4Magnets: First4Magnets A Division of Magnet Expert. http://www.first4magnets.com/. Accessed 18 August 2016
25. Faulhaber: Faulhaber Drive Systems. https://www.faulhaber.com/en/global/. Accessed 15 July 2016
26. MaxonMotor: DC Motors and Drive Systems by Maxon Motor. http://www.maxonmotor.com/maxon/view/content/index. Accessed 18 August 2016
27. Test Standard Labs, LLC: ABS Data sheet (2014). http://www.teststandard.com/data_sheets/ABS_Data_sheet.pdf. Accessed 28 December 2016
28. Toner Plastics: PLA 3D Filament Data Sheet, March 2016. http://toner-plastics.com/wp-content/uploads/2016/04/PLA-3D-Filament-Data-Sheet.pdf. Accessed 28 December 2016

User Centred Design of Rehabilitation Robots

Jesús Varela[1(✉)], Roque J. Saltaren[1], Lisandro J. Puglisi[1], Javier López[2], María Alvarez[2], and José Carlos Rodríguez[2]

[1] Centro de Automática y Robótica (CAR-UPM), José Gutierrez de Abascal 2, 28006 Madrid, Spain
jvsa@iberdrola.es
[2] Hospital Universitario Infanta Sofía. Servicio de Rehabilitación, Paseo de Europa, 34, 28703 San Sebastián de los Reyes, Madrid, Spain

Abstract. The design of modern rehabilitation robots must be user centred. Apart from safety, robustness and the classical technical aspects (kinematics, dynamics, etc.) other issues related with the patient should be considered in the design phase. The shoulder is a complex joint frequently simplified as spherical, but the alignment of the robot and the shoulder during the movements is relevant to deal with patients that often suffer weakness, pain, etc. Specially, for robotic exoskeletons given that they are "dressed" on the body. Aspects related to shoulder and upper-limb mobility are analysed. Clinical aspects such as the patient's disorder or capabilities and the treatment prescribed impact the design as well as the patient comfort and the possibility of being treated at home instead of in a hospital. All these aspects are reviewed through its application on the rehabilitation exoskeleton SOFI design.

1 Introduction

In 1995 Marc D. Taylor patented [1] a support for the upper-limb fixed to a chair in which the user could practice abductions and shoulder rotations with rehabilitation purposes. Since then, many rehabilitation robots [2–5] have been designed to place the patient in a chair near the robot for the treatment. In 1998, Mark E. Rosheim presented one of the first patents of robotic modern exoskeletons [6] which was used as master-element to control a slave-robot. Others have presented versions of the same concept [7,8] for different applications.

In 2009 a new product with the shape of a big joy-stick or a hand-stick to be used sit in wheels-chair, ReoTherapy [9], was presented by Motorika. The solution is connected to a computer game to motivate the user with interesting activities to move the upper-limb.

MacARM [10] was a structure with the form of a room with a stretcher inside. A set of cables going out the corners controlled a handler that moved the arm of the patient lying down on the litter. Another solution to accommodate the patient on a litter is described in [11] composed of two collaborative robots.

A few rehabilitation exoskeletons have been presented in scientific journals. RUPERT [12] is a pneumatic system with 4 Degrees of Freedom (DoF): one on

I. Chang et al. (eds.), *Advances in Automation and Robotics Research in Latin America*, Lecture Notes in Networks and Systems 13, DOI 10.1007/978-3-319-54377-2_9

the shoulder, two on the elbow and the last one on the wrist. A more complete version of exoskeleton can be found at [13,14]. Powerful actuators equipped with "harmonic drives" tries to solve the alignment problem of robot and shoulder using four actuators.

2 Classification of Rehabilitation Robots

In the available bibliography different classifications of robots can be found [15,16] using as criteria the type of actuators (electric, pneumatic, hydraulic, ...), functionality (increase of strength, rehabilitation, haptic interface, ...), kinematics type, part of the body to which it is applied, etc.

The existing rehabilitation robots could be classified by the functionality taking into account the external aspect, portability, type of exercises that can be done.

1. Industrial-fix robots: robots of high size, weight and cost of industrial aspect, usually installed on the wall or floor of the rehabilitation place. ARMin is one of the most famous robots of this type but there are others [1–5].
2. Cubic-fix type: The aspect is a cubic structure with cables going out from the corners towards a central element that is used to move the limb of the patient lying on a litter [10,11].
3. Joystick type robots: these solutions are usually a "handhold" to be manipulated by the patient. Frequently, they are associated to some kind of game-software making more pleasant the rehabilitation task [9,17].
4. Semi-portable robots: some of them are not too large and appeared fixed on a wheels-chair. This kind of systems is relatively portable considering the mobility of the chair [18,19].
5. Exoskeletons: Wearable robots to be dressed on the body. They are opening the door to home-rehabilitation [12–14].

The exoskeleton SOFI belongs to type 5 designed to be portable, light and to reduce the cost.

3 Design of a Rehabilitation Robot

The design of a rehabilitation robot starts analysing the requirements which basically comes from:

- General needs derived from the problem:
  - Solving the problem of the "shoulder multi-joint complex". The shoulder is usually treated as a spherical joint but it presents mobility clearly far from this simplified approach. Physicians present the shoulder as a complex composed of five joints: gleno-humeral, acromio-clavicular, externo-clavicular, escapulo-torax and sub-acromial joint.
  - The user is a "patient" that presents limitations, disorders, and probably pain in the shoulder and consequently will need careful manipulation, a certain comfort and easy dressing of the system.

- Specific requirements derived from the disorder to be treated. It will impose conditions and restrictions. In a clinical analysis performed in Hospital Universitario Infanta Sofía, some of them raised:
  - Specific "rehabilitation exercises" appropriated for each disorder to be treated.
  - Necessity of working in the area in which the patient feels pain.
  - Necessity of working out of the area in which the patient is able to move the upper-limb actively.
  - Capacity of measuring compensations in certain exercises.
- Other aspects that have to be taken into account:
  - Hiding mechanical parts, electronics and cables. The user should not be afraid of the robot.
  - Ergonomics and adaptability to different sizes of patients, low weight, etc.
  - Robustness. The system will be used by patients and must work without incidences.

Only a few of these aspects will be discussed here, exposing the most interesting conclusions applied to the exoskeleton SOFI design. In this case, it was intended to produce a portable solution to make possible that the patient took the robot to work at home.

The first selected application case was a Brachial plexus. In the birth process some neural connexions were broken and the control of certain muscles in the shoulder was lost. A muscle was removed from its original position in the shoulder zone to reinsert it for a different use. After surgery, the patient has a "new muscle" to move up the upper-limb, but the brain is not yet able to control this new element, and need to be trained for that. Therapists ask the patient to execute certain movements covering a given area or direction for brain education, and after that "strengthen movements" will be necessary. In the case of robot therapy some questions rise:

- How to reinforce the patient when the execution is right (maybe completing the movement, increasing the distance covered) or how to contribute to the improvement of patient control?
- How to generate the trajectory for the patient to follow it?
- How to execute the movement (smoothness, reasonable speed, patient safety and comfort,...)?
- What are the limits inside which the trajectory executed by the patient can be considered valid and when to correct it?

All this questions will receive an answer in the second part of the article. The article is structured as follows: First, the general requirements are addressed focusing on the analysis of the "shoulder complex" issue and how to implement a solution for it. Then, some specific requirements related to the implementation of the exercise are exposed.

4 General Requirements: Shoulder Complex Movement

As introduced before, the movement of the shoulder contributed to the upper-limb movement and needed to be understood and considered when implementing a rehabilitation robot. For this reason the limits of this movement in healthy people were examined. Then, the working space of the person was analysed to impose limits to the robot. Finally, a solution was proposed to follow the gleno-humeral movements appropriately.

4.1 Shoulder Movement Limits

An analysis of the gleno-humeral joint, on 37 healthy persons (Fig. 1) informed that in average the horizontal displacement (Fig. 1(a)) is 13.68 cm for men (std. deviation 2.048) and 12.21 cm for women (std. deviation 2.345). In vertical (Fig. 1(b)) the average displacement of GHJ for men is 9.36 (std. deviation 1.68) and for women is 7.24 (std. deviation 3.18). All these differences have been found significant using Student-t-test with Graph-prism.

Intuitively, with the height of the person the displacements should increase. The data analyzed has shown this in horizontal for men and women, but not so clearly in vertical.

The horizontal movement of GHJ is 12.65 cm (std. dev. 0.57) for people below 1.50 m high while it is 14.83 cm (std. dev. 0.87) for those higher than 1.80 m. The women groups show a similar tendency with movements of 11.79 cm for persons below 1.50 m high while 13.43 cm for women higher than 1.60 m. Note that groups of women higher than 1.70 m are not in present in the experiment. Nevertheless, only a few groups present significant differences compared among them (see Fig. 1(c) and (d)).

Comparing groups of the same height but different sex, men present in all groups larger excursions of the GHJ than women (12.65 vs 11.79 for persons below 1.50 m high; 13.65 vs 11.63 for persons between 1.50 and 1.60 m high; 14.45 vs 13.44 for the group 1.60 to 1.70 m high) but the t-test result not significant, probably due to the low number of persons in each group.

Reviewing the data related to vertical gleno-humeral displacement, it is more difficult to establish conclusions. As explained before, a significant difference has been found between men and women of around 2 cm, but the differences observed among the different height groups analyzed provide no significant t-test results. The difference between average displacements in men of the first and last groups is of 1 cm only. In women the difference is 0.5 cm.

The studies regarding the variability of horizontal and vertical displacement show that elder people suffer a reduction of this capacity while the age groups below 60 years old maintain more or less a normal range of motion.

All together suggest that there is a real difference in horizontal and vertical displacement between men and women and a certain variation with the height of persons, at least for the horizontal range of movements, but with an important variability (Fig. 2). Consequently, robotic systems must take this into account

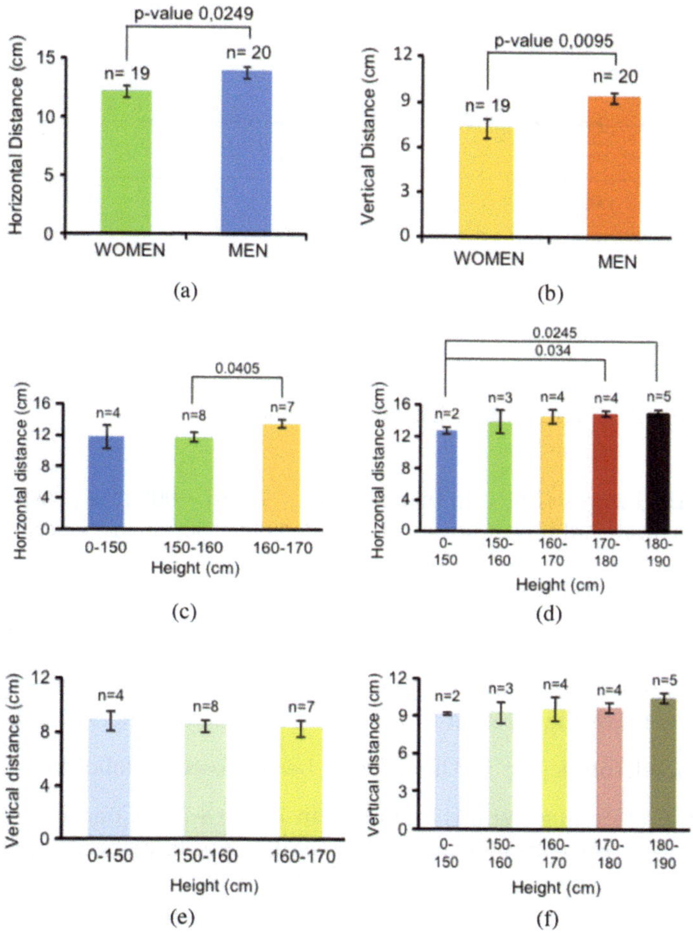

Fig. 1. Gleno-humeral horizontal and vertical displacement limits. (a) horizontal. (b) vertical. (c) Horizontal displacement by age groups for women. (d) Horizontal displacement by age groups men. (e) vertical displacement by age groups for women. (f) vertical displacement by age groups for men.

offering the possibility to adjust person by person the necessary excursion of the gleno-humeral joint.

4.2 Working Space of the Upper-Limb and Robot

The working space of the upper-limb can be described as follows:

- Area I: Main volume (Fig. 3 in green): from -30° to $+135^o$ approximately in the transversal plane. Sportsmen or normal people with external support (therapist, pressing against a wall,...) are able to overpass these limits.

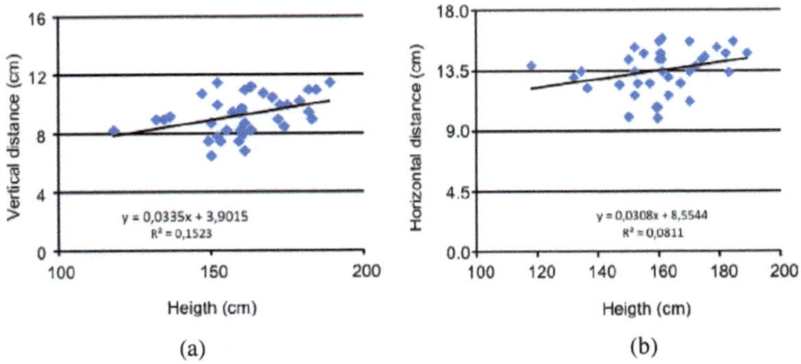

Fig. 2. Horizontal (a) and vertical (b) tendency of variation of the gleno-humeral excursion distance in function of the height of the persons.

- Area II: Backwards (Fig. 3 in black): from -30° to -90° it is possible to move the upper-limb holding it in an elevation below 80° using internal shoulder rotation.
- Area III: Back (Fig. 3): the forearm can be place behind the back and move there placing the humerus down and backwards. It is also possible to move the forearm behind the head placing the humerus up.
- Area IV: Frontwards (Fig. 3 in blue): After 135° in the transversal plane, it is possible to move the forearm to touch the opposite shoulder, the head, etc.

For the exoskeleton design the limits have been chosen as follows:

- Area I: Elevation in general is limited from 75° to $+75^o$. Concerning the movements in the transversal plane, the general limits have been adopted (from -30° to 135°). The reason to impose the upper limit ($+75^o$) is just safety for the patient. Many healthy people have problems to arrive there, so it was not really a necessity for the treatments but was a risk for the patients. It was different regarding the lower limit. The mechanical solution to cover all other volumes and also below -75° was complex and usually therapists do not work below -75° because this zone is normally well covered by patients. There was a known exception to this case that was addressed with other robot for safety reasons.
- Area II can be covered by the exoskeleton at any height, but no treatment was foreseen there, and there was a risk for the users in the upper part. Consequently it was mechanically blocked for safety reasons.
- Area III is important for certain treatments of patients analysed in the clinical studies. The mechanical design has been done to be able to cover these areas.
- Area IV is also covered given that no special risk is foreseen in this zone.

Other restrictions applied are:

- Elbow flexion between 0° and 150°. Mechanically it is easy to overpass these limits, but it is a risk for the patient and was not required for therapists and physicians.

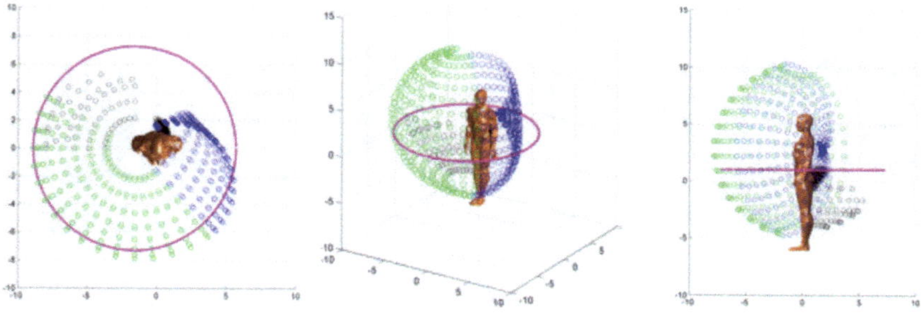

Fig. 3. Impact of the gleno-humeral joint in the volume covered by the right upper-limb. (a) top-view. (b) isometric view. (c) sagittal view.

- Shoulder rotation between -110° and $+110^o$. Usually human shoulder rotation with humerus in horizontal position and forearm perpendicular to it do not overpass -85° and $+110^o$. When humerus is below the horizontal plane the new limits are more closed to -90° and $+90^o$. A certain excess has been tolerated for special exercises.

In Fig. 3 the effect of shoulder joint is visible. The green area exceeds the limits of a circle on the frontal part, and a bit backwards.

4.3 Sliding Elbow Structure

While the humerus is progressing on the transversal plane from the horizontal position aligned with the shoulders line (both contained in the frontal plane) towards a more advanced position in the sagittal plane the GHJ moves frontwards. Something similar happened in the elevation movement. When the upper-limb goes up in the frontal plane, the GHJ also progress upwards.

In order to be able to follow the elevation of the GHJ during the upper-limb movements, the exoskeleton was equipped with a linear degree of freedom (Fig. 4 - J2). For the horizontal movement described of GHJ, a different solution was adopted. Instead of implementing a new degree of freedom around the shoulder, this complexity was transferred to the "elbow platform". A platform was designed to support the weight of the limb.

The first problem was to select a place that could be valid for patients of different ages and heights. The humerus length was different in each case and the appropriate place to install the support platform should be found. Installing the elbow at the end of this platform seemed to be a good idea, to avoid additional structures and weight. Making this platform able to slide over a guide it was possible to absorb the movements of the GHJ sliding the elbow along the guide, and removing the complexity from the shoulder to place it in the new element at the elbow.

Finally, it was decided to let this degree of freedom out of control. With an actuator the platform can be placed in the right place at any moment, but

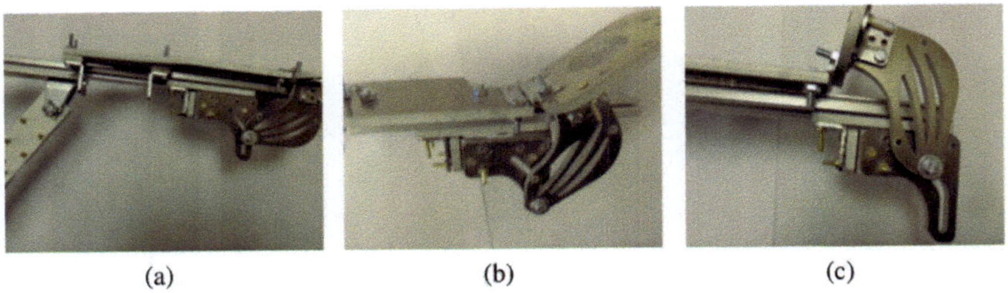

Fig. 4. Sliding elbow structure

allowing the patient to be free to move it, the alignment of shoulder joints and robot mechanical ones is complete. In fact this movement is sensed to detect abnormal postural compensations of the patient during the execution of the rehabilitation exercises.

4.4 General Kinematic Design

The exoskeleton SOFI is made of aluminium and plastic, with a short set of steel pieces. It is actuated by Maxon motors, and equipped with economic sensors. The misalignment problem in the shoulder was solved incorporating two elements: a "sliding elbow system" (patented) [20] which makes possible to transmit part of the complexity of the shoulder to the elbow area, and a sliding piece in the axilla that permits a fine regulation of the position of the rotational axis taking into account the displacements of the connexion between humerus and shoulder.

The exoskeleton had five actuated joints and six non-actuated ones. The harness that fixed the exoskeleton to the body was flexible and adjustable to a wide range of body sizes. To support the arm, there was a mechanical structure (Fig. 5) which was connected to the harness through an actuated joint (J1) that allowed moving the structure forward and backwards in the transversal plane. This actuator combined with the elevation one (J3) enabled the exoskeleton to execute the basic movement of the upper-limb on the three axes.

The third actuated joint (J2) was placed on the axilla. It is a sliding piece that moves the elevation axe (J5) up and down to follow the natural movements of the shoulder maintaining the arm in a comfortable position at any elevation posture. The following element that provides shoulder alignment was the "sliding elbow" (J7). This element, which was able to slide on the "arm structure", contained several sub-elements: the support for the arm and the elbow and forearm structures. This linear joint was not actuated and provided at the same time:

- The alignment of the shoulder with the arm when the GHJ displaced horizontally, independently of the angle of elevation of the arm.
- The self-adaptation for different body dimensions, and particularly for the size of the arm.

A sensor measured the linear displacement of this platform in relation to the support structure. The forearm structure also had sliding piece to adjust

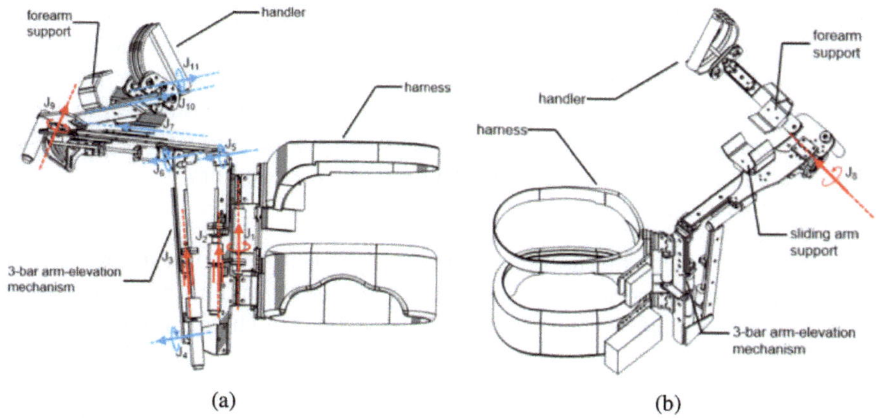

Fig. 5. Kinematic topology of the robotic rehabilitation exoskeleton SOFI

the handle to different lengths of forearm (J10). The handle adds an additional degree of freedom (J11) not actuated for forearm rotation. It contributes to accommodate comfortably the arm to different positions.

The elbow structure contains two actuators for the shoulder rotation (J9) and elbow flexion-extension (J8). The design allowed deep internal and external rotation to make possible the exercises previously described.

Safety was ensured through several means: a general stop-switch in an eventual case of emergency situation, mechanical limits controlled with end-switches, software security limits established in the electronics, software limits stablished by the therapists and doctors, and gear systems connected to worm screws which provide low speed movements while large forces using small actuators. Electric ones were selected instead of hydraulic also for security reasons among others. A hydraulic actuator was very strong and could hurt the patient in case of malfunction. Additionally, it produced noise; the maintenance was dirty and it needed space for the compressor and tubes. Pneumatic actuation presented similar problems of space and noise, as it can be seen at [21] and it was less controllable than hydraulic one.

4.5 Specific Requirements

In previous sections the need of specific requirements was stablished. Once the shoulder disorder to treat is selected (brachial plexus), it should be decided the protocol to follow to obtain the best possible progress for the patient. In this case, physicians and therapists propose a procedure to re-educate the brain in the control of the "transplanted" muscle.

1. They would establish a trajectory to be followed by the patient that would be programmed in the robot.
2. The robot would execute the trajectory twice to teach the patient about what it is expected from him.

3. The patient would try to do the movement by himself. If the patient executes correctly the exoskeleton followed the trajectory stablished by the patient. If the error in the trajectory is more than the limits stablished by therapists, the robot corrects the trajectory teaching the patient. The limits were wide at the beginning but were reduced more and more once the patient progressed. The alternative of stopping when the patient's trajectory is not good was discarded to avoid discouraging him.

The first need was a method to generate the trajectories to be followed by the patient and to introduce them in the robot in an easy way. This was easily solved with a system composed of two inertial units placed in the arm and forearm of the physician or therapist to register the movements executed by them.

The result can be observed on computer screen to verify if it is necessary to repeat the movement or not. Usually therapists register 4 similar trajectories and they select in the visualization phase the preferred one. Part of the verification process is to ask the exoskeleton to execute the movement without patient to ensure that the movement is fine and safe for the patient, but this require a pre-processing.

The trajectory is registered as a couple of four components vector (quaternions), because it is an efficient format from the transmission point of view. Each matrix of quaternions is converted to a coordinates x,y,z, taking into account if it is a right or left upper-limb.

The inverse kinematics of the robot is used to calculate the commands for the actuators. The sequence is analyzed to eliminate the high speeds and accelerations without producing important changes on the initial trajectory (Fig. 6).

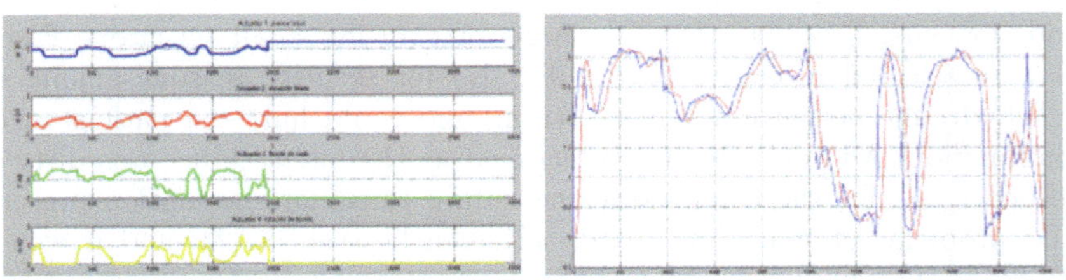

(a) (b)

Fig. 6. (a) Trajectory for the patient converted in a sequence of commands for actuators. (b) filtered result used to eliminate important changes in the acceleration Kinematic topology of the robotic rehabilitation exoskeleton SOFI

5 Conclusions

The design of rehabilitation robots offer many possibilities, but it involves dealing with patients; persons with physical disorders, frequently suffering pain and limitations different from the standard ones of healthy persons. This article deals about patients with shoulder disorders and the design of the rehabilitation exoskeleton SOFI.

The main general concerns when dealing with shoulders are the alignment of the robot with the human joint to avoid discomfort or pain and consequently the limits of the movements of the GHJ.

A study about the vertical and horizontal mobility of the GHJ has been presented establishing these ranges the standard deviation and the percentage that this suppose in the global upper-limb movement. Based on these results the alignment problem has been analysed and a solution has been proposed. An additional degree of freedom has been included in the shoulder zone while the other has been transferred to the upper-limb support platform that slides over a guide parallel to the humerus. The solution has been object of a patent under the name of "sliding elbow".

A simulation analysis of the working space, based on the results obtained from the clinical analysis about GHJ motion limits, of human shoulders provided a first approach to the safe and unsafe volumes. These frontiers between the safe and unsafe zones could be modified depending on the shoulder disorder to be treated and the exercises proposed by therapists.

In the design of the rehabilitation robots it is so relevant the robot design as the "rehabilitation exercise" design. The best method to improve the performance of the patient is a good exercise design. The case of a "brachial plexus" has been analysed proposing a protocol to treat it, based on tasks that are easy at the beginning and increase the complexity gradually, but trying at any moment to propose achievable targets to encourage the patient with positive reinforcements.

The exercise designed for the study-case selected involves the use of movement trajectories adapted for a particular patient. The problem and calculations (inverse kinematics, etc.) can be simplified using appropriated sensors to generate the trajectory and introducing minor restrictions in the robot design.

As explained before, safety must be considered in every design phase. Hardware limits, mechanical gears, actuators selection as well as software limits should be stablished to protect the patient and the hardware. The movement speed and the accelerations when moving the limbs of ill persons must be carefully stablished. On top of this, the doctor should have the capability to limit the working volume or even exceed the limits in certain cases. Finally, the robot should be able to evaluate the capabilities of the patient every day before starting each session for additional safety.

Other aspects as robustness, ergonomics and comfort should also be considered from the design phase.

References

1. Taylor, M.D.: US 5,417,643. Continuous passive motion exercise device. Danninger Medical Technology Inc., Columbus Ohio
2. Quintinskie Jr., J.J.: US 6,007,500. Shoulder, rotator cuff, and elbow stretching machine

3. Zemlyakov, V., McDonough, P.: US 20030115954 A1. Upper extremity exoskeleton structure and method
4. Nef, T., Mihelj, M., Kiefer, G., Perndl, C., Muller, R., Riener, R.: ARMin- exoskeleton for arm therapy in stroke patients. In: IEEE 10th International Conference on Rehabilitation Robotics, Noordwijk, pp. 68–74 (2007), doi:10.1109/ICORR.. 4428408
5. Tsai, B.C., Wang, W.W., Hsu, L.C., Fu, L.C., Lai, J.S.: An articulated rehabilitation robot for upper limb physiotherapy and training. In: IEEE/RSJ International Conference on Intelligent Robots and Systems, Taipei, pp. 1470–1475 (2010)
6. Rosheim, M.E.: US 5,845,540. Robotic Manipulator. Ross-Hime Designs Inc.
7. Kim, M.S.; Lee, S.Y.; Lee, C.W.: US 6,301,526 B1. Master device having force reflection function. Institute of Science and Technology, Seoul (KR)
8. Schiele, A.; Visentin, G.: E03291149. EPO E03291149. Exoesqueleto para un brazo humano, Particularmente para aplicaciones espaciales
9. http://motorika.com/product-1/
10. Mayhew, D., Bachrach, B., Rymer, W.Z., Beer, R.F.: Development of the MACARM–a novel cable robot for upper limb neurorehabilitation. In: 9th International Conference Rehabilitation Robotics, ICORR 2005, pp. 299–302 (2005), doi:10.1109/ICORR.2005.1501106
11. Morales, R., Badesa, F.J., Domenech, L.M., Garca-Aracil, N., Sabater, J.M., Menchn, M., Fernandez, E.: Design and control of a rehabilitation robot driven by pneumatic swivel modules. In: 3rd IEEE RAS EMBS International Conference Biomedical Robotics and Biomechatronics, Tokyo, pp. 566–571 (2010), doi:10.1109/ BIOROB.2010.5626919
12. Sugar, T.G., He, J., Koeneman, E.J., Herman, R., Huang, H., Shultz, R.S., Herring, D.E., Wanberg, J., Balasubramanian, S., Swenson, P., Ward, J.A.: Design and control of RUPERT: a device for robotic upper extremity repetitive therapy. IEEE Trans. Neu. Sys. Reh. Eng. **15**(3), 336–346 (2007), doi:10.1109/TNSRE. 2007.903903
13. Carignan, C.R., Liszka, M.S.: US 7,862,524 B2. Portable arm exoskeleton for shoulder rehabilitation
14. Carignan, C., Liszka, M.: Design of an arm exoskeleton with scapula motion for shoulder rehabilitation. In: 12th International Conference on Advanced Robotics, Seattle, WA, pp. 524–531 (2005), doi:10.1109/ICAR.1507459
15. Guizzo, E., Goldstein, H.: The rise of the body bots [robotic exoskeletons]. IEEE Spectr. **42**(10), 50–56 (2005). doi:10.1109/MSPEC.2005.1515961
16. Gopura, R., Kiguchi, K.: Mechanical designs of active upperlimb exoskeleton robots: state-of-the-art and design difficulties. In: IEEE International Conference Rehabilitation Robotics, Kyoto, pp. 178–187 (2009). doi:10.1109/ICORR.5209630
17. Qiu, Q., Ramirez, D.A., Saleh, S., Fluet, G.G., Parikh, H.D., Kelly, D., Adamovich, S.V.: The new jersey institute of technology robot-assisted virtual rehabilitation (NJIT-RAVR) system for children with cerebral palsy: a feasibility study. J. NeuroEngineering Rehab. **6**(1), 40. doi:10.1186/1743-0003-6-40
18. Vitiello, N., Lenzi, T., Roccella, S., Rossi, S.M.M., Cattin, E., Giovacchini, F., Vecchi, F., Carrozza, M.C.: NEUROExos: a powered elbow exoskeleton for physical rehabilitation. IEEE Trans. Rob. **29**(1), 220–235 (2013). doi:10.1109/TRO.2012. 2211492

19. Housman, S.J., Le, V., Rahman, T., Sanchez, R.J., Reinkensmeyer, D.J.: Arm-training with T-WREX after chronic stroke: preliminary results of a randomized controlled trial. In: IEEE 10th International Conference Rehabilitation Robotics, Noordwijk, pp. 562–568 (2007). doi:10.1109/ICORR.4428481
20. Varela Sanz, J.; Saltaren Pazmio, R.: ES 2544890 B2. Exoesqueleto robotizado con soporte de codo deslizante autoajustable para brazo humano
21. Ueda, J., Ming, D., Krishnamoorthy, V., Shinohara, M., Ogasawara, T.: Individual muscle control using an exoskeleton robot for muscle function testing. IEEE Trans. Neu. Sys. Reh. Eng. **18**(4), 339–350 (2010). doi:10.1109/TNSRE.2010.2047116

Towards Tracking a Semi-autonomous, Pneumatic Colonoscope Robot

Bradley Woosley[1(✉)], Prithviraj Dasgupta[1], Hossein Dehghani[2], Ross Welch[2], José Baca[1], Carl Nelson[2], Benjamin Terry[2], and Dmitry Oleynikov[3]

[1] Computer Science Department, University of Nebraska at Omaha, Omaha, USA
{bwoosley,pdasgupta,jbacagarcia}@unomaha.edu
[2] Mechanical and Materials Engineering Department, University of Nebraska-Lincoln, Lincoln, USA
{cnelson5,bterry2}@unl.edu
[3] Department of Surgery, University of Nebraska Medical Center, Omaha, USA

Abstract. We describe a semi-autonomous tracking technique based on a Bayesian prediction mechanism for a pneumatic colonoscopy robot. One of the principal problems in using a robot colonoscope is to determine its position or displacement after it has been inserted into the colon. To address this problem for our robot, we have developed a temporal Bayesian framework that uses the observations of the air flow rate and pressure in the tube guiding the robot to predict the current location of the robot. Our experimental results show that the predicted location and velocity of the robot are accurate over most of the time-steps, with a maximum error of 11.5 cm (about 2.5 times the length of the robot's head) occurring only 7% of the time.

1 Introduction

Robotic colonoscopies have been experiencing a growing demand over the last decade in the hope for non-invasive diagnostic and therapeutic colonoscopy. The American Cancer Society estimates a high incidence rate of colorectal cancer, with more than 14,000 new cases each year in the United States. Early diagnosis significantly reduces the risk of developing colorectal cancer, but the pain and discomfort experienced during colonoscopy usually discourages patients, delaying the diagnoses. To address this problem, our paper uses a novel, pneumatic robot [5] that diminishes the shear forces between the colonoscope (robot tip) and the colon wall. It is important for the doctor to know the location of the colonoscope inside the colon to be able to perform acurate biopsies, and for making correct diagnoses. With a traditional colonoscope, this is accomplished via length markers on the outer surface of the scope. However, placing such markers might not be feasible on all types of colonoscopy robots. For example, the robot discussed in this paper uses an initially spooled, flexible and expandable tubing for inserting the robot tip inside the colon, which is not amenable to markings. To address this problem, we propose a tracking method that predicts the velocity and displacement of colonoscope robot tip after it has been inserted

I. Chang et al. (eds.), *Advances in Automation and Robotics Research in Latin America*, Lecture Notes in Networks and Systems 13, DOI 10.1007/978-3-319-54377-2_10

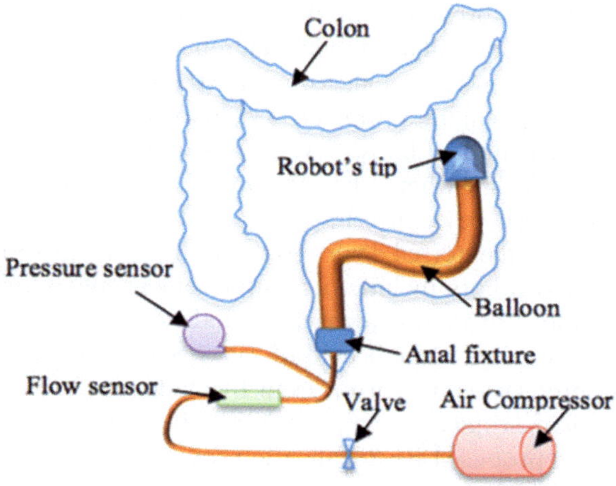

Fig. 1. A general schematic of our proposed technique for tracking the colonoscope robot.

into the colon. A general schematic of our proposed robot tracking technique is shown in Fig. 1. We have presented experimental results of tracking inside a Syndaver colon and from insertion of our proposed robotic colonoscope within a swine intestine along with predicted data from our tracking algorithm. Our results show that the tracking algorithm can predict the robot's displacement with a maximum error of 4.5 cm ($<=$ length of the robot's head) for over 50% of the time-steps, and with an error between 7.5–11.5 cm (1.5–2.5 times the length of the robot's head) for about 35% of the total time-steps, when the robot has proceeded towards the end of the colon.

2 Related Work

Robotic colonoscopy has been an active area of research in the past decade and a comprehensive overview of the existing technologies for colorectal endoscopy is available in [10]. Many researchers have proposed novel designs for colonoscope robots that enables to maneuver efficiently within the constrained environment of the human intestine. Qiurini *et al.* proposed a 12-legged micro-capsule endoscope [12], while a microcapsule robot with a legged anchoring mechanism was described in [6]. To prevent the robot's motion from getting impeded by intestinal walls and bends, a colonoscope robot with a 3-DOF maneuverable tip that can be actuated pneumatically was proposed in [3]. Other novel robot designs and acutation mechanisms include an inchworm robot with hollow body and steering including camera and light source [8], peristaltic crawling robot [1], and wireless anchoring and extending microrobot [9]. In addition, several robotic technologies for endoscopy have been commercialized including Endotics (Caltech), Invendoscope (Germany), Aer-o-scope and Colonosight (UK). In contrast to the above works, the design presented in this paper is unique in the way that it locomotes based on balloon inflation, while eliminating the relative motion, and hence the friction, between the colonoscope and the tissue.

One of the most challenging problems of colonoscope robots is how to track their position inside the human body. Several researchers have proposed wireless capsule endoscopes (WCE) where the location of the capsule inside the body is tracked using magnetic resonance imaging (MRI) [7]. However, MRI involves considerable costs and an external device to track the robot. Alternative tracking techniques for in-vivo robots include computer tomography, X-ray fluoroscopy [2] and 3D ultrasound. But their drawbacks include low image quality, distortion from rigid instruments and a small field of view. Several researchers have proposed using computer vision algorithm from live video captured from a camera mounted on the colonoscope's tip to provide reliable tracking for the robot [4]. In [11], authors have proposed RF-localization for determining the position of a WCE, while fusing RF data with camera images has been reported to provide more accurate localization for a WCE inside the human small intestine.

3 Colonoscope Robot

The pneumatic robot is composed of a 3D-printed tip, an anal fixture and latex tubing. Figure 2 shows the robot's tip composed of a clamshell-type structure, a sealing mechanism, a shaft and two bearings. The sealing mechanism is made of two shafts and each shaft has three bearings, as shown in Fig. 3(a). The shafts are configured to compress the tube, preventing air passage to the tip (Fig. 3(b)). Admitting the compressed air into the tip may cause the tip or tubing rupture. The tubing is spooled around a shaft, a small portion of which is placed between the robot's tip and the anal fixture (Fig. 3(c)). Therefore, the robot's tip is propelled as the pressurized air inflates the tube (Fig. 3(d)), unspooling more tubing from the robot tip. Figure 4 shows the robot's tip inserted in an excised swine colon; the anal fixture serves as an anchor while the tip advances. The outer diameter and length of the robot's tip are 3.175 and 4.5 cm, respectively. The outer diameter is within an acceptable range for colon insertion. The tube length is about 49 cm, which allows the robot's tip to reach the cecum given that the tube increases to three times its original length as it inflates.

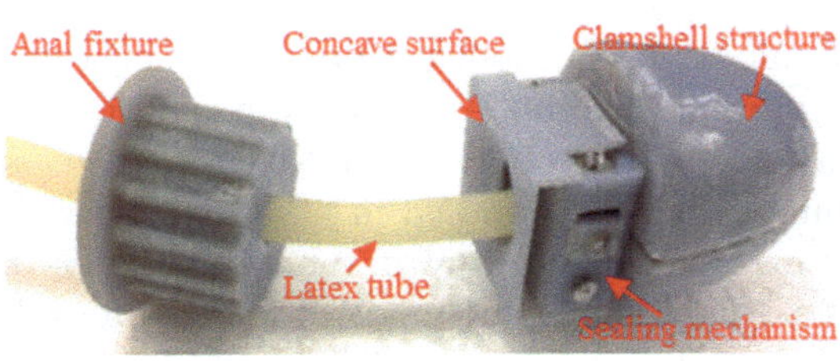

Fig. 2. Design of the colonoscope robot.

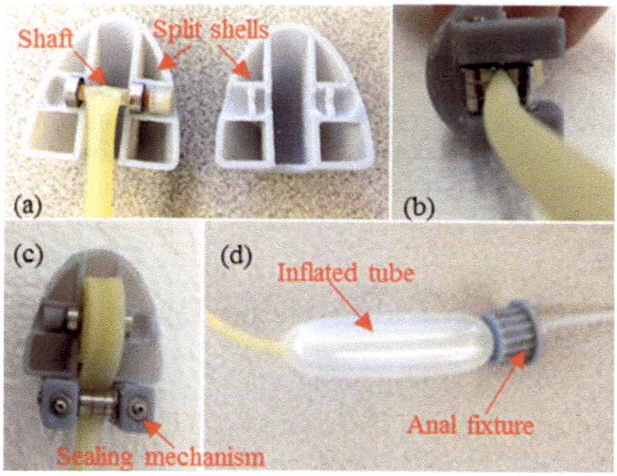

Fig. 3. Components of the colonoscope robot: (a) clamshell structure, (b) sealing mechanism, (c) packed tube around the shaft, (d) anal fixture.

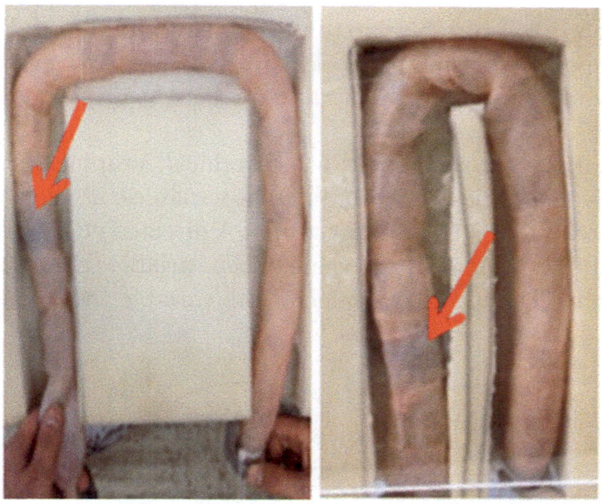

Fig. 4. Photographs of the robot inserted into an excised pig intestine. The red arrows indicate the robot's tip.

This design reduces the shear force on the colon. The net shear force on an object is $\int \mu p dA$, where p is the pressure applied. For a constant pressure and coefficient of friction, the shear force becomes proportional to the area. In this design, only the head of the robot moves, which is 4.5 cm long, compared to a typical colonoscope, where the whole scope moves, and is close to 1 m in length.

A pneumatic circuit provides compressed air for the robot. This circuit is equipped with an air compressor, a pressure regulator, a flow sensor, a digital pressure sensor, a solenoid valve and an electrical switch. To run a test, the pressure of the compressor and the flow rate are set within a range of 10–20 psi and 2–5 standard cubic feet per hour (SCFH) respectively. The maximum pressure that the balloon can tolerate is 20 psi (the range of failure depends on the temperature and condition of the tube); therefore, the pressure and flow are adjusted manually during the test to provide the best performance and avoid balloon failure. This is especially critical when the robots tip encounters an obstacle or the colon bends. More details on the mechanics and functionality of the design can be found in [5].

4 Robot Position Tracking Using Dynamic Bayesian Network

Let x_t denote the displacement of the robot's tip and V_t denote its velocity during time-step t following the robot's insertion. Also, let FR_t and P_t denote the two variables, air flow rate and air pressure in the robot's tube, respectively, at time-step t; corresponding observed values of these variables are $FRmeter_t$ and $Pmeter_t$. Our objective is to predict the displacement and velocity of the robot's tip upto T_{pred} time-steps into the future, $x_{t+1:t+T_{pred}}$ and $V_{t+1:t+T_{pred}}$, given the readings of the observed variables up to the current time step, $FR_{0:t}$ and $P_{0:t}$. We have used a temporal Bayesian network [13] to model the dependencies between the observed and unobserved variables and predict the displacement and velocity of the robot.

In a temporal Bayesian network, a hierarchical, graphical structure is used to capture the dependencies between different variables, as shown in Fig. 5. Each variable is represented as a node in the graph. A directed edge between two nodes indicates causality - a change in the value of the variable (node) at the beginning of the edge effects a change in the value of the variable (node) at the end of the

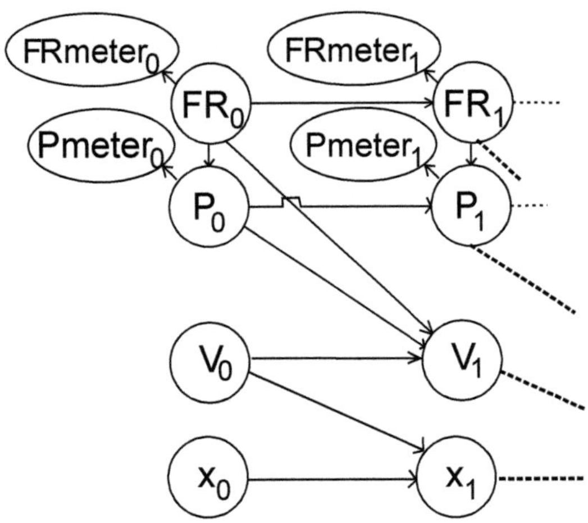

Fig. 5. Temporal causal network used to predict the displacement and velocity of the colonoscope from the observed variables flow rate (FRmeter) and air pressure (Pmeter).

edge; the variables at the edge's beginning and end are called the parent and child nodes respectively. If a child node has multiple parent nodes, it means that the variable in the child node is affected by all the parent variables (nodes). For example, in Fig. 5, variable V_1 is affected by variables V_0, P_0 and FR_0. The temporal aspect of the Bayesian network implies that causality between variables occurs over time. Following conventional temporal Bayesian networks [13], we have assumed that our temporal Bayesian network involves first-order Markov dependencies, that is, a variable's value in the current time-step depends on its value during the preceding time-step only and not on its values from earlier time-steps. For example in Fig. 5, the robot's current displacement is affected by its displacement and velocity in the previous time-step only. Similarly, the velocity at the current time-step is determined from its previous time-step's velocity, and the previous time-step's air pressure and flow rate. Finally, the air pressure and flow rate are also affected by their previous time-step values, as shown in Fig. 5.

To enable fast prediction using the temporal Bayesian network, we discretize the analog variables in the network. The displacement of the robot at time t is discretized using the equation $dX_t = \left\lfloor \frac{\Delta x}{x_{thr}} \right\rfloor$, where $\Delta x = x_t - x_{t-1}$, denotes the change in the position of the robot inside the colon and x_{thr} is a distance threshold that is used to represent 1 unit of movement by the robot. Using this definition, $dX_t \in \{0, 1, 2\}$. We then discretize the continuous variables V, P, FR into discrete variables V_t, P_t, FR_t for time-step t, based on respective threshold values, $V_{thr}, P_{thr}, FR_{thr}$, while using the following equations.

$$V_t = \begin{cases} \text{stop} : v = 0 \\ \text{slow} : 0 < v \le V_{thr} \\ \text{fast} : v > V_{thr} \end{cases}$$

$$P_t = \begin{cases} \text{low} : p < P_{thr} \\ \text{hi} : p \ge P_{thr} \end{cases}$$

$$FR_t = \begin{cases} \text{low} : fr < FR_{thr} \\ \text{hi} : fr \ge V_{thr} \end{cases}$$

To quantify the effect between different variables (nodes) in a temporal Bayesian network, conditional probability tables (CPTs) are used for each node in the network. A CPT of a node gives the probability of each discrete value of the variable in the node, conditioned by the combined values of the parent variables of that node. For our causal network in Fig. 5, the CPT values for the different nodes were generated with data from a single run of the real robot through a swine intestine. For calculating the probability value inside a CPT of a node (FR_t, P_t, V_t, or dX_t) we determined the ratio of the number of time-steps in which the node's variable had the desired value and the total number of instances where the variable occurred, while keeping the values of its parents fixed. For example, for the CPT of node P_t, that has P_{t-1} as its parent node, we calculated $P(P_t = low|P_{t-1} = hi)$ as $\frac{no.\,of\,time-steps\,at\,which\,P_t=low \text{ and } P_{t-1}=hi}{no.\,of\,time-steps\,at\,which\,P_{t-1}=hi}$. The probability values are displayed in Tables 1, 2, 3, 4, and 5.

Table 1. Conditional probability table for temporal component of FR.

FR_{t-1}	$P(FR_t = low \mid FR_{t-1})$	$P(FR_t = hi \mid FR_{t-1})$
low	0.7727272727	0.2272727273
hi	0.0842911877	0.9157088123

Table 2. Conditional probability table for temporal component of P.

P_{t-1}	$P(P_t = low \mid P_{t-1})$	$P(P_t = hi \mid P_{t-1})$
low	0.8543046358	0.1456953642
hi	0.0772727273	0.9227272727

Finally, the CPT values and the causality relationships from the causal network are used in conjunction with the observed values of the pressure meter and the flow rate meter readings at each time-step, to calculate the most likely change in the robot's position (or displacement) and its most likely velocity. We have used a particle filtering algorithm [13] for this calculation. As time continues, the cumulative values of the robot's displacement provide the most likely distance the robot has traveled inside the colon.

5 Experimental Results

To verify our approach, we first performed a series of experiments running the robot inside a Syndaver colon, shown in Fig. 6. We ran 11 experiments, using 5 of the runs to train our model, and tested the model using the remaining 6 experiments[1]. The robot was placed inside the Syndaver colon, then the air pressure inside the robot's balloon is increased using an air flow regulator connected to an air compressor. In earlier test experiments, the balloon was found to rupture when the pressure inside it exceeded 20 psi. To maintain a suitable pressure in the balloon that pushes the robot's head but does not risk rupture, the pressure was maintained between $10 - 20$ psi. As the robot proceeds through the colon, the balloon pressure varies depending on the air flow rate and the resistance that the robot experiences. During our experiments, if the pressure approached

Table 3. Conditional probability table for dependency of pressure P on flow rate FR at time-step t.

FR_t	$P(P_t = low \mid FR_t)$	$P(P_t = hi \mid FR_t)$
low	0.2429906542	0.7570093458
hi	0.4545454545	0.5454545455

[1] Videos of the experiments can be found at https://sites.google.com/site/inversproject/.

Table 4. Conditional probability table for change in position dx for velocity V at time-step t.

V_t	$P(dX_t = 0 \mid V_t)$	$P(dX_t = 1 \mid V_t)$	$P(dx_t = 2 \mid V_t)$
stop	1.0	0.0	0.0
slow	0.2513089005	0.7486910995	0.0
fast	0.0	0.6732026144	0.3267973856

Table 5. Conditional probability table for change in position dx for velocity V at time-step t.

FR_{t-1}	P_{t-1}	V_{t-1}	$P(V_t = stop \mid FR_{t-1}, P_{t-1}, V_{t-1})$	$P(V_t = slow \mid FR_{t-1}, P_{t-1}, V_{t-1})$	$P(V_t = fast \mid FR_{t-1}, P_{t-1}, V_{t-1})$
low	*low*	*stop*	1.0	0.0	0.0
low	*low*	*slow*	0.25	0.75	0.0
low	*low*	*fast*	0.0	0.0	1.0
low	*hi*	*stop*	0.7142857143	0.2857142857	0.0
low	*hi*	*slow*	0.0	0.6829268293	0.3170731707
low	*hi*	*fast*	0.0	0.3870967742	0.6129032258
hi	*low*	*stop*	0.0	0.0	1.0
hi	*low*	*slow*	0.0	0.8214285714	0.1785714286
hi	*low*	*fast*	0.0	0.4411764706	0.5588235294
hi	*hi*	*stop*	0.0	1.0	0.0
hi	*hi*	*slow*	0.0	0.6557377049	0.3442622951
hi	*hi*	*fast*	0.0	0.2564102564	0.7435897436

20 psi, the air flow was immediately adjusted manually to reduce the pressure to within the allowable range. We first recorded the observed variables, $FRmeter$ and $Pmeter$, at discrete time-steps. Each time-step for our experiment was considered 0.1 sec in duration. For our experiments we set the distance moved by the robot head that is considered as one unit of movement, $x_{thr} = 1.5$ cm, which corresponds to about one-third of the length of the robot's head. The values of the other thresholds V_{Thr}, P_{Thr}, and FR_{Thr} were set to the average of the velocity, pressure and flow rate values over the duration of the training data. We tracked the ground truth location of the robot using an overhead camera and tracking software. For the Syndaver experiments, the thresholds were $V_{Thr} = 1.4415469568$ cm/s, $P_{Thr} = 8.5292857707$ psi, and $FR_{Thr} = 4.495146746$ SCFH. Figure 7(top) shows the average error in tracking the robot's position across the 6 testing runs. As can be seen, the error is always less then 10 cm, and for most of the run is close to 5 cm, which is approximately the size of the robot. Figure 7(bottom), shows the predicted position and actual position for the robot in a single run through the Syndaver colon. As can be seen from the graph, the tracked position is very

Fig. 6. First test environment using a clear, straight tube.

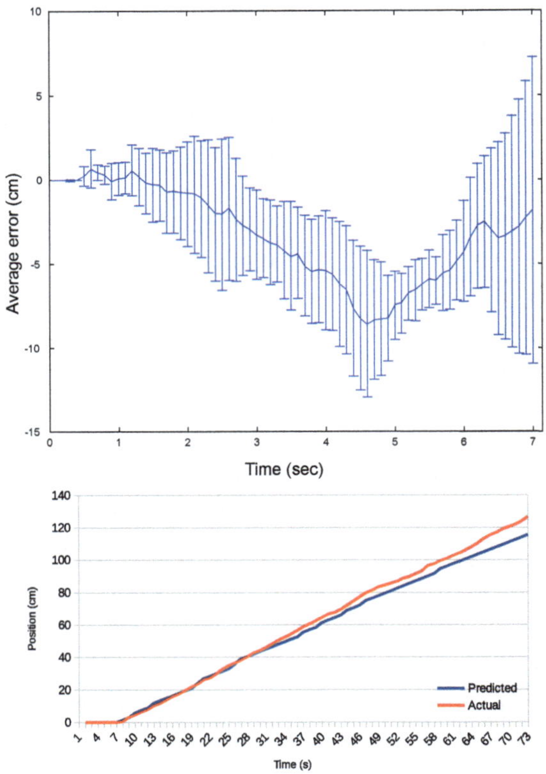

Fig. 7. (top) Average error over 6 runs of the robot in a Syndaver colon. (bottom) Example of one run in a Syndaver colon, showing predicted distance and actual distance traveled by the robot.

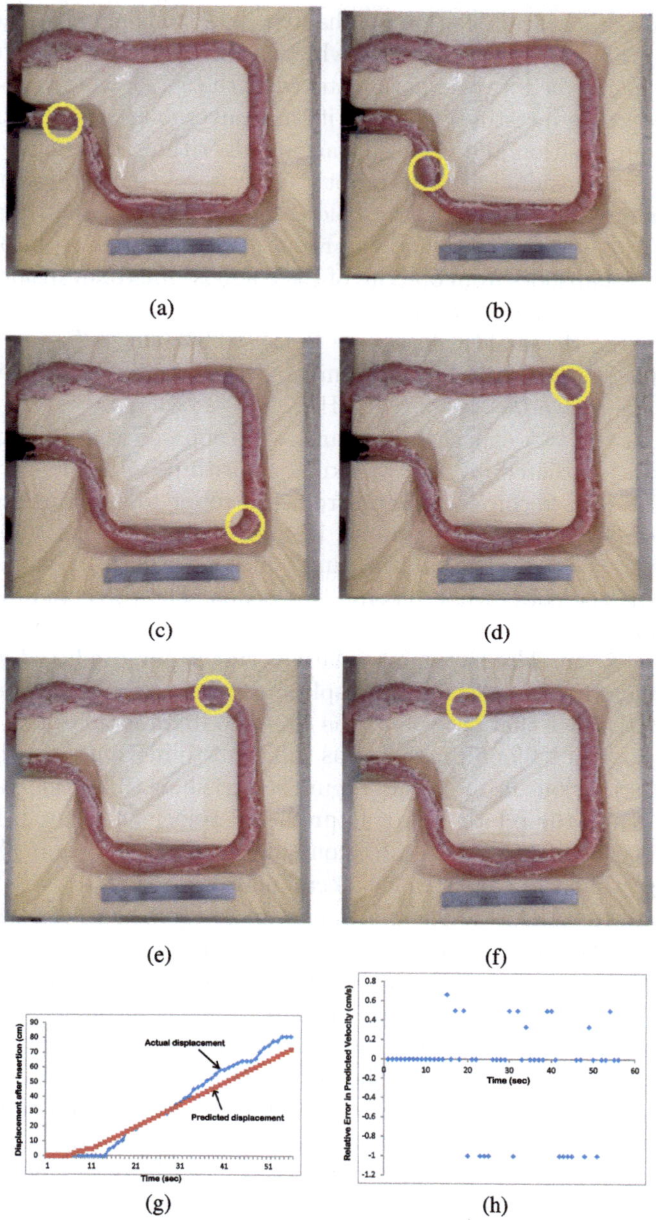

Fig. 8. (a–f) Snapshots from run 2 of the colonoscope robot within an excised pig intestine; the yellow circle shows the current position of the robot. (g) Actual and predicted displacement data and (h) Relative error in predicted velocity from run 2 of the colonoscope robot.

close to the robot's actual position for the first half. It is only near the middle of the run when the error begins to grow, which is also visible in Fig. 7(top).

We then performed two experiments with the robot colonoscope moving inside excised swine intestines to verify its maneuverability and to test the performance of our position and velocity tracking algorithm. The excised intestines were cleaned and insufflated. The robot's tip was covered with a lubricant to facilitate smooth motion inside the colon. The colon was then placed inside a foam-core frame for support and the robot's tip was inserted into the colon. Figure 8 shows snapshots from one run of the colonoscope robot inside an excised swine intestine with the position of the robot marked with a yellow circle. For experiment 1, the threshold values were $V_{Thr} = 0.875$ cm/s, $P_{Thr} = 15$ psi and $FR_{THr} = 3.5$ SCFH, while for experiment 2 they were, $V_{Thr} = 0.875$ cm/s, $P_{Thr} = 15.2$ psi and $FR_{THr} = 2.5$ SCFH.

For each of the two experimental runs, we compared the displacement and velocity values predicted by our tracking algorithm with the corresponding ground truth values recorded directly from the robot moving inside the colon. The results of one experiment is shown in Fig. 8(g–h). In run 1, the prediction algorithm was used for 10 seconds during which the robot traveled a distance of 15 cm inside the colon while passing two bends. The predicted displacement is mostly commensurate with the actual displacement. The maximum displacement error is 1.5 cm. The maximum relative error in the predicted velocity is 1 cm/s. The relatively lower errors in displacement and velocity in run 1 can be attributed to the fact that the robot was closer to its insertion point during the run and the applied balloon pressure was sufficient to move it smoothly through bends and obstructions in the colon. Figure 8(g–h) show the actual vs. predicted displacements and the relative error in predicted velocity for run 2. In this run, the robot proceeded further into the colon almost up to the cecum. We observed from Fig. 8(g) that the displacement accumulates an error towards the end of the run when the robot is in the latter part of the transverse colon and in the ascending colon. We believe this is because in these parts the robot has proceeded further into the colon; although the robot continues to progress into the colon, the pressure inside the robot's balloon increases when it encounters obstructions, which results in a stop and jump motion instead of uniform motion. The relative error in the predicted velocity also shows that the velocity after about half-way through the experiment gives more errors. Overall, we see from run 2 that the error in the robot's displacement is between $7.5 - 11.5$ cm ($1.5 - 2.5$ times the length of the robot's head) for only about 35% of the time-steps when the robot has proceeded towards the end of the colon.

6 Discussions and Future Work

In this paper we have presented our work on a novel, pneumatically controlled colonoscope robot along with an algorithm that automatically tracks the robot's displacement and velocity inside the colon. We have successfully demonstrated the navigation of the robot within swine intestines along with results of the

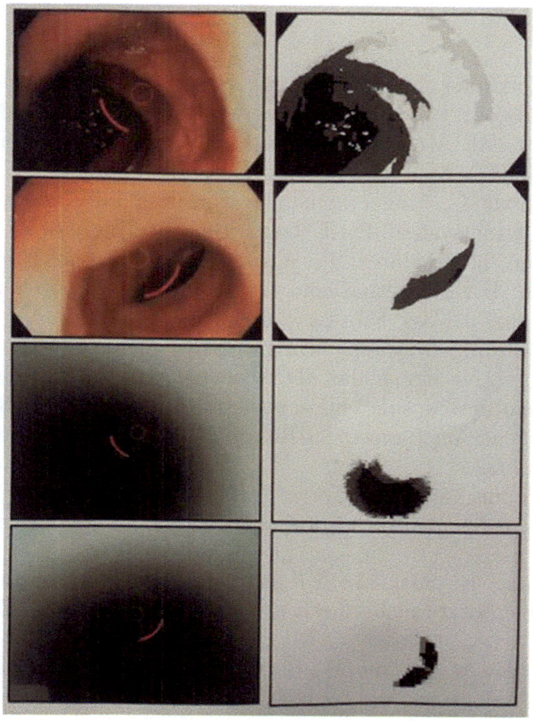

Fig. 9. Video images from a conventional endoscopy procedure. The solution from the image processing algorithm is the direction that the robot's tip has to take with respect to the center of the image. The direction is calculated in real-time and displayed with a red line on the image.

tracking algorithm. Currently, we have ongoing efforts to add a miniature wireless camera to the robots tip and use the image captured by it as an additional input. We plan to use image-fusion techniques to aid locating the robot and keeping track of it by combining air pressure, air flow rate and the solution given by the image processing algorithm. This algorithm will identify the inner walls of the colon and predict the direction that the robots tip should take, as show in Fig. 9. The robots tip will become a steering mechanism capable of guiding the entire robot and minimizing the friction generated during displacement. Particularly, during critical moments where the robot reaches the colons bends. This new addition will definitely improve the results from our tracking mechanism presented in this work.

Another direction we are investigating is to close the feedback loop so that the predicted error in displacement and velocity of the robot's tip is provided to the control mechanism to automatically adjust the air flow into the robot's balloon. Our work presented here is our first step towards semi-autonomous maneuver of a pneumatically controlled robot and we envisage that understanding and addressing the issues presented herein will help in solving the open problems in robotic colonoscopy and make colonoscopies more amenable to patients.

References

1. Adachi, K., Yokojima, M., Hidaka, Y., Nakamura, T.: Development of multi-stage type endoscopic robot based on peristaltic crawling for inspecting the small intestine. In: 2011 IEEE/ASME International Conference on Advanced Intelligent Mechatronics (AIM), pp. 904–909, July 2011
2. Carpi, F., Pappone, C.: Magnetic maneuvering of endoscopic capsules by means of a robotic navigation system. IEEE Trans. Biomed. Eng. **56**(5), 1482–1490 (2009)
3. Chen, G., Pham, M., Redarce, T.: A semi-autonomous micro-robotic system for colonoscopy. In: IEEE International Conference on Robotics and Biomimetics, ROBIO 2008, pp. 703–708, February 2009
4. Ciuti, G., Visentini-Scarzanella, M., Dore, A., Menciassi, A., Dario, P., Yang, G.-Z.: Intra-operative monocular 3D reconstruction for image-guided navigation in active locomotion capsule endoscopy. In: 2012 4th IEEE RAS EMBS International Conference on Biomedical Robotics and Biomechatronics (BioRob), pp. 768–774, June 2012
5. Dehghani, H., Pourghodrat, S., Terry, B., Nelson, C., Oleynikov, D., Dasgupta, P.: Semi-autonomous locomotion for a diagnostic endoscopy device. J. Med. Devices (2015)
6. Glass, P., Cheung, E., Sitti, M.: A legged anchoring mechanism for capsule endoscopes using micropatterned adhesives. IEEE Trans. Biomed. Eng. **55**(12), 2759–2767 (2008)
7. Keller, H., Juloski, A., Kawano, H., Bechtold, M., Kimura, A., Takizawa, H., Kuth, R.: Method for navigation and control of a magnetically guided capsule endoscope in the human stomach. In: 4th IEEE RAS EMBS International Conference on Biomedical Robotics and Biomechatronics (BioRob), pp. 859–865 (2012)
8. Kim, B., Lim, H.Y., Park, J.H., Park, J.-O.: Inchworm-like colonoscopic robot with hollow body and steering device. JSME Int. J. Ser. C **49**, 205–212 (2006)
9. Lin, W., Shi, Y., Jia, Z., Yan, G.: Design of a wireless anchoring, extending micro-robot system for gastrointestinal tract. Int. J. Med. Robot. Comput. Assist. Surg. **9**, 167–179 (2013)
10. Obstein, K., Valdastri, P.: Advanced endoscopic technologies for colorectal cancer screening. World J. Gastroenterol. **19**(4), 431–439 (2013)
11. Pahlavan, K., Ye, Y., Khan, U., Fu, R.: RF localization inside human body: enabling micro-robotic navigation for medical applications. In: 2011 International Conference on Localization and GNSS (ICL-GNSS), pp. 133–139 (2011)
12. Quirini, M., Webster, R., Menciassi, A., Dario, P.: Design of a pill-sized 12-legged endoscopic capsule robot. In: IEEE International Conference on Robotics and Automation, (ICRA), pp. 1856–1862 (2007)
13. Russell, S., Norvig, P.: Artificial Intelligence: A Modern Approach. Prentice Hall, Englewood Cliffs (2009)

Neural Networks for FDI on the First Actuator of a Two-Link Planar Manipulator

Jesús A. Esquivel(✉), Juan A. Díaz, Isela Carrera, and Héctor Moreno

Facultad de Ingeniería Mecánica y Eléctrica, UAdeC, Monclova, Mexico
{jesus.esquivel,iselacarrera,h_moreno}@uadec.edu.mx, jads2100@hotmail.com

Abstract. This paper shows an approach to use a Neural Network trained by the classic backpropagation algorithm for solving the problem of fault detection and isolation (FDI) of simple mechanisms subject to failures in actuators. The approach taken was to reserve the term of the projection of the tuning algorithm used for keeping bounded the weight, and use it at the time of the fault. Works like Vemuri *et.al.* [12], where faults are focused in the inertia matrix and the isolation technique does not show clearly the results it aims, were the inspiration for this research. Here the fault is modelled as a torque suddenly bounded at first actuator and a neural network of two layers is used with an adaptive law whose projection operation is a reserved degree of freedom for keeping the system under control.

1 Introduction

The problems of fault detection and isolation (FDI), have attracted a lot of attention since their incipient days from the Ph.D. results of Beard [3] and Jones [9]. The goal for solving the problem is to be able to isolate the particular input meaning a fault from the other signals: control inputs, disturbances or other faults. These works aims solutions by means of a memoryless feedback, while [11] and [2] have addressed the problem in a more complete framework as is the geometric approach and aims to solve the problem by a *dynamic* feedback.

This work inspired in [12] for fault detection and isolation on robotic system and on their approach due to the lack of affinity to the mainstream of FDI's works. Their solution has not great difference from what is a control problem of a robotic manipulators under effects of perturbations, see [10]. It is considered only faults on the commands and how the system must detect their occurrence, as well as to isolate them from the other input signals for not affecting the system behavior. When the fault rises there must be a kind of observer, *residual generator*, for this single fault, and insensitive to the rest. In other words, the residual generators maps inputs, the faults, over non interactive subspaces over which evolute their dynamics. These non interactive spaces are found under the study of *unobservable subspaces* which leads to a *quotient*, observable, subspaces, giving the necessary and sufficient conditions for the construction of asymptotic observers of each of the quotient subsystem, so getting the desired filter. In [7] is presented a survey of the various model-based fault detection, isolation and

I. Chang et al. (eds.), *Advances in Automation and Robotics Research in Latin America*,
Lecture Notes in Networks and Systems 13, DOI 10.1007/978-3-319-54377-2_11

reconfiguration (FDIR) methods. Their approach of FDI, correspond to the generation of residuals that are robust to noise, unknown disturbance, and model uncertainties, as well as various statistical techniques of testing the residuals for abrupt changes, or faults. The performance of a fault detection algorithm is usually measured in terms of the tradeoffs between the false alarm rate and the mean detection delay, [7]. In terms of the residual characteristics, in [6] is shown that the various observer-based methods, such as eigenstructure assignment, fault detection filters and unknown input observers give identicals residuals with that of an equivalent parity relation method. With this last approach the transfer function or state-space models are transformed to yield directional or structural residual vectors directly [4,5]. It is a complex task in the robust residual generation techniques to generate residuals insensitive to noise and uncertainties, and at the same time have very good sensitiveness to fault. These techniques include full-state observer-based methods, unknown input observers, parity relations approach, optimization-based, the Kalman filter-based, the system identification, artificial intelligence approaches among others, see [7]. In [13] is proposed a FDI approach with recurrent neural network-based observers for simultaneously detecting, isolating and identifying the severity of actuator faults in presence of disturbances and uncertainties in the model and sensor measurements. The neural network weights are updated based on a modified *dynamic backpropagation* scheme. The algorithm of Backpropagation, or a variation of it, is a recurrent strategy for updating the weights of the artificial neural networks (ANN) employed for solving the problem of FDI as is the case of [1] where a synergism of ANN with Fuzzy theory used a Takagi-Sugeno Model in discrete time for providing the data to train each time one of the neuronal models.

2 Robot Dynamics and Filtered Error [10]

The robot arm has dynamics

$$M(q)\ddot{q} + V_m(q, \dot{q})\dot{q} + G(q) = \tau \tag{1}$$

where $q, \dot{q}, \ddot{q} \in \mathbb{R}^n$ are vectors of joint positions, velocities and accelerations, respectively, $\tau \in \mathbb{R}^n$ is the input torque vector, $M(q) \in \mathbb{R}^{n\times n}$ is the inertia matrix, $V_m(q, \dot{q}) \in \mathbb{R}^n$ is a matrix containing the centripetal and Coriolis terms and $G(q) \in \mathbb{R}^n$ is the gravity vector.

To make the robot follow a prescribed desired trajectory, [10], $q_d(t)$, define the tracking error $e(t)$ and filtered tracking error $r(t)$ by

$$e = q_d - q \tag{2}$$

$$r = \dot{e} + \Lambda e \tag{3}$$

with Λ a positive definite design parameter matrix and $q_d(t)$ the desired trajectory. The robot dynamics are expressed in terms of this filtered error as

$$M\dot{r} = -V_m r + f(x) - \tau \tag{4}$$

where the nonlinear function $f(x)$ is defined as

$$f(x) = M(q)(\ddot{q}_d + \Lambda\dot{e}) + V_m(q,\dot{q})(\dot{q}_d + \Lambda e) + G(q) \tag{5}$$

vector x contains all the time signals needed to compute $f(\cdot)$ and may be defined for instance as $x \equiv [e^T, \dot{e}^T, q_d^T, \dot{q}_d^T, \ddot{q}_d^T]^T$.

Approximation-Based Controllers. When $f(\cdot)$ changes for a reason, e.g., faults, an estimate $\hat{f}(\cdot)$ is used instead and a general sort of approximation-based-controller by setting

$$\tau = \hat{f} + K_v r - v(t) \tag{6}$$

with

$$K_v r = K_v \dot{e} + K_v \Lambda e$$

as an outer PD tracking loop, and $v(t)$ an auxiliary signal to provide robustness in the face of disturbances and modelling errors. The estimate $\hat{f}(\cdot)$ and the robustifying $v(t)$ is defined with neural networks to estimate the fault and accommodate the control to keep the desired behavior, e.g., tracking control, see Fig. 1. The controller design problem is to select the estimate $\hat{f}$ and the robust term $v(t)$ in control law so that the dynamics error is stable and like [10], where the goal is keep only bounded the tracking error, the goal of the control of the two-link manipulator is to track the desired trajectory in spite of a fault detected.

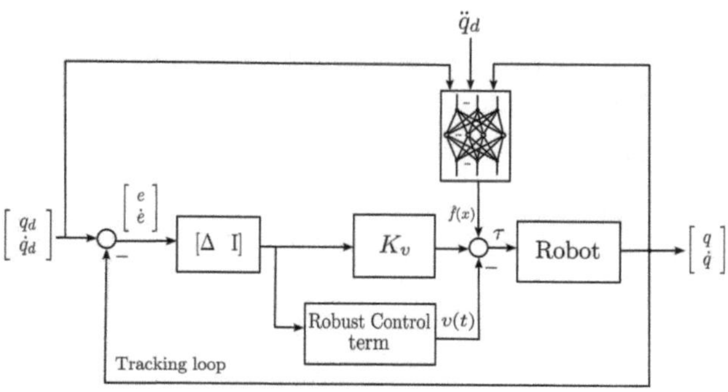

Fig. 1. Filtered error approximation-based controller

3 Artificial Neural Networks

The ANN scheme used in this work is of two layers. In a generalized form its model is

$$y_i = \sigma\left(\sum_{l=1}^{L} w_{il}\,\sigma\left(\sum_{j=1}^{n} v_{lj}x_j + v_{l0}\right) + w_{i0}\right) \tag{7}$$
$$i = 1, 2, \ldots, m$$

The thresholds can be embedded in the sum doing $x_0 = 1$. Then we can define the following quantities

$$y_i = \sigma\left(\sum_{l=0}^{L} w_{il} z_l\right) \tag{8}$$

$$z_l = \sigma\left(\sum_{j=0}^{n} v_{lj} x_j\right) \tag{9}$$

where $z_0 = 1$ too.

Backpropagation [10]. Backpropagation weight tuning is a gradient descent algorithm, so the weights in the two layers are updated according to

$$w_{il} = w_{il} - \eta \frac{\partial E}{\partial w_{il}} \tag{10}$$

$$v_{lj} = v_{lj} - \eta \frac{\partial E}{\partial v_{lj}} \tag{11}$$

with E as a prescribed cost function. Normally is selected as

$$E = \frac{1}{2} e^T e = \frac{1}{2} \sum_{i=1}^{m} e_i^2 \tag{12}$$

$$e_i = Y_i - y_i \tag{13}$$

The learning rates η in the two layers can be selected different or *ad hoc* for each case. The backward recursion for backpropagated errors:

$$e_i = Y_i - y_i \qquad i = 1, 2, \ldots, m \tag{14}$$

$$\delta_i^2 = y_i(1 - y_i)e_i \tag{15}$$

$$\delta_l^i = z_l(1 - z_l) \sum_{i=1}^{m} w_{il} \delta_i^2 \qquad l = 1, 2, \ldots, L \tag{16}$$

These quantities are used for update the weights and thresholds,

$$w_{il} = w_{il} + \eta z_l \delta_i^2, \qquad i = 1, 2, \ldots, m \tag{17}$$

$$v_{lj} = v_{lj} + \eta X_j \delta_l^1 \qquad l = 1, 2, \ldots, L \tag{18}$$

$$j = 0, 1, \ldots, n$$

In order to keep the boundedness of the weights in a continuous online backpropagation strategy so that the control input $\tau(t)$, as well as the weights $\hat{V}$ and $\hat{W}$, remains bounded, the proposed controller in Table/Theorem 4.3.2 of [10] guarantees the boundedness of the control input by a small value with large gains in K_v. The control is given by

$$\tau = \hat{W}^T \sigma(\hat{V}^T x) + K_v r - v, \tag{19}$$

where the robustifying signal, in [10], is

$$v(t) = -K_z(||\hat{Z}||_F + Z_B)r \tag{20}$$

The matrix Z of all the NN weights is defined as

$$Z \equiv \begin{bmatrix} W & 0 \\ 0 & V \end{bmatrix}$$

with Z_B known and $||\cdot||_F$ the Frobenius norm defined next.

Definition 1 (Frobenius Norm). Given a matrix $A = [a_{ij}]$, the *Frobenius norm* is defined as the root of the sum of the squares of all elements:

$$||A||_F^2 \equiv \sum a_{ij}^2 = tr(A^T A),$$

with $tr(\cdot)$ the matrix *trace.*

The tuning algorithms for the weights/thresholds are an unsupervised version of backpropagation through time; it is not necessary to know the ideal plant output are given by

$$\dot{\hat{W}} = F\hat{\sigma}r^T - F\hat{\sigma}'\hat{V}^T x r^T - \kappa F||r||\hat{W} \tag{21}$$

$$\dot{\hat{V}} = Gx\left(\hat{\sigma}'^T\hat{W}r\right)^T - \kappa G||r||\hat{V} \tag{22}$$

The matrices F and G are positive definite, $\kappa > 0$ is a small design parameter. According to [10] ten hidden-layer neurons suffices.

Assumption 1 (Bounded Ideal Target NN Weights). *On any compact subset of $\mathbb{R}^n$, the ideal NN weights are bounded so that*

$$||Z||_F \leq Z_B$$

It is used the sigmoidal functions as activation function of the NN for implementing the tuning algorithms due that helps implement the required Jacobian $\hat{\sigma}'$,*i.e.*,

$$\sigma'(z) = diag\{\sigma(z)\}(I - diag\{\sigma(z)\})$$

so that

$$\hat{\sigma}'^T\hat{W}r = diag\{\sigma(\hat{V}^T x)\}[I - diag\{\sigma(\hat{V}^T x)\}]\hat{W}r$$

4 Fault Detection with Neural Network

This section is based on Vemuri, *et.al* [12] where the approach for fault detection is similar to the foregoing shown in this paper. The kind of manipulators are

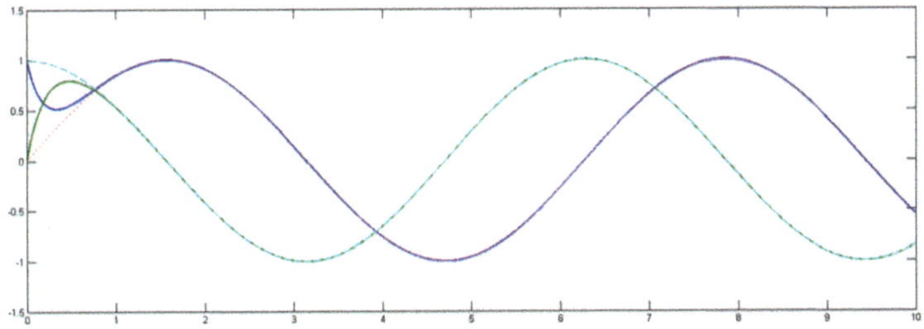

Fig. 2. Neural control without faults

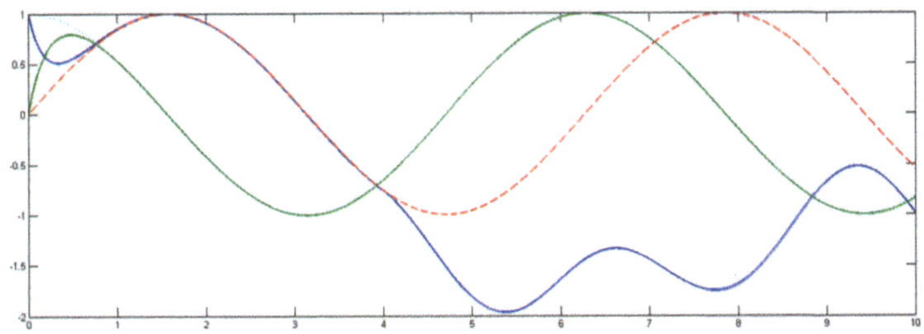

Fig. 3. Neural control with fault at 4 s

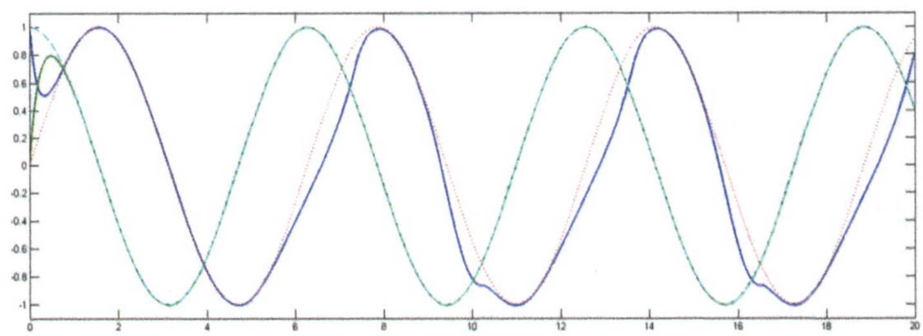

Fig. 4. Neural control with accommodation

modelled by (1). The failure is supposed in the inertia and is modelled relying every part of the robot model, that is,

$$\psi(q,\dot{q},\tau) = [\tilde{M}^{-1}(q) - M^{-1}][\tau - V_m(q,\dot{q})\dot{q} - G(q)] \tag{23}$$

Due to sigmoidal neural networks are good enough for aproximating and analyzing nonlinear models, as pointed out by [12,14], they are used here.

Consider a sigmoidal neural network whose input-output characteristics are described by

$$y = \hat{\psi}(q,\dot{q},\tau;\hat{\theta}) \tag{24}$$

where $y \in \mathbb{R}^n$ is the output of the network and $\hat{\theta} \in \mathbb{R}^p$ represents the adjustable weights of the network. Let the weights $\hat{\theta}(0) = \hat{\theta}_0$ of the selected neural network be initialized such that

$$\hat{\psi}(q, \dot{q}, \tau; \hat{\theta}_0) = 0, \qquad \forall q, \dot{q}, \tau \tag{25}$$

corresponding to the no-failure situation. This can be achieved by initializing the weights at the output to zero. starting from these initial conditions, the main objective is to adjust the parameter $\hat{\theta}(t)$ at each time t so that $\hat{\psi}(q, \dot{q}, \tau; \hat{\theta})$ approximates the unknown function $\beta(t - T)\psi(q, \dot{q}, \tau)$.

In [12] is used a neural network with updating law

$$\dot{\hat{\theta}} = \Gamma Z^T \varepsilon - \chi^* \Gamma \frac{\hat{\theta}\hat{\theta}^T}{\hat{\theta}^T \Gamma \hat{\theta}} \Gamma Z^T \varepsilon, \quad \hat{\theta}(0) = \hat{\theta}_0$$

where χ^* denotes the indicator function if the weights $\hat{\theta}$ are confined or not in a hypersphere of size M, and is used in a projection operation to avoid *parameter drift* that may arise like the standard adaptive laws in the presence of modeling uncertainties [8].

The accommodation by the reconfiguration, also referred as self-correction, of the control law is achieved by

$$\tau_r = \tau - M(q)\hat{\phi}(q, \dot{q}; \hat{\theta}) \tag{26}$$

where τ is the nominal control law and τ_r is the reconfigured control law. Here the control action sent as τ is compared to the expression $M(q)\ddot{q} + V_m(q, \dot{q})\dot{q} + G(q)$ so that the difference is due to the error given by the actuators. This approach point at exactly the actuator in trouble. The intrincacies of this class of fault are precisely for the underactuated state in which the robot drops and the difficulties for sending a command through this via. Our approach take action on τ adding the projection term directly in the tuning algorithm when a fault appears.

5 Example

The two-link planar manipulator in (Fig. 5) has dynamics given by

$$M(q) = \begin{bmatrix} \alpha + \beta + 2\eta \cos q_2 & \beta + \eta \cos q_2 \\ \beta + \eta \cos q_2 & \beta \end{bmatrix} \tag{27}$$

$$V_m(q, \dot{q}) = \begin{bmatrix} -\eta \dot{q}_2 \sin q_2 & -\eta(\dot{q}_1 + \dot{q}_2) \sin q_2 \\ \eta \dot{q}_1 \sin q_2 & 0 \end{bmatrix} \tag{28}$$

$$G(q) = \begin{bmatrix} \alpha e_1 \cos q_1 + \eta e_1 \cos(q_1 + q_2) \\ \eta e_1 \cos(q_1 + q_2) \end{bmatrix} \tag{29}$$

where $\alpha = (m_1 + m_2)a_1^2$, $\beta = m_2 a_2^2$, $\eta = m_2 a_1 a_2$, $e_1 = g/a_1$. Here we took the same parameters as [10] for the manipulator which are $a_1 = a_2 = 1$ m, $m_1 =$

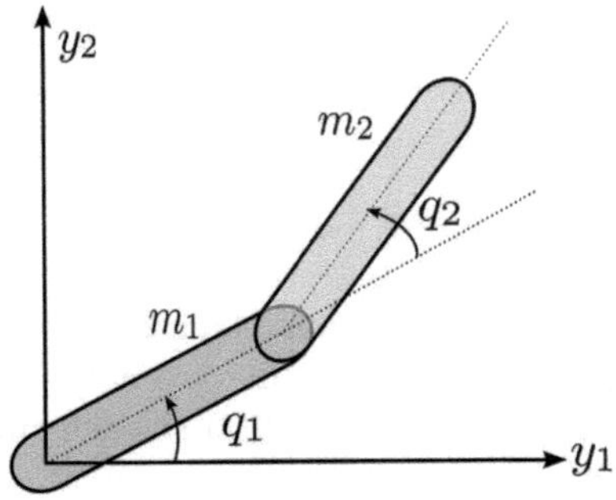

Fig. 5. Two-link planar elbow arm

0.8 kg, $m_2 = 2.3$ kg. The desired trajectory was selected as $q_d(t) = \{\sin t, \cos t\}$. The controller parameters were selected as $K_v = 20 * I_2$, where I_n is eye(n), the identity matrix of $\mathbb{R}^n$ in MATLAB, $K_z = 10 * I_2$, $\Lambda = 5 * I_2$, $F = 10 * I_{10}$, $G = 5 * I_{10}$, $\kappa = 0.1$ and $Z_B = 1000$. With initial conditions as $q_1(0) = 1$ rad, $q_2(0) = 0$ rad, $\dot{q}_1(0) = 0$ rad/s and $\dot{q}_2(0) = 0$ rad/s, $W = 0_{10\times 2}$ and $V = 0_{10\times 10}$.

In the Fig. 2 is shown the result of the controller without faults, filtered error and neural networks trained with augmented backpropagation.

A failure was simulated at the torque of the first link where was enforced to be bounded by 10, *i.e.*, $|\tau_1| \leq 10$ after 4 s. The numerical algorithm had troubles for going on and break out by the singularities of the matrices W and V, so the matrix gains of the tuning algorithms were modified to $F = 0.005 * \text{eye}(10)$, $G = 0.005 * \text{eye}(10)$ and $K_z = 2 * \text{eye}(2)$. The results are shown at Fig. 3.

One of the major challenges in designing intelligent robotic systems after detecting the occurrence of a fault is to provide a way to accommodate or self-correct itself the system, see [12,15]. In this work the detection of the fault is achieved simply checking the difference between the torque desired and the model of the robot, *i.e.*,

$$||\tau - M(q)\ddot{q} - N(q, \dot{q})|| > \delta$$

for a given $\delta > 0$, in our case $\delta = 50$. Once detected the fault starts the reaction of the controller for the accommodation. The action law (see Fig. 4) used was a modification of (30) and (31) shown at the following equations,

$$\dot{\hat{W}} = F\hat{\sigma}r^T - \text{sw} * F\hat{\sigma}'\hat{V}^T x r^T - \kappa F||r||\hat{W} \tag{30}$$

$$\dot{\hat{V}} = \text{sw} * Gx\left(\hat{\sigma}'^T \hat{W} r\right)^T - \kappa G||r||\hat{V} \tag{31}$$

where sw was defined as

$$\text{sw} = \text{Heaviside}(||\tau - M(q)\ddot{q} - N(q, \dot{q})|| - \delta) * 50$$

which is a non-normalized version of the projection term given in [12].

6 Conclusion

The investigation of this work was about a fault detection and accommodation with artificial neural networks of a fault given in the first actuator of a two-link planar robot which was set in the form of a bound in its torque so affecting this way the behavior of the mechanism. The approach for the fault detection was simply to take the difference between the torque sent to the actuators and its complement of the mathematical model of the manipulator. And the accommodation or self-correcting action of the controller was a change of the augmented tuning backpropagation algorithm which uses a term to keep the weights within a bounded region; in this case was not used during the normal operation and activated when the fault was detected.

References

1. Anzurez-Marin, J., Cuevas-Silva, O., Pitalúa-Díaz, N.: The fault diagnosis problem: residual generators design using neural networks in a two-tanks interconnected system. In: Electronics, Robotics and Automotive Mechanics Conference (2009)
2. De Persis, C., Isidori, A.: A geometric approach to nonlinear fault detection and isolation. IEEE Trans. Autom. Control **46**(6), 853–865 (2001)
3. Beard, R.V.: Failure accomodation in linear systems through self-reorganization. Ph.D. dissertation, Massachusetts Institute Technology, Cambridge (1971)
4. Czajkowski, A., Patan, K.: Design of predective fault tolerant control by the means of state space neural networks. In: 24th Mediterranean Conference on Control and Automation, Athens, Greece, June 2016
5. Czajkowski, A., Luzar, M., Witczak, M.: Robust multi-model fault detection and isolation with a state-space neural network. In: 24th Mediterranean Conference on Control and Automation, Athens, Greece, June 2016
6. Gertler, J.: All linear methods are equal and extendible to (some) nonlinearities. Int. J. Robust Nonlinear Control **12**, 629–648 (2002)
7. Hwang, I., Kim, S., Kim, Y., Seah, C.E.: A survey of fault detection, isolation, and reconfiguration methods. IEEE Trans. Control Syst. Technol. **18**(3), 636–653 (2010)
8. Ioannou, P.A., Sun, J.: Stable and Robust Adaptive Control. Prentice-Hall, Englewood Cliffs (1995)
9. Jones, H.L.: Failure detection in linear systems. Ph.D. dissertation, Massachusetts Institute Technology, Cambridge (1973)
10. Lewis, F.L., Jagannathan, S., Yeşildirek, A.: Neural Network Control of Robot Manipulators and Nonlinear Systems. Taylor & Francis, London (1999)
11. Massoumnia, M.A.: A geometric approach to the synthesis of failure detection filters. IEEE Trans. Autom. Control **31**, 839–846 (1986)
12. Vemuri, A.T., Polycarpou, M.M., Diakourtis, S.A.: Neural network based fault detection in robotic manipulators. IEEE Trans. Robot. Autom. **14**(2), 342–348 (1998)
13. Talebi, H.A., Khorasani, K.: A neural network-based multiplicative actuator fault detection and isolation of nonlinear systems. IEEE Trans. Control Syst. Technol. **21**(3), 842–851 (2013)

14. White, D.A., Sofge, D.A.: Handbook of Intelligent Control: Neural, Fuzzy and Adaptive Approaches. Van Nostrand Reinhold, New York (1993)
15. Yen, G.G.: Reconfigurable learning control in large space structures. IEEE Trans. Control Syst. Technol. **2**, 362–370 (1994)

Experimental Identification of Lu-Gre Friction Model in an Hydraulic Actuator

Lisandro J. Puglisi(✉), Roque J. Saltaren, and Cecilia E. Garcia Cena

Centro de Automática y Robótica (UPM-CSIC),
Jose Gutierrez de Abascal 2, 28006 Madrid, Spain
lisandro.puglisi@alumnos.upm.es

Abstract. In this work is presented the experimental identification of friction effects defined by the parameters of the LuGre model. The parameters are found by means of two experiments. The first one is performed with motions at constant velocity and the second one is performed under controlled force. These experiments allow to find separately the set of parameters that govern the steady state and the pre-sliding regime, respectively.

1 Introduction

Friction is an inevitable non-linear phenomenon that occurs in all kinds of mechanical system. It appears as reaction forces at the physical interface between two surfaces in contact that depends on contact geometry and topology, material of the bodies, relative motion and presence of lubrication [4]. In a hydraulic actuator the interaction between the rod, seals, the piston o-rings, and the viscous effects of the hydraulic fluid all generate friction.

It plays a major role in control systems and it is generally an impediment for servo control resulting in steady state errors and instabilities due to the appearance of limit cycles [8,10,13,17].

In Fig. 1 a typical friction effect in an hydraulic actuator (HA) is presented. The suddenly burst of oscillation occurs at low velocity motion under controlled force excitation. The curve at the top of the figure shows the sinusoidal excitation with an amplitude of $10\,N$ at $0.04\,Hz$ applied to the HA. As it can be observed, the force exerted by the piston (middle curve) falls into oscillations at $t \approx 17\,s$, and the position of the HA decays (see curve at the bottom of the figure).

This particular behavior, named stick-slip motion is caused by the friction force at zero velocity, which is higher than at a small nonzero velocity. When motion starts, the friction force decreases rapidly and the accelerating force is therefore large. The motion is jerky and switches between periods of sticking and slipping, resulting in poor control performance, and can also give rise to vibrations and noise [14].

In order to deal with friction effects, there are two different approaches. The first one, is to try to diminish these effects by designing the system and/or operate in regions where the nonlinear behavior are minimum. As in [16] where

I. Chang et al. (eds.), *Advances in Automation and Robotics Research in Latin America*,
Lecture Notes in Networks and Systems 13, DOI 10.1007/978-3-319-54377-2_12

an extra motor is added in order to provide a permanent rotational velocity to a linear actuator avoiding the Stribeck region between the rod and the piston. The other alternative, is to identified the effect and compensate it within the control loop. This last approach requires of the understanding of the effect and further modeling of friction.

Despite the persistent efforts of many researchers there is still no model that captures all the aspects of friction into one formulation. Many models based on experimental observation have been proposed. A review of frictions models can be found in [6,15], and a discussion of the properties of some dynamical models are presented in [3], physics-motivated models and empirically motivated models for friction's dynamics are presented and discussed in [1].

In particular, the LuGre model [9] presents a good compromise between simplicity and reproduction of complex friction behavior. It has few parameters and represents a good choice to express friction at zero velocity and gross motion.

Based on the afore mentioned, this work tackles the experimental identification of the friction effects acting in a hydraulic actuator considering the LuGre model. Therefore, the two step off-line identification methodology presented in [12] is implemented in this work. First, constant velocity tests are performed to

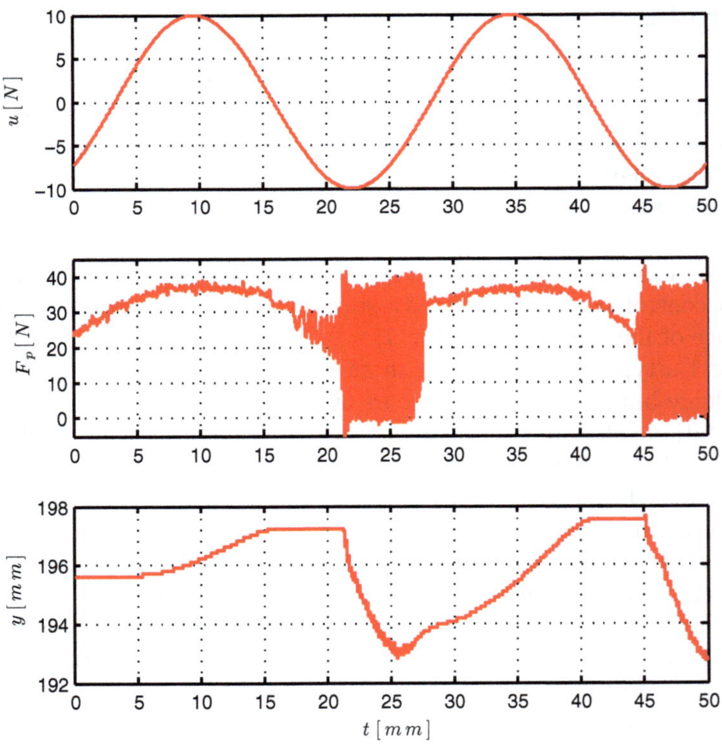

Fig. 1. Oscillations caused by friction.

identify the *static velocity to friction map*. Following in a second step, a simplified friction model is derived and motions at low speed are performed in order to identify the governing parameters of the *stick-slip* phenomena.

2 Hydraulic Actuator

The hydraulic actuator (HA) under analysis is presented in Fig. 2. Its main parts are the servo-valve and the double effect cylinder. Additionally, the position and force are measured by a linear digital encoder and a load cell at the distal end of the piston rod, respectively. The dynamic model and its experimental identification is fully described in [19].

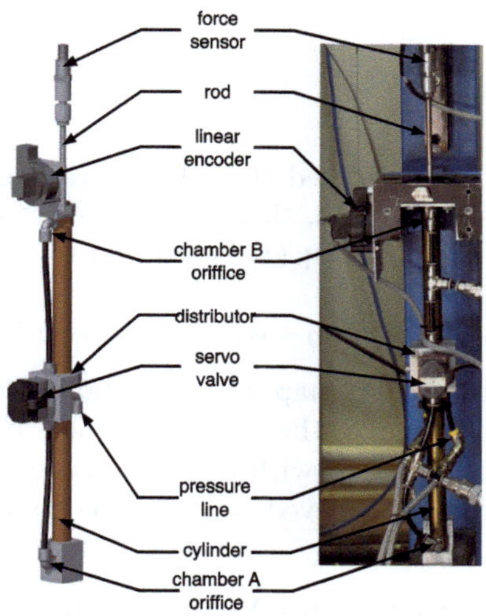

Fig. 2. Schematic diagram of the servo hydraulic actuator circuit.

3 LuGre Model Friction

The LuGre model [9], is an extension of the dynamic Dahl model, but it captures the Stribeck effect and describes the stick-slip motion. Moreover, the LuGre model has passivity properties that are useful for designing friction compensator that give asymptotically stable closed-loop systems. A comprehensive description of the model and its properties can be found in [14].

The standard LuGre model is described by the following parametrization:

$$F_f = \sigma_0 z + \sigma_1 \frac{dz}{dt} + \sigma_2 \dot{y} \tag{1}$$

$$\frac{dz}{dt} = \dot{y} - \sigma_0 \frac{|\dot{y}|}{g(\dot{y})} z \tag{2}$$

$$g(\dot{y}) = \alpha_0 + \alpha_1 e^{-(\dot{y}/v_s)^2} \tag{3}$$

where F_f is the friction force, $\dot{y}$ is the relative velocity between the surfaces in contact, and z is the internal friction state that can be interpreted as the average bristle deflection. The LuGre model reproduces the spring-like behavior for small displacement, where σ_0 is the stiffness and σ_1 is the microdamping. The macrodamping (i.e. viscous friction) is governed by σ_2. The velocity dependent function $g(\dot{y})$, provides the Coulomb friction (α_0) and Stribeck effect (α_1 and v_s).

4 Experimental Identification of LuGre Model Friction

The parameters that compose the dynamical model can be found experimentally by considering the different regimes of friction and the set of parameter associated to them [2,11].

4.1 Steady State Regime

At constant velocity, it is considered that the interaction between the contact surfaces are in steady state regime, therefore the micro-damping effect can be neglected (i.e. $\dot{z} = 0$), therefore Eq. (1) is reduced to the classic static friction model:

$$F_{ss}(\dot{y}) = \alpha_0 + \alpha_1 e^{-\dot{y}/{v_s}^2} + \sigma_2 \dot{y}. \tag{4}$$

The static velocity to friction map, can be obtained by establishing a series of constant velocity movements of the piston and measuring the friction force.

Given the integrating behavior with regard to the position of the HA, constant velocity motions can be achieved by providing steps inputs in open loop.

4.1.1 Experiment #1: Constant Velocity Motions.

- **Objective**
  Finding the parameters that govern the behavior of friction in steady state regime: α_0, α_1, σ_2, v_s.
- **Description of the experiment**
  The HA is submitted to step inputs in open loop driving external loads of 0 Kg, 3 Kg and 6 Kg. More than 50 tests with different step amplitudes within the range of $[-u_{min}, u_{max}]$ of the servo-valve for each external load are performed.
- **Data processing**
  - The velocity and acceleration of the piston are obtained from its position by means of $1st$ and $2nd$ order finite difference, respectively. The signals are filtered with a $4th$ order Butterworth low-pass filter with a cut off frequency at 200 Hz. It must be highlighted that finite difference has an acceptable performance for off-line processing, but not for real time requirements (see [18]).
  - The friction force can not be directly measured, therefore equating the dynamic equation of motion of the HA (see [19]), it is found the following relation:

$$F_{fp} = F_P - m_T\, \ddot{y}(t) - m_T\, g - F_{fg}, \tag{5}$$

where F_P is the force exerted by the piston given by: $F_p = A_p(P_A - \alpha P_B)$, P_A and P_B are the pressure at the chambers A and B of the piston respectively. m_T is the external load, F_{fg} is the friction force of the linear guide, $\ddot{y}$ is the piston's acceleration and g is the acceleration of gravity.

- It is assumed that the transient response to step input can be neglected after $0.5\,s$. Hence, the static velocity to friction map is generated with the set of data of $F_{fp}(t)$ and $\dot{y}(t)$ that belongs to $t \varepsilon [0.5,\,1]\,s$.
- The parameters of $F_{ss}(\dot{y})$ are estimated by solving a non linear curve-fitting problem in least-squares sense implementing the *lsqcurvefit* function of Matlab®. All the values of F_{fp} and $\dot{y}$ obtained during the tests are gathered into two matrices and used as input data of this function. Lower and upper bounds for each parameters are defined according to the maximum and minimum values observed during the tests.

- **Results**

1. A typical step response of the HA in open loop is presented in Fig. 3. As it can be observed, a constant voltage at the input of the servo-valve produces a motion with constant velocity. The velocity curve, shows that transients can be neglected for $t > 0.5\,s$. For the interval $0.5\,s \leqslant t \leqslant 1\,s$, it can be considered that $F_{fp}(t)$ and $\dot{y}(t)$ remain constant.
2. The static velocity to friction map of the HA for different external loads found is depicted in Fig. 4(a). The red, green and cyan circles correspond to the average value of the F_{fp} for a given constant velocity of the rod moving an external load of $0\,Kg$, $3\,Kg$ and $6\,Kg$, respectively. As it can be observed, the mapping presents an asymmetrical behavior and it is more sensible to external load changes in motions with negative velocities. Furthermore, for gross motion with positive velocities (i.e. $\dot{y} > v_s$), it can be said that there is no significant difference between the curves for different loads, which is not the case for $\dot{y} < 0$. However, it can also be observed that there is a tendency of the curves to converge to a line for $\dot{y} < -150\,mm/s]$.
3. The non linear curve fitting of the data obtained during the tests to the friction model (4) is presented in Fig. 4(b), superposed to the experimental data presented as grey dots. It can be observed that the fitted curve, preserves the asymmetrical characteristic of the curves found during experimentation.
4. The resulting parameters from the non linear curve fitting of the friction model in steady regime are summed up in Table 1. The $^+$ sign, indicates that the parameter corresponds to the half plane of positive velocity, while $^-$ indicates that the parameter corresponds to the half plane of negative velocity. This asymmetry of the parameters found regarding the sign of the velocity, it is assumed to be a consequence to the fact that the experimentation was performed against the action of gravity.

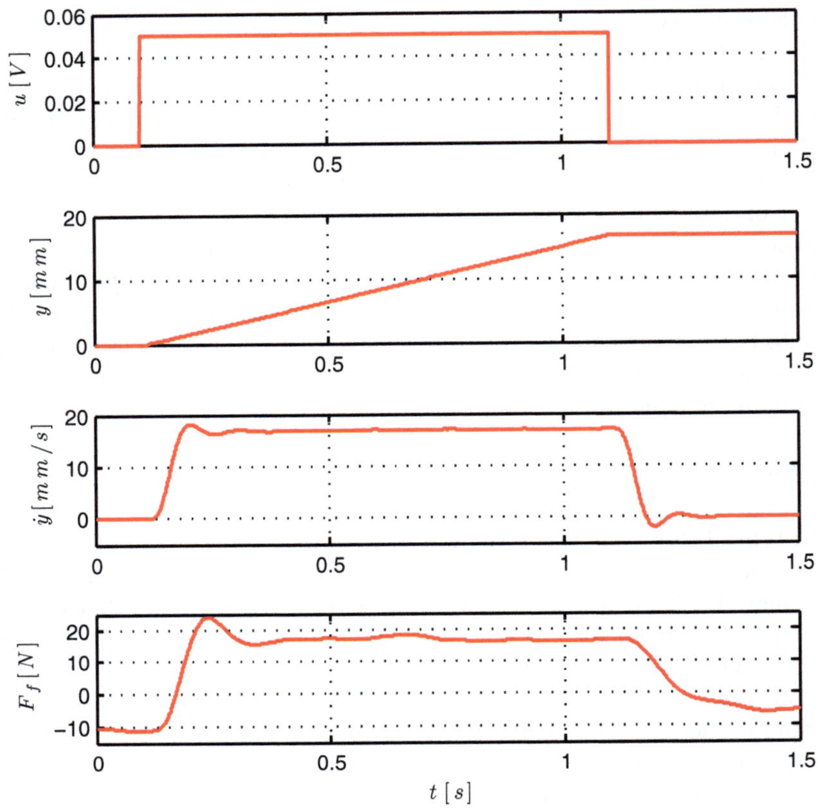

Fig. 3. Step response in open loop.

4.2 Pre-sliding Regime

In pre-sliding regime the friction force is due to elastic and plastic deformations of the asperity contacts. The interface between two contact surfaces can be viewed as an interface composed by springs. Therefore, if a force is applied the springs are extended but generating an opposing force, i.e. the friction force. If the extension becomes too large the springs snap and sliding occurs.

The pre-sliding regime occurs in motions at very low velocity (i.e. $\dot{y} \approx 0$). Then, linearizing Eq. (1) for $z = 0$ and $\dot{y} = 0$, the friction model is approached to the following expressions (see [14] for the detailed maths and assumptions considered for the linearization).

$$\delta F_f = \sigma_0 \delta z + (\sigma_0 + \sigma_1)\delta \dot{y}, \tag{6}$$

$$\frac{d}{dt}[\delta z] = \delta \dot{y}. \tag{7}$$

Considering that $z \approx 0$ and $\dot{y} \approx 0$, it can be assumed that $\dot{y} \approx \dot{z}$, then:

$$F_f \approx \sigma_0 y + (\sigma_1 + \sigma_2)\dot{y}. \tag{8}$$

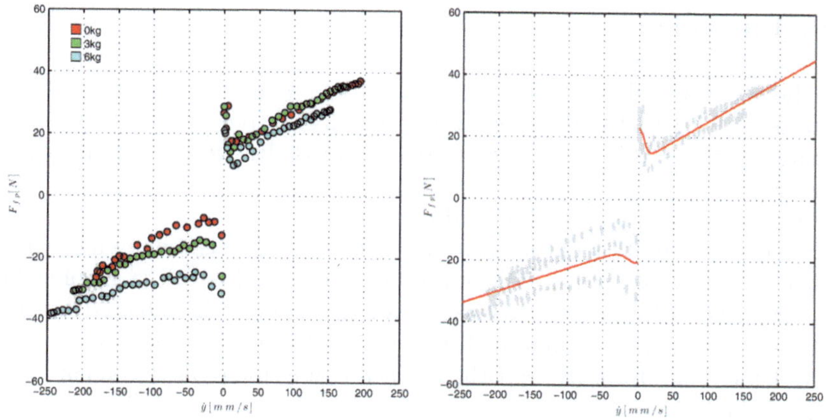

(a) (b)

Fig. 4. Velocity to friction map of the HA in steady state regime for different external loads.

Table 1. Estimated parameters for the friction model in steady state regime.

Parameter	Value	Unit
$\alpha_0{}^+$	12.20	N
$\alpha_1{}^+$	10.64	N
$v_s{}^+$	10.00	mm/s
$\sigma_2{}^+$	0.13	Ns/mm
$\alpha_0{}^-$	−15.00	N
$\alpha_1{}^-$	−5.39	N
$v_s{}^-$	−20.00	mm/s
$\sigma_2{}^-$	0.07	Ns/mm

Hence, the elastic characteristic of friction can be explored experimentally, by applying an external force F smaller than the stiction force (i.e. $F < F_s$) to the HA at rest and measuring its displacement.

4.2.1 Experiment #2: Force Control Response.

- **Objective**
  Finding the parameters that define the pre-sliding behavior in the HA: σ_0 and σ_1.
- **Description**
  The HA is submitted to controlled force tests, providing small motions at very low velocities. Controlled force is achieved by feeding back the force exerted by the piston (i.e. F_p) and implementing a proportional controller. A sinusoid input at 0.04 Hz, with 30 different amplitudes below static friction is applied to the HA. Tests for loads of 0 Kg, 3 Kg, and 6 Kg are performed.

- **Data Processing**
  - The velocity and acceleration of the piston are derived from its position by means of $1st$ and $2nd$ order finite difference, respectively. The signals are filtered with a $4th$ order Butterworth low-pass filter with a cut off frequency at $200Hz$.
  - Friction force is obtained according to Eq. (5).
  - Since, the resolution of the position sensor is not precise enough, the numerical identification of Eq. (8) can not be processed. Instead, a graphical estimation is performed. Therefore, in all the tests, it is detected the forces that produce the minimum distance that the position sensor is capable of detect (i.e. $0.1\,mm$), and the corresponding force-displacement chart is generated.
  - It is assumed that at small displacement the linearized friction model can be simplified into:

$$F_f \approx \sigma_0 y, \tag{9}$$

therefore, at low velocities friction is governed by σ_0 and it is obtained graphically as the slope of the force in the force-displacement chart.

  - The effects of the parameter σ_1 are critical at micro-scale, however in mechanical systems where the sensor resolution is within the millimeter scale, the effect of σ_1 is minor, and its main role is to damp the linearized equation in the pre-sliding regime rather than to finely match the data [5]. Therefore, replacing Eq. (8) in the motion's dynamics of the HA, the pre-sliding regime results into:

$$m_T\ddot{y} = u - (\sigma_0 y + (\sigma_1 + \sigma_2)\dot{y}), \tag{10}$$

where $u = F_p - m_T g$. Defining the linear approximated map $G(s) : y \rightarrow u$ as follows:

$$G(s) = \frac{1}{m_T s^2 + s(\sigma_1 + \sigma_2) + \sigma_0}. \tag{11}$$

It is imposed a damping ration $\zeta = 1$ for Eq. (11) in order to obtained a well-behaved stick-slip transition. Therefore, σ_1 will be given by:

$$\sigma_1 = 2\zeta\sqrt{\sigma_0 m_T} - \sigma_2. \tag{12}$$

- **Results**
  1. In Fig. 5 is presented the curves obtained during one of the tests carried on with an external load of $6\,Kg$. A set point of $60\,N$ is established in order to counteract the weight of the external load. From this result it can be observed that the response in position does not present the same behavior for positive and negative displacements. Furthermore, in positive displacements there is a bigger force component that opposes the motion. This result is clearly depicted in the friction force presented in the chart at the bottom. As it can be seen, the friction force rises up to almost $30\,N$ in the positive direction of motion, but it hardly reaches the $-10\,N$ in the opposite direction.

Table 2. Pre-sliding parameters estimation.

Parameter	Value	Unit
${\sigma_0}^+$	238.9709	N/mm
${\sigma_1}^+$	1.8944	Ns/mm
${\sigma_0}^-$	81.5808	N/mm
${\sigma_1}^-$	1.0911	Ns/mm

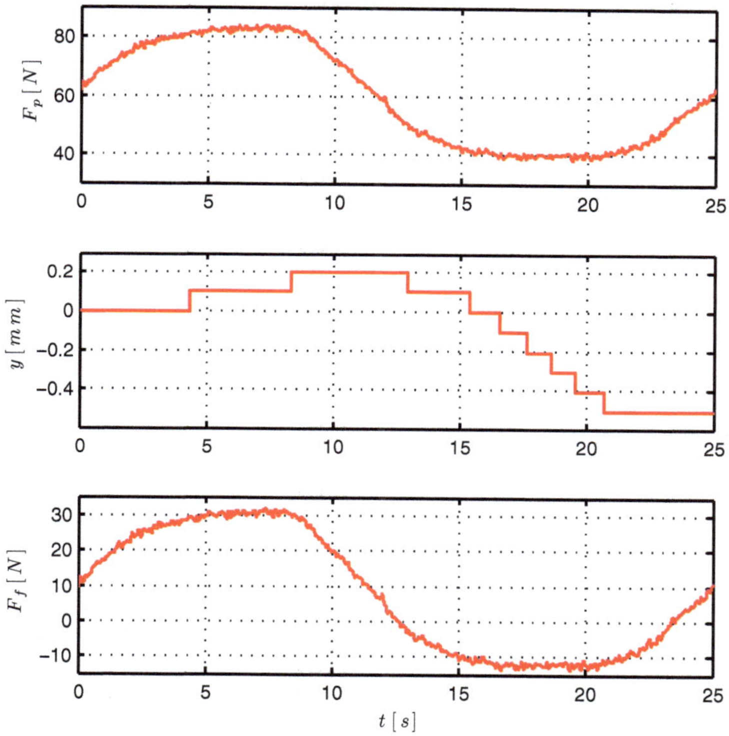

Fig. 5. Curves obtained from a force control test. Top: force generated exerted by the piston. Middle: position response. Bottom: calculated friction force.

2. The results of the tests are condensed in the chart graph presented in Fig. 6(a). As it can be seen, the elastic property of the pre-sliding regime conserves its characteristic even though the tests are carried on with different external loads. The asymmetry of the position response to force excursions can be also observed.
3. The linear fitting to the slope of the Force-displacement chart graph is presented in Fig. 6(b), remarking the asymmetry in the slope for positive and negative motions.

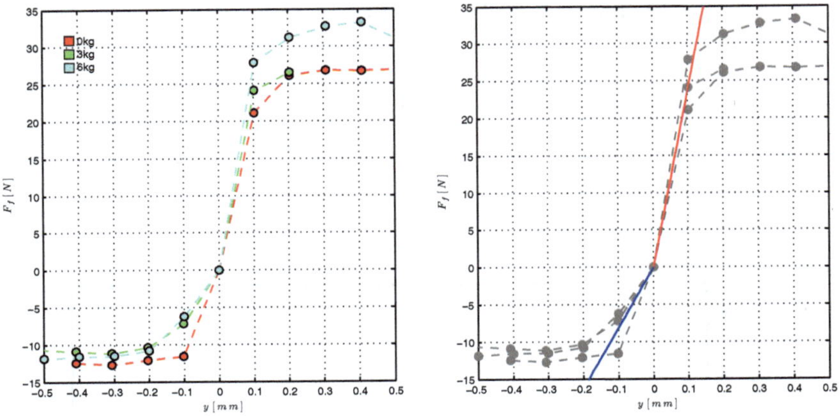

(a) (b)

Fig. 6. Force-displacement chart graph in pre-sliding regime.

4. The parameters that govern the pre-sliding regime are summed up in Table 2. The parameter σ_1, was calculated considering an external load of $3\,Kg$, in order to guaranty passivity of the model, i.e. $\zeta < \frac{\sigma_2}{2\sqrt{\sigma_0 m_T}}(\frac{\alpha_0}{\alpha_1} + 1)$ (see [5,7] for further details).

5 Conclusions

Friction effects are perhaps one of the most undesirable nonlinearities always present in any dynamic system, leading to instability, limit cycles and offset error.

Friction has several states and regimes that depends of the materials and relative velocity of the contact surface. The models found in the literature are as complex as the regimes that they pretend to recreate.

One of the critical regimes of friction is the *stick-slip* effect which are generated near zero velocity. This effect gives origin to limit cycles.

Therefore, for precision control it is needed a model that could represent this effects, such as the Lu-Gre model, that provides almost all the characteristics and regimes of friction: static friction, stick-slip, stribeck effect, viscous friction.

The identification of this model, can only be performed by steps, and identifying each of the parameters associated to each regime. For the identification of some parameters it is required to have high precision instrumentations. In the particular case of the parameters that govern the stick-slip effects, it is required position sensor with micrometer resolution in order to really perceive the spring like behavior of this regime.

Acknowledgements. The authors would like to thank the financial support of Spanish Government CICYT Project Ref. DPI2014-57220-C2-1-P, DPI2013-49527-EXP Universidad Politcnica de Madrid Project Ref. AL14-PID-15 and also to Comunidad de Madrid who supports the project ROBOCITY2030-III P2013/MIT-2748.

References

1. Al-Bender, F., Swevers, J.: Characterization of friction force dynamics. IEEE Control Syst. Mag. **28**(6), 64–81 (2008)
2. Altpeter, F.: Friction modeling, identification and compensation. Ph. D, École Polytechnique Fédérale de Lausanne (1999)
3. Armstrong, B., Chen, Q.: The Z-properties chart. IEEE Control Syst. Mag. **28**(5), 79–89 (2008)
4. Armstrong-Hélouvry, B., Dupont, P., Canudas-de Wit, C.: A survey of models, analysis tools and compensation methods for the control of machines with friction. Automatica **30**(7), 1083–1138 (1994)
5. Johanastrom, K., Canudas-de Wit, C.: Revisiting the LuGre friction model. IEEE Control Syst. Mag. **28**(6), 101–114 (2008)
6. Awrejcewicz, J., Olejnik, P.: Analysis of dynamic systems with various friction laws. Appl. Mech. Rev. **58**(6), 389 (2005)
7. Barahanov, N., Ortega, R.: Necessary and sufficient conditions for passivity of the LuGre friction model. IEEE Trans. Autom. Control **45**(4), 830–832 (2000)
8. Bombled, Q., Verlinden, O.: Dynamic simulation of six-legged robots with a focus on joint friction. Multibody Syst. Dyn. **28**, 395–417 (2012)
9. Canudas de Wit, C., Olsson, H., Astrom, K.J., Lischinsky, P.: A new model for control of systems with friction. IEEE Trans. Autom. Control **40**(3), 419–425 (1995)
10. Canudas-de Wit, C., Rubio, F.R., Corchero, M.A.: D-OSKIL: a new mechanism for controlling stick-slip oscillations in oil well drillstrings. IEEE Trans. Control Syst. Technol. **16**(6), 1177–1191 (2008)
11. Harnoy, A., Friedland, B., Cohn, S.: Modeling and measuring friction effects. IEEE Control Syst. Mag. **28**(6), 82–91 (2008)
12. Lischinsky, P., Canudas-de Wit, C., Morel, G.: Friction compensation for an industrial hydraulic robot. IEEE Control Syst. Mag. **19**(1), 25–32 (1999)
13. Olsson, H., Astrom, K.J.: Friction generated limit cycles. IEEE Trans. Control Syst. Technol. **9**(4), 629–636 (2001)
14. Olsson, H.: Control Systems with friction. Ph. D thesis, Lund Institute of Technology, June 1996
15. Olsson, H., Astrom, K.J., Canudas-de Wit, C., Gafvert, M., Lischinsky, P.: Friction models and friction compensantion. Eur. J. Control **4**(3), 176–195 (1998)
16. Owen, W.S., Croft, E.A.: The reduction of stick-slip friction in hydraulic actuators. IEEE/ASME Trans. Mechatron. **8**(3), 362–371 (2003)
17. Padthe, A., Drincic, B., Rizos, D., Fassois, S., Bernstein, D.: Duhem modeling of friction-induced hysteresis. IEEE Control Syst. Mag. **28**(5), 90–107 (2008)
18. Puglisi, Lisandro J., Saltaren, Roque J., Garcia, C.: On the velocity and accelertation estimation from discrete time-position sensors. Control Eng. Appl. Inf. **17**, 30–40 (2015)
19. Puglisi, L.J., Saltaren, R.J., Garcia, C., Banfield, I.A.: Robustness analysis of a PI controller for a hydraulic actuator. Control Eng. Pract. **43**, 94–108 (2015)

Comparative Analysis Between Fuzzy Logic Control, LQR Control with Kalman Filter and PID Control for a Two Wheeled Inverted Pendulum

Karen Viviana Chate García(✉), Oscar Eduardo Prado Ramírez(✉), and Carlos Felipe Rengifo Rodas(✉)

University of Cauca, Street 5 No. 4-70, Popayán, Colombia
{kchate,oscarprado,caferen}@unicauca.edu.co

Abstract. The aim of this article is to make a comparison between the performance of a fuzzy logic controller (FLC) and two linear controllers: a classical PID and a LQR with Kalman filter when they are applied to the two-wheeled mobile inverted pendulum robot *InstaBot SRAT-2*. The research is focused on determining the best controller for the mobile platform using three different performance indexes: *(i)* the standard deviation of the tilt angle, *(ii)* the root mean square value of the signal sent to the electrical motors and *(iii)* the region of convergence of the tilt angle. From the experimental test was observed that the LQR with Kalman filter controller presents a lower energy consumption and lower standard deviation of the error with respect to the PID controller; however the fuzzy controller presents a greater region of convergence than the two linear controllers. From the performance indexes it was concluded that fuzzy control is best suited for the mobile robot *InstaBot SRAT-2*.

1 Introduction

A mobile inverted pendulum robot is an unstable mechanical system with non-linear dynamics similar to a classical inverted pendulum [4], whereby it is used as a classic academic example to introduce the bases of system dynamics and feedback control [10]. In addition, this kind of robots has been used in multiple fields like agriculture, medicine and transportation [3,5] since they are autonomous, flexible and small, which increases its usefulness in narrow or dangerous working spaces [17].

In general, such vehicles are affected by external disturbances, unmodeled dynamics, errors in the estimated parameters and noise associated to the measurements, among others [14]. The different techniques of control proposed for these systems require of a good understanding of the mathematical model which can be obtained by Lagrange equations [4,9,10,13], Newton-Euler equations [1], Gibbs-Appell's movement equations, [3,16] or by Kane's method [4].

In order to achieve the control objective of the two wheeled inverted pendulum linear and non-linear control strategies has been applied. Among linear

I. Chang et al. (eds.), *Advances in Automation and Robotics Research in Latin America*, Lecture Notes in Networks and Systems 13, DOI 10.1007/978-3-319-54377-2_13

controllers we found the PID [4,15], LQR [1,9,13,21] and LQG [16]. These linear controllers present a limited operation range that depends on the linearization process, which limits the behavior of the robot. Therefore, it has been applied nonlinear control techniques that allows to achieve desired outcomes overcoming these restrictions. The literature presents non-linear control techniques such as fuzzy logic controllers (FLC) [18–20], neural networks [6–8], predictive control JIT [11], non-linear $H\infty$ controller [14], among others. In the same way, the development and implementation of hybrid controllers are highlighted, where two or three different control techniques are used to achieve the stabilization of the robot as shown in [10,17].

Despite of the great amount of research devoted to the stabilization of the two wheeled inverted pendulum, few articles address the performance comparison between the different control strategies. In [4], the three implemented controllers (FLC, LQR and PID) are compared from five different points of view: (i) rising time, (ii) settling time, (iii) overshoot, (iv) peak current supplied to the motors and (iv) linear displacement of the robot's platform obtained before steady state (tilt angle near to zero). In all tests the robot was initialized with a tilt angle of 24°. In [12], the authors compare by simulation a FLC and a PID controllers with respect to: (i) rising time, (ii) settling time, (iii) overshoot and (iv) steady state error with respect to the linear position. The current paper proposes a performance comparison considering two statistical measures, which are the standard deviation of the tilt angle and the root mean square value of the control law, and the region of convergence for the initial tilt angle.

The paper is organized as follows: In Sect. 2, describes the dynamic modeling of mobile pendular robot *InstaBot SRAT-2* using the *Lagrange* formalism. Section 3 presents the three implemented controllers used on the mobile platform. In Sect. 4, the performance indexes are defined as quantitative selection criteria to evaluate the performance of each controller. In Sect. 5, the time-response curves and performance indexes form implemented controller in the pendular mobile robot are shown and finally in Sect. 6, the conclusion of the paper is presented.

2 Dynamic Modelling of InstaBot SRAT-2

In Fig. 1 it is presented a graphic diagram of the mobile inverted pendulum robot in two dimensions.

The robot parameters are presented in the following Table 1:

A picture of the two wheeled inverted pendulum *InstaBot SRAT-2* is shown in Fig. 2.

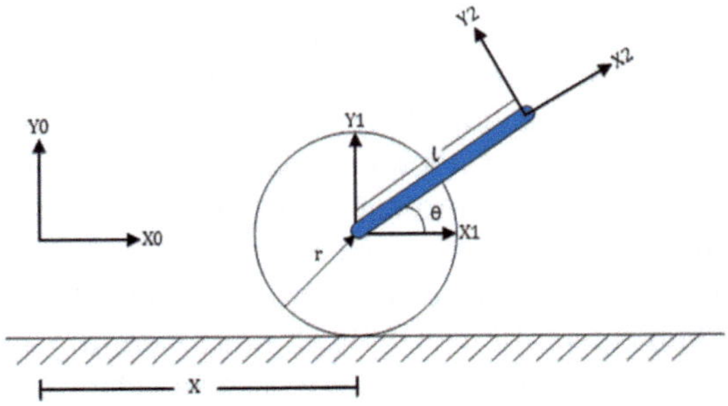

Fig. 1. Diagram of the mobile inverted pendulum robot in two dimensions.

Table 1. Parameters of the mobile inverted pendulum robot *InstaBot SRAT-2*.

Parameter	Symbol	Value	Unit
Pendulum body length	l	0.08530	m
Wheel radius	r	0.03310	m
Wheel mass	M	0.05304	Kg
Pendulum mass	m	0.39277	Kg
Wheel inertia momentum	J_1	5.8111×10^{-5}	$Kg \times m^2$
Pendulum inertia momentum	J_2	3.1754×10^{-4}	$Kg \times m^2$
Gravity	g	9.8	m/s^2

By using the *Lagrange* formalism the following mathematical model is obtained:

$$\underbrace{\begin{bmatrix} M+m+\frac{J_1}{r^2} & -ml\sin\theta \\ -ml\sin\theta & ml^2+J_2 \end{bmatrix}}_{I(q)} \underbrace{\begin{bmatrix} \ddot{x} \\ \ddot{\theta} \end{bmatrix}}_{\ddot{q}} + \underbrace{\begin{bmatrix} -ml\dot{\theta}^2\cos\theta \\ 0 \end{bmatrix}}_{H(q,\dot{q})} +$$

$$\underbrace{\begin{bmatrix} 0 \\ -mgl\cos\theta \end{bmatrix}}_{G(q)} = \underbrace{\begin{bmatrix} \frac{1}{r} \\ 0 \end{bmatrix}}_{E} \varGamma \tag{1}$$

$I(q)$ being the inertia matrix, $H(q, \dot{q})$ the vector of centrifugal and Coriolis forces and $G(q)$ the vector of gravitational forces. Model (1) can be rewritten as a state-space model composed by four differential equations:

$$\dot{x} = \begin{bmatrix} \dot{x_1} \\ \dot{x_2} \\ \dot{x_3} \\ \dot{x_4} \end{bmatrix}, \quad F = \begin{bmatrix} x_3 \\ x_4 \\ I^{-1}(x_2)\left(E\varGamma - H(x_2,x_4) - G(x_2)\right) \end{bmatrix} \tag{2}$$

Fig. 2. Mobile inverted pendulum *InstaBot SRAT-2*. Front view.

where the state vector is defined as:

$$x_1 = x \qquad x_2 = \theta - \frac{\pi}{2}$$
$$x_3 = \dot{x} \qquad x_4 = \dot{\theta}$$

To design a linear controller, the model described by (2) must be first linearized at the equilibrium point defined by:

$$\begin{aligned} \bar{x}_1 &= 0 \qquad \bar{x}_2 = 0 \\ \bar{x}_3 &= 0 \qquad \bar{x}_4 = 0 \end{aligned} \tag{3}$$

From the linearization process the following model is obtained:

$$\dot{x} = A\,x + B\,u \tag{4}$$

with A and B given by:

$$A = \begin{bmatrix} 0 & 0 & 1 & 0 \\ 0 & 0 & 0 & 1 \\ 0 & -\frac{g\,l^2\,m^2\,r^2}{d_n} & 0 & 0 \\ 0 & -\frac{g\,l\,m\,c_n}{d_n} & 0 & 0 \end{bmatrix}$$

$$B = \begin{bmatrix} 0 \\ 0 \\ \frac{r\,(m\,l^2 + J_2)}{d_n} \\ \frac{l\,m\,r}{d_n} \end{bmatrix}$$

and the constants c_n and d_n defined as:

$$c_n = J_1 + M\,r^2 + m\,r^2$$
$$d_n = J_1\,J_2 + J_2\,M\,r^2 + J_1\,l^2\,m + J_2\,m\,r^2 + M\,l^2\,m\,r^2$$

The matrices A and B define a linear state-space representation of the mobile inverted pendulum robot around the equilibrium point given by (3).

3 Controllers Design

This section describes the design and implementation of the control strategies (PID, LQR with Kalman Filter and Fuzzy) used in the mobile pendular robot *InstaBot SRAT-2* to achieve its stabilization.

3.1 PID Controller

This linear controller is based on the tilt angle error, which is defined as the difference between the set point (0°) and the measured tilt angle. The control law is defined by the Eq. (5).

$$u(t) = K_p\left(e(t) + \frac{1}{T_i}\int_0^t e(\tau)d\tau + T_d\frac{de(t)}{dt}\right) \tag{5}$$

Equation (5) is composed by the following three terms: proportional, integral and derivative. The proportional action is a gain value (K_p) multiplied by the error function. With a high value of K_p the steady state error decreases and accelerates the response of the system to the defined set point, however it can also raise the oscillations within the output signal [2]. On the other hand, the integral action eliminates the steady state error through the iterative accumulation of past errors. Finally, the derivative action reduces the oscillations in the output, resulting from the proportional and integral action. With the derivative gain value, the stability of the closed response is improved [2]. Based on this, the controller parameters that allowed to obtain the best result were $K_p = 47.5$, $K_i = 0.05$ and $K_d = 0.2$.

3.2 LQR Controller

The LQR controller design is based on the linear state space model described by (4), and the control law is defined as:

$$u = -Kx \tag{6}$$

where K is the state feedback gain and x the state vector. With the Eq. (6), the closed loop state-space model of the system is described below:

$$\dot{x} = (A - BK)x \tag{7}$$

The LQR controller aim is therefore to locate each of the eigenvalues of the matrix $A - BK$ in the left half plane s in such a way that the dynamic of the system is stable and it minimizes the quadratic cost function:

$$J = \frac{1}{2}\int_0^\infty x^T(t)\,W\,x(t) + u^T(t)\,R\,u(t)dt \tag{8}$$

W being a positive semidefinite matrix of 4×4 and R a positive scalar.

$$W = \begin{bmatrix} 1 & 0 & 0 & 0 \\ 0 & 190 & 0 & 0 \\ 0 & 0 & 1 & 0 \\ 0 & 0 & 0 & 0.7 \end{bmatrix}, \qquad R = 0.1 \tag{9}$$

Using the linear dynamic model (4) a position controller for the mobile inverted pendulum robot was achieved by adjusting the matrices W and R, which are related to the states and the inputs of system respectively. The controller gain K calculated with those matrices is:

$$K = \begin{bmatrix} 3.1623 & 44.1759 & 6.2140 & 0.3223 \end{bmatrix}$$

3.2.1 Tilt Angle Estimation with a Kalman Filter

The mathematical process performed by the Kalman filter to obtain the new state is based on a prediction and a correction mechanism, where through gain compensation between the previous estimate and the current observation, it is possible to achieve a convergence to the real states of the system.

The Kalman filter implementation proposed performs a sensor fusion between an accelerometer and a gyroscope. The data measured with the accelerometer are related to the angular orientation and the data delivered by the gyroscope are angular velocities. According to that, the following mathematical model is defined:

$$\begin{aligned} \dot{x} &= u \\ z &= x \end{aligned} \tag{10}$$

where u is the gyroscope measure, z is the orientation estimated with the accelerometer, and x is the estimated tilt angle. The discretization of the continuous time fusion sensor model (10) results in the following Kalman equations:

- Prediction

$$\begin{aligned} \hat{x}_k^- &= \hat{x}_{k-1} + h\,u_k \\ \sigma_k^- &= \sigma_{k-1} + Q \end{aligned} \tag{11}$$

- Update

$$\begin{aligned} G_k &= \sigma_k^-(\sigma_k^- + \rho)^{-1} \\ \hat{x}_k &= \hat{x}_k^- + G_k(z_k - \hat{x}_k^-) \\ \sigma_k &= (1 - G_k)\sigma_k^- \end{aligned} \tag{12}$$

For this particular system, u_k is the angular velocity measured with the gyroscope, z_k is the orientation estimated from the measurements delivered by the accelerometer and $\hat{x}_k$ is the estimated orientation. The Kalman filter implementation was performed using the *KalmanFilter*[1] from *TKJElectronics*.

[1] *Kalman filter* library. Available at https://github.com/TKJElectronics/KalmanFilter.

3.3 Fuzzy Control

Fuzzy logic presents an alternative to the classical reasoning and set theory, introducing the degrees of truth in a proposition and the degree of membership of a set as real values in the range [0, 1], this is done in order to achieve a more realistic approximation of human reasoning.

The fuzzy PD controller designed for the robot *InstaBot SRAT-2* uses as inputs the current value of the error signal (*Error*) and its time derivative (*ErrorChange*). For fuzzy PD controller we define 3 fuzzy sets for the variable (*Error*): *Negative error (N), Zero error (Z), Positive error (P)* and 3 fuzzy sets for the variable (*ErrorChange*): *Negative Change (NC), Zero Change (ZC), Positive Change (PC).* For the control law, we define 4 fuzzy sets: *Large Negative (LN), Small Negative (SN), No Control Law (NCL), Small Positive (SP), Large Positive* (LP), being *No Control Law* a triangular set and the other sets being of type *singleton.* The designed fuzzy PD controller has a base of 9 rules shown in Table 2.

Table 2. Rules matrix for the fuzzy PD controller.

	PC	ZC	NC
P	LP	SP	NCL
Z	SP	NCL	SN
N	NCL	SN	LN

A surface diagram of the control law obtained by using the above mentioned fuzzy sets and the rules shown in Table 2 is presented in Fig. 3.

4 Performance Indexes

The following indexes were used to compare the three controllers described in last section:

1. Standard deviation of the error

$$\varepsilon_e = \sqrt{\frac{1}{N}\sum_{k=1}^{N}\left[\theta_d(k) - \theta(k)\right]^2} \tag{13}$$

   Being ε_e the standard deviation of the error signal, N the number of samples, $\theta^d(k) \triangleq 0$ the desired tilt angle and $\theta(k)$ the measured tilt angle.
2. Root mean square of the control law

$$\varepsilon_u = \sqrt{\frac{1}{N}\sum_{k=1}^{N} u^2(k)} \tag{14}$$

   where ε_u is the root mean square value of the control effort $u(k)$.

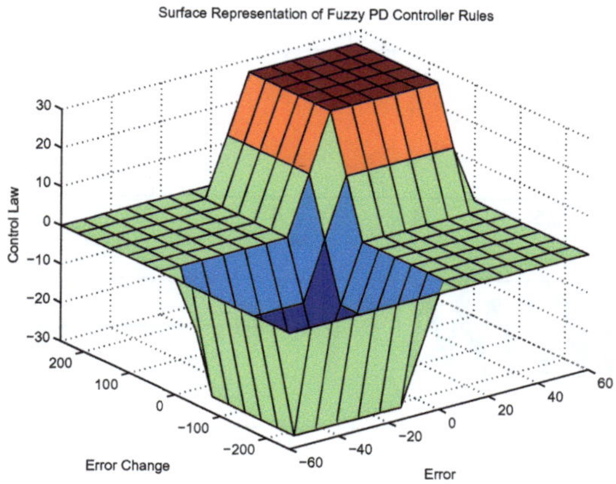

Fig. 3. Fuzzy PD controller surface diagram applied to the robot *InstaBot SRAT-2*. Input variables: *Error* (axis x), *Error Change* (axis y), output variable: *Control Law* (axis z).

3. The region of convergence of the tilt angle was obtained by initializing the robot with different degrees of inclination and determining whether it falls or it converges to the equilibrium position. The initial tilt angle was gradually increased until the controllers were not able to return to the equilibrium position $\theta = 0$.

For each controller the first two indexes were calculated multiple times (N = 11), each time using a different data set. The comparison between the three controllers was made by using the average value of each index.

5 Experimental Results

The results obtained from the implementation of each controller are illustrated in Figs. 4 and 5. Figure 4 presents the tilt angle as a function of time and Fig. 5 the control law

Table 3 presents the average value of the performance indexes (ε_e and ε_u) obtained from the eleven test developed for each controller.

Tables 4 and 5 show the ratios of the performance indexes of the linear controllers and the fuzzy control.

Figure 6 presents the critical convergence angles for the three implemented controllers. In the case of the PID controller, it has a poor performance due to the integral component which caused the control law to overact by the presence of a steady state error for a long period of time. LQR and fuzzy PD controllers did not present this inconvenience due to the absence of cumulative components in its control law.

Table 6 shows the range of initial tilt angles for which the control convergences to the desired equilibrium point.

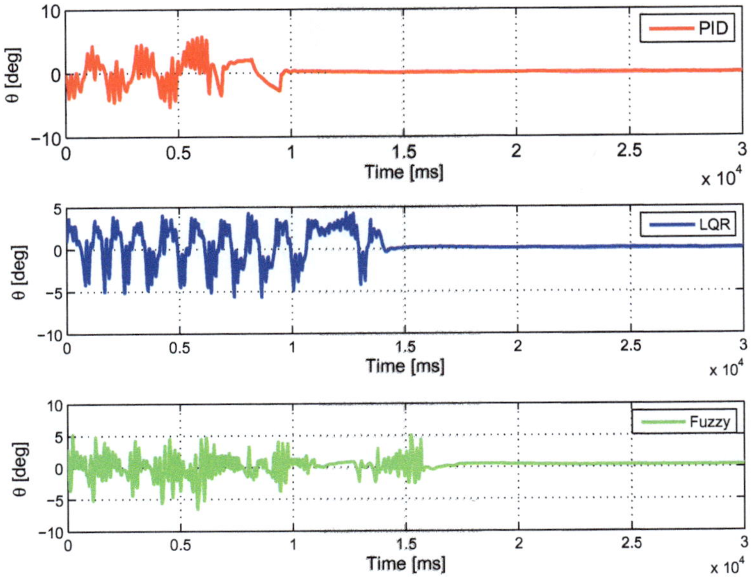

Fig. 4. Time response of the implemented controllers (tilt angle) *(i)* PID , *(ii)* LQR with Kalman filter, *(iii)* fuzzy PD.

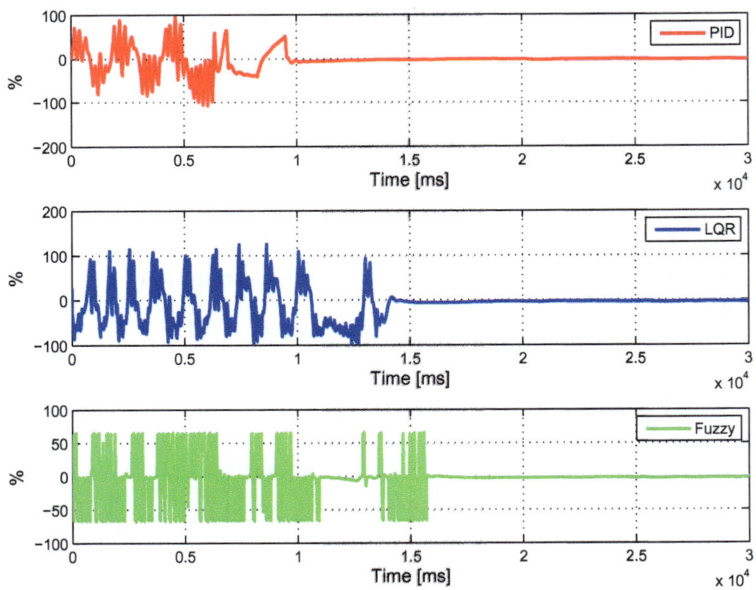

Fig. 5. Control effort obtained from the implemented controllers *(i)* PID , *(ii)* LQR with Kalman filter, *(iii)* fuzzy PD.

Table 3. Comparison of the average of standard deviations obtained from the eleven experiments realized with each controller.

Controller	$\varepsilon_e()$	$\varepsilon_u(\%)$
PID	1.95	37.7719
LQR	1.1559	26.4835
Fuzzy	0.7689	12.4778

Table 4. Comparison of the ratios of standard deviations of error in the implemented controllers.

$\varepsilon_{ePID}/\varepsilon_{eFuzzy}$	$\varepsilon_{eLQR}/\varepsilon_{eFuzzy}$	Least error
2.5361	1.5033	LQR

Table 5. Comparison of the ratios of standard deviations of control effort of the implemented controllers.

$\varepsilon_{uPID}/\varepsilon_{uFuzzy}$	$\varepsilon_{uLQR}/\varepsilon_{uFuzzy}$	Least control effort
3.0271	2.1224	LQR

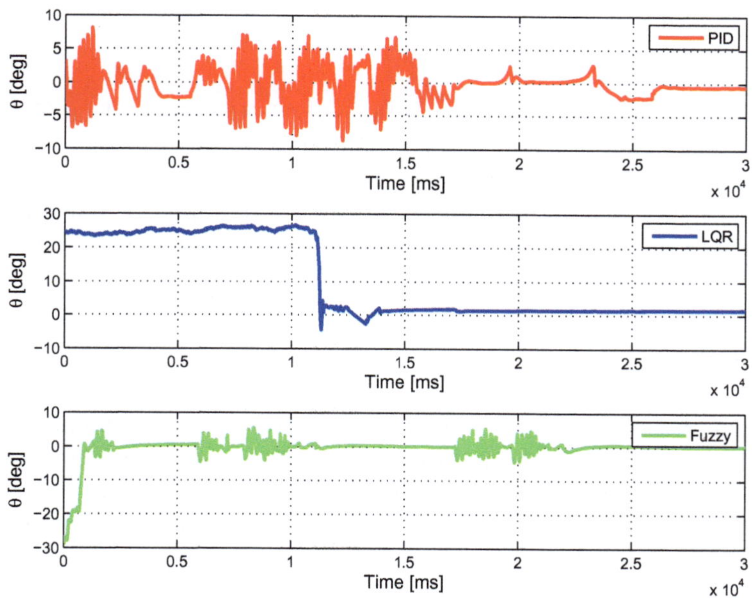

Fig. 6. Critical convergence angle of the implemented controllers *(i)* PID, *(ii)* LQR with Kalman filter, *(iii)* fuzzy PD.

Table 6. Comparison of the convergence regions found in the implemented controllers.

PID	LQR	$Fuzzy$	Most convergence region
$[-4, 4]$	$[-24.4, 24.4]$	$[-28, 25]$	Fuzzy

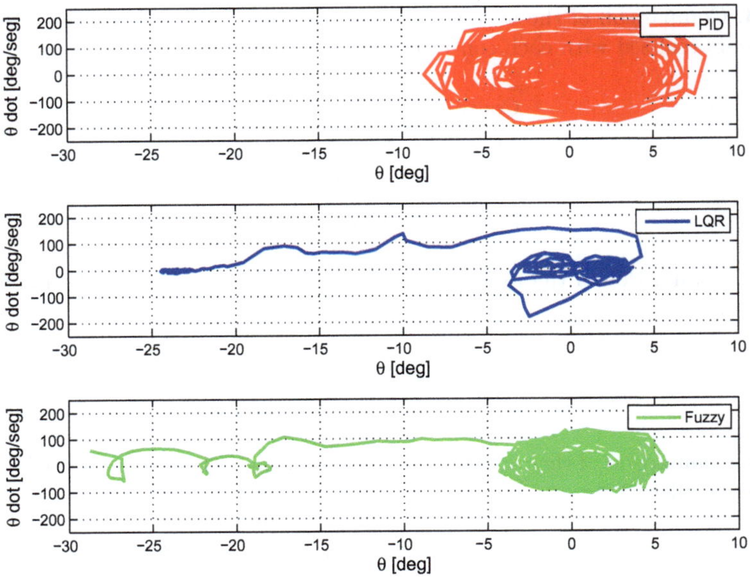

Fig. 7. Convergence towards the equilibrium point ($\theta = 0, \dot{\theta} = 0$) of the implemented controllers *(i)* PID, *(ii)* LQR with Kalman filter, *(iii)* fuzzy PD.

Figure 7 presents the trajectory of the robot in the phase plane ($\theta, \dot{\theta}$) for an initial tilt angle close the critical angle of Table 6.

It is observed that the PID convergence region is the smallest from the three implemented controllers, due to the control law overflow produced by the integral component. The LQR and fuzzy controllers present a wide range of initial values for which the controller can return to the desired set point.

6 Conclusion

In this paper the dynamical model of a mobile inverted pendulum robot was obtained by using the *Lagrange* formalism. For this robot three controller, two linear (PID and LQR) and one nonlinear (Fuzzy-PD) were implemented. The experimental results obtained for each controller were compared by using three performance indexes. Two of them respectively depending on the standard deviation of the tilt angle and on the control effort. The third index is the range of initial tilt angles converging to the equilibrium point. With these three indexes it was verified that the fuzzy controller presents a smaller error deviation and less energy consumption, in addition it presents a greater region of attraction to the equilibrium point. The Kalman filter design was applied to improve the estimation of the robot orientation angle for the LQR controller, allowing it to have a better result of the deviation error with respect to the PID controller.

References

1. Akeson, J., Blomdell, A., Braun, R.: Design and control of yaip- an inverted pendulum on two wheels robot. In: Proceedings of the 2006 IEEE International Conference on Control Applications, Munich, October 2006
2. Astrom, K.J., Hagglund, T.: Control PID Avanzado. Pearson Prentice Hall, Upper Saddle River (2009)
3. Azimi, M., Koofigar, H.: Model predictive control for a two wheeled self balancing robot. In: Proceeding of the lnternational Conference on Robotics and Mechatronics (RSIIISM), Tehran, February 2013
4. Bature, A.A., Buyamin, S., Ahmad, M.N., Muhammad, M.: A comparison of controllers for ballancing two wheeled inverted pendulum robot. Int. J. Mech. Mechatron. Eng. IJMME - IJENS **14**(3), 62–68 (2014)
5. Brisilla, R.M., Sankaranarayanan, V.: Nonlinear control of mobile inverted pendulum. Rob. Auton. Syst. **70**, 145–155 (2015)
6. Chih-Hui Chiu, You-Wei Lin, and Chun-Hsien Lin. Real-time control of a wheeled inverted pendulum based on an intelligent model free controller. Mechatronics, 21(3):523–533, 2011.
7. Chiu, Chih-Hui, Peng, Ya-Fu, Lin, You-Wei: Intelligent backstepping control for wheeled inverted pendulum. Exp. Syst. Appl. **38**(4), 3364–3371 (2011)
8. Chiu, C.-H., Tsai, W.-R., Chou, M.-H., Peng, Y.-F.: Two - wheeled robot control based on self-tuning output recurrent CMAC. In: Proceedings of the International MultiConference of Engineers and Computer Scientists, Hong Kong, vol. 1, March 2009
9. Jamil, O., Jamil, M., Ayaz, Y., Ahmad, K.: Modeling, control of a two-wheeled self-balancing robot. In: International Conference on Robotics and Emerging Allied Technologies in Engineering (iCREATE 2014), pp. 191–199, April 2014
10. Mahler, B., Haase, J.: Mathematical model and control strategy of a two-wheeled self-balancing robot. In: 39th Annual Conference of the IEEE Industrial Electronics Society, (IECON), pp. 4198–4203, November 2013
11. Nakpong, N., Yamamoto, S.: Just-in-time predictive control for a two-wheeled robot. In: 10th International Conference on ICT and Knowledge Engineering (ICT Knowledge Engineering 2012), pp. 95–98, November 2012
12. Nasir, A.N.K., Ahmad, M.A., Ghazali, R., Pakheri, N.S.: Performance comparison between fuzzy logic controller FLC and PID controller for a highly nonlinear two-wheels balancing robot. In: First International Conference on Informatics and Computational Intelligence (ICI), pp. 176–181, December 2011
13. Peng, K., Ruan, X., Zuo, G.: Dynamic model and balancing control for two-wheeled self-balancing mobile robot on the slopes. In: Proceedings of the 10th World Congress on Intelligent Control and Automation, Beijing, July 2012
14. Raffo, Guilherme V., Ortega, Manuel G., Madero, Vicente, Rubio, Francisco R.: Two-wheeled self-balanced pendulum workspace improvement via underactuated robust nonlinear control. Control Eng. Pract. **44**, 231–242 (2015)
15. Ren, Tsai-Jiun, Chen, Tien-Chi, Chen, Chun-Jung: Motion control for a two-wheeled vehicle using a self-tuning PID controller. Control Eng. Pract. **16**(3), 365–375 (2008)
16. Sun, F., Yu, Z., Yang, H.: A design for two-wheeled self-balancing robot based on kalman filter and LQR. In: International Conference on Mechatronics and Control (ICMC), Jinzhou, July 2014

17. Sun, L., Gan, J.: Researching of two-wheeled self-balancing robot base on LQR combined with PID. In: 2nd International Workshop on Intelligent Systems and Applications (ISA 2010), pp. 1–5, May 2010
18. Wang, W.-J., Huang, C.-H.: Model-based fuzzy control application to a self-balancing two-wheeled inverted pendulum. In: IEEE Control Applications, (CCA) and Intelligent Control, (ISIC), pp. 1158–1163, July 2009
19. Wen, Y.-H., Lin, Y.-S., Leu, Y.-G.: Design and implementation of the balance of two-wheeled robots. In: International Conference on Advanced Robotics and Intelligent Systems (ARIS 2013), pp. 130–134, May 2013
20. Wu, J., Zhang, W.: Design of fuzzy logic controller for two-wheeled self-balancing robot. In: 6th International Forum on Strategic Technology (IFOST 2011), vol. 2, pp. 1266–1270, August 2011
21. Xu, C., Li, M., Pan, F.: The system design and lqr control of a two-wheels self-balancing mobile robot. In: International Conference on Electrical and Control Engineering (ICECE), pp. 2786–2789, September 2011

Manufacturing Control Architecture for FMS with AGV: A State-of-the-Art

Sergio R. González[1(✉)], Ivan Mondragón[1], Gabriel Zambrano[1], Wilson Hernandez[1], and Henry Montaña[2]

[1] Department of Industrial Engineering, Pontificia Universidad Javeriana, Bogotá D.C, Colombia
{sgonzalezb,imondragon,gzambrano,hernandez-w}@javeriana.edu.co
[2] Department of Electronics Engineering, Universidad Distrital Francisco José de Caldas, Bogotá D.C, Colombia
hmontana@gmail.com

Abstract. In environments of constant development, the production system is going more complex, focusing on new tendencies about continuous improvement for technologies in manufacturing allows to reduce times and costs of production. The implementation of new Material Handling Systems (MHS) in manufacturing line, allows decrease times for the process of transport. The MHS is utilized principally in the production system for repetitive tasks (i.e., internal and external transport for raw material and goods). Through effective implementation of an MHS, it reduces damages to the materials and risks for workers and the same time increases the efficiency of the operation. In the manufacturing, transport costs is associate to different aspects such as reduced reactivity, recovery system failures, inflexibility, low autonomy and limitation of classical architectures of a MHS, all of this is due to reduced capacities of interaction between control systems (i.e., MHS control and Flexible Manufacturing System (FMS) control). For this reason, the costs of material handling can be reduced through integrated control architectures. In these circumstances, the challenge is to develop manufacturing control architecture for FMS and Automatic Guided Vehicle (AGV) with reactivity to the environment changes, scalability, robustness against the occurrence of disturbances, easier integration of manufacturing resources, and autonomy and intelligence capabilities. Although specific research in this topic has achieved a number of great successes, the general framework for the development on architectural level has not been defined by the community. This paper focuses on the overview over principal development in control architecture literature for FMS, AGV and FMS-AGV, in order to overcome of different aspects of transport and the limitations of classical hierarchical architectures.

1 Introduction

The shrinking product life cycles, globalization, mass customization, market volatility, changing nature of industrial requirements are some of the challenge of

I. Chang et al. (eds.), *Advances in Automation and Robotics Research in Latin America*, Lecture Notes in Networks and Systems 13, DOI 10.1007/978-3-319-54377-2_14

accomplishing a global optimal performance in current manufacturing systems [1]. A manufacturing system is *"a collection of integrated equipment and human resources, whose function is to perform one or more processing and/or assembly operations on a starting raw material, part, or set of parts"*, (Fig. 1) [5]. A manufacturing system requires various types of inputs (e.g., energy, raw materials, tools, equipment, labor, market information and product design) and inevitably non-desired outputs are generated (e.g., waste and scrap). Manufacturing system involves highly adaptive, reactive and fault-tolerant manufacturing control [4].

In this context, control is *"concerned with managing and controlling the physical activities in the factory aiming to execute the routing plan provided by the manufacturing planning activity"* [2]. In the detailed view of control, each level of the whole manufacturing system is seen as a set of controllers that execute the control function in a *"hierarchical"* level [9]. The control properties allow to produce high quality parts with reduced duty cycles and costs. All of this despite disturbances as tools, equipment and material failures [7,8]. According to industrial requirements it is priority developing an intelligent control in manufacturing systems with properties of flexibility and quick reconfiguration for new manufacturing process.

The flexibility in manufacturing is the *"capability to adapt rapid and frequent changes in flow of materials and parts"* [6]. Therefore, flexibility of a manufacturing system is dependent upon its components (machines, MHS, etc.), capabilities, interconnections, and the mode of operation and control [4]. The flexibility should be inherent to the control architecture (e.g., centralized and decentralized control) being possible to adapt with a minimum effort in the programing. Control architecture is a structure model for the FMS that determines interrelationships and establishes mechanisms among control components (e.g. machines, transport system and equipment control). Depending of such structure, its allow controller coordinated the execution of control decisional for transform raw materials in goods [4,9]. Analyzing a single problem is possible by determining the performance of architecture control. This, under statics conditions while con-

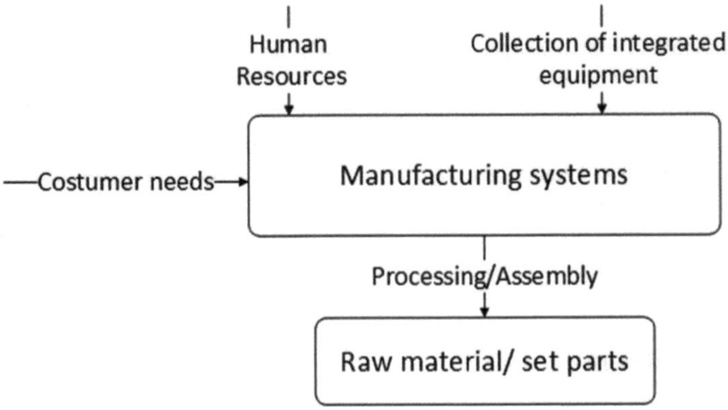

Fig. 1. Manufacturing System.

trol the system is activate. This brings evidence that the centralized control of production system do not accomplish flexibility requirement, manifesting the inadequate the architecture centralized control [4,7,8].

It depends on factors such as quantity of components and the automation level that manufacturing systems can be classified in three groups. The first group is characterized by manual manufacturing with low product variety or similar characteristics. The second group consists in multiples machines, where the manufacturing operations and material transfer between machines are manual proceses. The last one, integrate automatic machines processing with automatic material handling, are known as Flexible Manufacturing Systems (FMS). Roughly speaking a FMS is a manufacturing system in which there is some amount of flexibility that allows react in case of changes. It has recently gained increasing attention becomes an important issue for a growing range of industries.

2 Flexible Manufacturing System (FMS)

Many types of manufacturing systems are currently implemented, including assembly lines, batch production and Flexible Manufacturing Systems (FMS). A FMS is a manufacturing system with a high degree of flexibility, defined as *"...an automated, mid-volume, central computer controlled manufacturing system..."* [10]. Generally speaking, FMS are one of the systems that combine productivity-efficiency of transfer lines and flexibility to react in case of changes, reducing or eliminating problems in manufacturing industries [4]. Flexible Manufacturing Systems (FMS) have become very popular due to the production low level costs and their high levels of productivity [3].

From a systemic point of view, a hierarchical FMS is composed of three subsystems: production system, a material handling system (MHS), and a hierarchical computer system for control purposes [1]. The control system in FMS deals with three types of control decisions in real time: sequencing (i.e., product launching order in the FMS), machine routing (i.e., machine selection among alternative machines for the same manufacturing operation) and material handling (i.e., route selection among the alternative transfer paths allowed by the transportation system that connects the machines) [11]. The final effect on the performance in a FMS is attained if control decisions are appropriately taken by the control system [12]. In FMS, such control decisions are closely related to different types of flexibilities, hence, flexibility and FMS control Architectures are attached for a good performance level [4].

2.1 FMS Control Architectures

In current FMS (Fig. 2), there are two configuration commonly used in classical manufacturing control system [1,13,14]. The control architecture defines the blueprint for the design and construction of FMS control [4]. According to [15], the traditional architectures of the manufacturing system common implemented are: centralized and hierarchical. It is a centralized where a single control simultaneously controls all subsystems (e.g. master-slave configuration). The centralized control architectures can be found in the lowest levels of hierarchical and hierarchical architectures. It is especially in basic process and low complexity where it can be driven by unique central computer.

The second one is a hierarchical structure where each subsystem is controlled separately by one controller based on local information [16]. These subsystems are controller from central control. The capacity of decision-making is distributed to entities (i.e., decisional entities (DE)) or subsystems with below decision in the process [4,15].

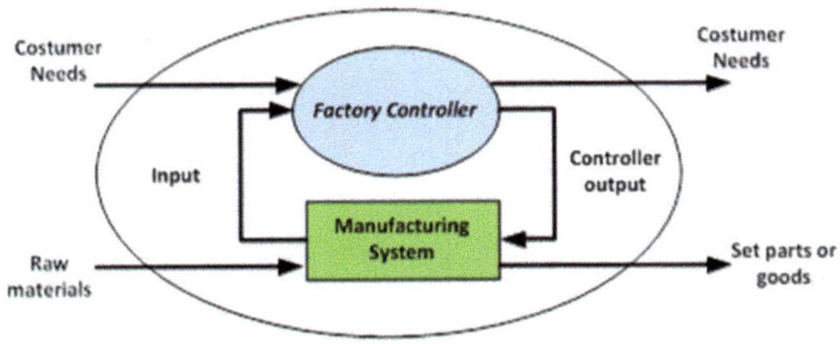

Fig. 2. Block diagram of control manufacturing system, based on [15].

Recent works have integrated dynamic functions on FMS control architectures, specifying how data and control are organized according to the structural and behavioral characteristics that define the elements (e.g. attributes, structure composition and operational) of the control system [1], and how the individual components interact with each other. Such architectures introduced heterarchical relationships in order to respond to changes or perturbations (i.e., reactivity), fault tolerance, scalability, among others [8,17].

In the decentralized control, the process can be considered as a unit functional. In this is allocated an element of control according to system requirements or process. Each one unit is interconnected to the other through a complex system of communication for information exchange [5]. The existence of different unit dont involves complete detection of all process, only one the fail unit [18].

In the early 1970's according to [15], the first class of distributed control in the manufacturing is known Computer Integrated Manufacturing (CIM). This method of manufacturing depends of process of close loop in real time. CIM is a method of manufacturing complete controlled by a computer. This depends of sub process of closed loop control. This approximation is lead to Manufacturing Resource Planning (MRP2) and Enterprise Resource Planning (ERP).

2.1.1 Fully Hierarchical Control Architectures

The perception of hierarchical (Fig. 3) in the complex structure of the organization of an industrial process is taking as base for the coordination of the systems

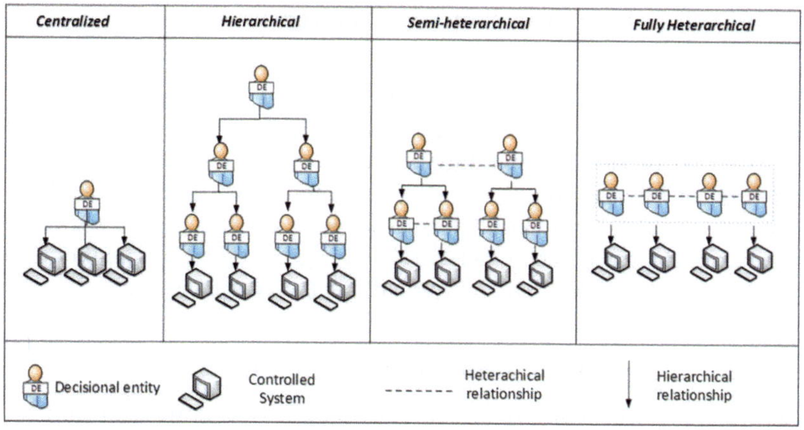

Fig. 3. Architectures of control in the manufacturing, based on [4].

between subordinates and entities of control. Interacting through a constant flow of information in both directions [19]. This is realized by a division of assignment of control and supervision in one level or more of controllers for reduce the complexity of a system centralized unique. This allows a data exchange with a better synchronization with a minimum of disturbances [4].

Typically at the top of the hierarchical, there is no single decisional entity that leads the decisions making of architecture, (Fig. 4). This is responsible for the overall efficacy of the planning horizontal. The planning horizontal is gradually less in the lowest level of the hierarchical [4]. Compared to the centralized architecture, hierarchical allows a gradual increase of control. This resulting in a reduction in development time software allowing to limit (i.e., response times near real time) the complexity of system [4].

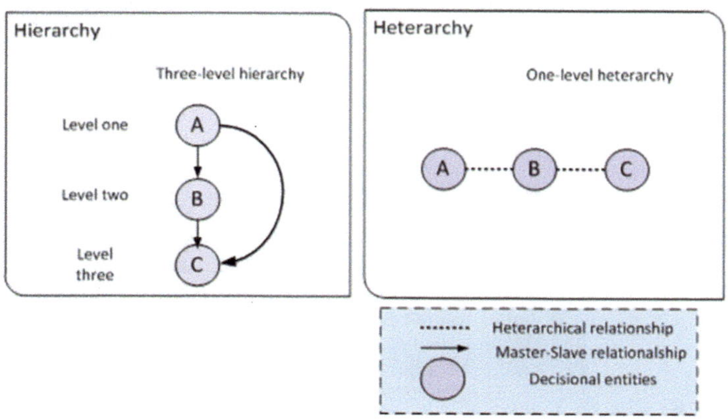

Fig. 4. Hierarchical and heterarchical architectures, based on [15].

2.1.2 Fully-Heterarchical Control Architectures

In control architectures with only one level of hierarchical (i.e., heterarchical architecture), is representing though a distribution of responsibilities of each one decisional entities. The heterarchical structure is based in a descentralized control, (Fig. 4). In this, is eliminating the need of a online control and minimal retention of information. This allows eliminate the use of a data base. The critical response times are handled locally without affecting other entities Adaptive capacity of the architecture heterarchical is guaranteed by the independence between decisional entities and their equal right access to resource [4].

2.1.3 Semi-heterarchical Control Architectures

Semi-hetarchical control combines the advantages of hierarchical and heterarchical architectures avoiding the disadvantage of each one, (Fig. 4). So involve multilevel relations with a low level of autonomy compared with pure structures heterarchicas [4]. According to [20] the systems semi-heterarchical can be represented by a control system. It is designed to perform real time a list of predetermined tasks operated by an active resource (AR). All of this taking account each resource allocation and routing possible [21]. So a semi-hetarchical system can be controlled by a dynamic assignation process (DAP) and dynamic routing process (DRP). These structures can be observed in (Fig. 4). The input to DRP is composed by a pair of nodes (ns,nd), where (ns), is the source node of resources and (nd) is a final node of resources that associate to one or more products in an instant time(t). The output of a DRP is a real time optimized of transport by the routing products. This information will used for a DAP to improve the assignations.

The overall structure of a DAP and a DRP is administrate as a architecture heterarchical unsupervised, (Fig. 5). The relationship between a DAP and DRP is then considerate dependent herarchical [20, 21].

In semi-heterarchical control two points are important of stand out. The first one refers to the hierarchical in heterarchical system. This helps in the prediction of behaviors of the control systems. The second one, the hierarchical benefits to

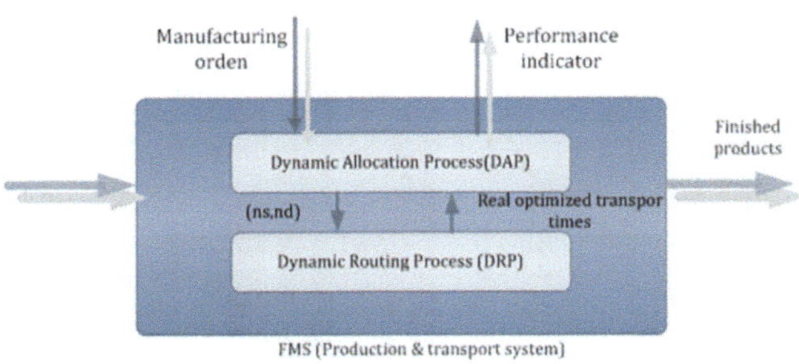

Fig. 5. Semi-Heterachical Control structure, based on [15].

the hetarchical architectures in the migration of the industrial application totally hierarchical to a focusing more descentralized [4].

3 Material Handling Systems (MHS)

FMS require a qualified MHS to transport material (i.e., raw materials, partially manufactured products and goods) safely and with low cost throughout the manufacturing and distribution process. MHS is an important area in a FMS because more than 80% of time is spent in waiting queues or in transportation [22]. Conventional solutions of MHS are based on forklift, industrial trucks, belt, roller and vertical conveyors, elevators, material handling robots and AGV. MHS face some limitations such as the occurrence of bottlenecks, deadlocks, local optimization and low efficiency [6,14].

Automated Guide Vehicle (AGV) defined as *"a material handling system that uses independently operated, self-propelled vehicles guided along defined pathways in the facility floor"* [22] have gained new interest to transport materials between workstations (e.g. load and unload points, machines of processing material) and can be used for the resolution of bottlenecks due to reconfigurable setup that allows AGVs to create new routes. In addition, AGVs are flexible and have the capability to make their own decisions and cooperate with other AGVs [23]. AGVs are employed to maintain flexibility (e.g. diversity of vehicle types, route simplification between processes within complicated networks and the ability to program and retrofit with new tooling to deal with diverse industrial needs), space utilization and efficiency of production and transport [24].

Transport processes required a detailed scheduling for the efficient organization similar to production. In this case, a schedule allocates jobs to available transport vehicles (e.g. trucks, AGV, conveyor, etc.), determines routes, pick-up and drop-off points (P/D) and due dates. The main objective is to minimize the total transportation time required to fulfill the requested jobs [25]. The transportation scheduling problem is decomposed into two sub-problems: AGV allocation and routing.

The industrial use of AGV has grown due to its great potential on the performance of manufacturing environments (e.g. distribution centers, transshipment terminals, warehousing systems, production plants and FMS) being most frequently implemented where they operate alongside humans [14,24]. In this situation, autonomy is vital for the safety of human workers and effective operation of the system. Autonomy in robotics can be defined as: *"within a rational behavior, by the effectiveness and robustness of a robot in carrying out tasks in different and well-known environments"* [24]. Autonomous controls have the ability for self-governance in the performance of control functions (e.g. tracking, regulation and the ability to tolerate failures). A MHS based on an autonomous AGV is known as Flexible Manufacturing Handling System (FMHS) due to its intrinsic ability to accommodate rapid and frequent changes in work-flow [6].

Different approaches to solve problems to optimize the control of AGV systems in FMS were found in the literature: scheduling transport, dispatching of loads, production planning and design of facilities, all related with operational decision, where the main objective for control is to satisfy demands for transportation tasks as soon as possible without conflicts (e.g. deadlocks and collisions) [14,26]. Often, the efficient control and coordination of these decisions is taken to solve different problems, such as: reduce material handling cost, in-process inventories and overall operational cost [14].

The performance measures for evaluating the AGV control system module are: the number of deadlock situations (number of problems that require operator intervention), dispatching rate (i.e. the number of dispatches per hour), quantify the performance of algorithm (i.e. average waiting time of ready parts and the average orders queue length) and AGV utilization (i.e. AGV empty travel rate and AGV idle time) [27]. Embedding autonomous controller structure into AGV releases the higher level production management systems from routing of parts and materials besides provide high level adaptation to changes in the plant and environment [24,28].

There are a number of issues that frequently arise in a typical AGV system. These can occur during the conception, designing, implementation, or operational stages [29]. The AGVs systems design includes problems like: flowpath design, deadlock prevention, conflict-free routing, capacity, fleet size, jobs, traffic management, determination of pick-up and drop-off points (P/D), number and location of points, idle points number and location, battery management, fault management, navigation and guidance and system management (i.e. method of system control used to dictate system operation) [30].

4 Control Architecture of an Autonomous AGV

The control structure (i.e., control architecture) of an autonomous AGV is a framework in which the processes are carried out (e.g. sensing, control, errors detection and recovery, path planning, tasks planning and monitoring of events) during the execution of a particular job. This defines how these should be integrated to get the desired results through decisional capacity [31]. The development of control architecture heavily depends on the environment, jobs, and hardware components [6].

However, due to the inherent complexity of the manufacturing environment, traditional control architecture systems still do not exhibit the capability of adaptation and evolution in terms of production control [6]. In fact, the centralized and hierarchical control approaches present good production optimization but a weak response to change. More, many existing warehouses deploy AGVs use a centralized or hierarchical control paradigm that is integrated with the rest of the material handling systems [32]. With the intervention of a central controller, AGVs require various kinds of guidance for navigation, communication media for transmission of information among AGVs and well-organized jobs definition generated during the system planning stage. With these approaches to material handling, these AGVs cannot be regarded as fully autonomous [6,32].

Autonomous problem arise in that is constrained by rules that are imposed by the strictly specified task execution routines for each resource. Traditional

AGV require costly and rigid changes to infrastructure, typically these utilize a limited drivepath using predetermined routes (i.e., closed loop in one direction), which are frequently demarcated by striping the floor in some manner or by using buried cables or chemical stripes painted on the plant floor. In this case one of the most important problems of deploying AGVs is that the environment around them is not static. These cannot adapt to a changing environment, and are not safe for collaboration with warehouse personnel. AGVs also have limited interaction with the workstations [33].

5 Literature Review

In this section, it is reviewed literature of architecture control for robot, AGV, FMS, and FMS integrated with AGV. The authors provided a list of references this field of control that present either a dedicated or distributed component to the production control.

[34] presented a hierarchical queueing network approach to determine the number of AGVs. Three main issues emerge: track layout, the number of AGVs required and operational transportation control.

[35] proposed a distributed control architecture KAMROs Multi Agent Robot Architecture (KAMARA) that are responsible to overcome coordination problems caused by the independent task execution of systems.

[36] proposed control architecture for autonomous vehicle driving in a dynamic and uncertain traffic environment. The architecture is composed of three levels; the operational level, tactical level, and meta-tactical level, which is the feature of the architecture. The proposed architecture was tested on a highway driving simulator in various traffic scenarios; simulation results show the feasibility of the architecture.

[37] investigated the control system for a robot system with a certain degree of autonomy and complexity. The main specification and design requirements are: Reactivity to the environment, intelligent behavior, multiple sensor integration, resolving of multiple goals, robustness, reliability, programmability, modularity, flexibility, expandability, adaptability, global reasoning.

[38] described an integrated architecture allowing a mobile robot to plan its tasks taking into account temporal and domain constraints to perform corresponding actions and to control their execution in real time while being reactive to possible events. The general architecture is composed of three levels a decision level an execution level and a functional level. The authors proposed a control structure of an autonomous robot must have both decision making and reactive capabilities.

[38] proposed a generic architecture for autonomous robots. The architectural concepts have been justified with respect to the properties required in an autonomous robot. Autonomy in a rational behavior can be evaluated by the robots efficiency and robustness in carrying out various tasks in a partially known environment. The main properties for autonomous robots such as programmability reactivity adaptability or evolutiveness.

[40] discussed how intelligent decision-making is performed for the top Decision Layer of CLARAty architecture for robotic autonomy. This layer provides support for the new trend in planning and executive systems. This layer interfaces with a Functional Layer that provides robot behaviors and control. The interface between these two layers is flexible so that different instantiations of the architecture can use different levels of Decision Layer and Functional layer capabilities.

[41] introduced an architecture that integrates shop floor agents for scheduling, cell control, transportation, and material management. This work introduced a multi-agent system architecture that controls different aspects of a manufacturing environment.

[7] evaluates the performance of alternative control architectures for manufacturing production. The authors evaluated the effect of modifying reactive-based control architecture to incorporate partial hierarchies of agents and planning capabilities. The authors to pose the principal question: what is the most appropriate control architecture for a given system has led industrial and academic researchers to develop a spectrum of decentralised control architectures ranging from hierarchical to non-hierarchical structures? They investigated how increases in planning horizon affect the performance of initially reactive control architecture.

[42] investigated a multi-agent system to architecture control an automated manufacturing environment. The architecture includes functions at the manufacturing cell level, materials handling and transport level, and factory scheduling level. The authors focus attention on the functions of the agents of the transport system, which is composed of a set of AGVs. Agent is an autonomous, computational entity that can be viewed as perceiving its environment and acting upon it. Agents are event-driven objects that can be integrated in automated manufacturing environments to control certain tasks.

[43] investigated decentralizing control of AGVs based on quality requirements such as flexibility and openness. The AGV control system is structured as a multi-agent system. Presented an overview of the agent-based architecture of the AGV system.

[31] investigated on control architecture for autonomous underwater vehicle (AUV). The architecture is organized in three layers: mission layer, task layer and execution layer. The test with real vehicle have been done to validate the architecture. The autonomous control architecture should have well planning or re-planning ability as well as reactive ability to the changing of the external environment.

[44] propose a robotic control architectures. The control architecture defines abilities that should be integrated to develop an autonomous navigation. This could be classified into three categories: Deliberative (Centralized) navigation, Reactive (Behaviour-based) navigation and hybrid (Deliberative - Reactive) navigation.

[45] investigated a hybrid systems framework to behavior control of nonholonomic AGV. This framework has the 3-layered hierarchical structure containing

a hybrid automata of the motion control as the middle process. The hybrid automata has three states, stop, line path following and circle path following.

[46] development an architecture for controlling autonomous mobile robots. The proposed architecture is composed of modules integrates deliberation with a standard planner, execution, monitoring and replanning. The authors present results from experiments that were conducted with the robot Pioneer P3DX.

[47] presents the control system architecture of the autonomous vehicle, called Intelligent Pioneer. The authors investigated the path tracking and stability of motion to effectively navigate in unknown environments. In this approach, a two degreeof freedom dynamic model is developed to formulate the pathtracking problem in state space format.

[48] present a classification scheme that provides a structured mechanism for organizing the relevant information about the design of the AGVS from a control perspective. It allows the system designer to determine how design decisions will impact the control complexity.

[49] investigated the path planning and coordination of multiple Automated Guided Vehicles (AGVs) in an automated warehouse. This paper deals with decentralized coordination of Automated Guided Vehicles (AGVs). The authors propose a hierarchical traffic control algorithm, that implements path planning on a two layer architecture. Describe a coordination strategy for a fleet of AGVs, through an architecture based on a two-layer approach. They treated the planning and the path optimization as a common entity. The path planning is split on the two layers in order to simplify the problem.

[50] propose an architecture for a control system of an autonomous robot as well as an architecture for a multi-robot system in which the robots cooperate in order to accomplish client's tasks. The solution is based on the SOA paradigm and an ontology as a way of representing an environment.

[51] presents the principal components needed in a functional architecture for autonomous driving. They proposed on the division of the architecture into layers, and reasoning on the distribution of the architectural elements across these layers.

[52] proposed a control framework in which a controller is developed for FMS scheduling. This control approach is based on reducing the planning horizon leads to a more stable environment.

[26] discussed the literature related to design and control issues of AGV systems at manufacturing, distribution, transshipment and transportation systems. This paper research perspectives in the design and control of AGV systems in distribution, transshipment and transportation systems.

[1] introduced a framework that includes a governance mechanism in control system architectures that dynamically steers the autonomy of decision-making between predictive and reactive approaches. This paper focuses on architecture of control of FMS, the authors propose a developing a framework that includes a governance mechanism in control system architecture (CSA) that dynamically guide of autonomy of decision making between predictive and reactive approaches. The contribution of this paper is related to the strategy of control

architecture applied in FMS. Although this article focuses only on FMS, it can be a start point to evaluate this framework in AGV.

[53] studied a modeling and traffic control problem of the automated guided vehicle (AGV) system in a container terminal. A set of traffic rules is proposed to ensure the completion of all jobs with the absence of vehicle deadlocks and collisions. Moreover, these rules can be realized almost decentralized requiring little intervention from a central controller.

[8] this article presents a classification of architectures of FMS in the state-of-the-art. These architectures can be grouped in three classes (I, II and III). This paper focuses on the class II hybrid control architecture (i.e., hierarchical and heterarchical) found in the domain of manufacturing scheduling. They propose a dynamic class II architecture called ORCA (dynamic Architecture for an Optimized and Reactive Control). ORCA was applied to a real flexible Manufacturing System (FMS) to prove the applicability of this architecture in an industrial environment. The authors proposed a dynamic architecture, that switching between two functioning modes: normal mode (i.e., the entity is controlled hierarchically) and disrupted mode (i.e., the entity is controlled herarchically).

[23] this paper propose a coordinated control of AGVs in an FMS in areas of group behaviours like formation control, path following maintaining a formation pattern (marching), and collision avoidance between robots or static obstacles. This article proposes a hybrid architecture which objective is to design control architecture that coordinates the AGVs and the process tasks, using Petri Networks (PN). This is based in two levels: in the high level propose a models of FMS using PN, in the low level the control select an adequate AGV, and control laws for process task. Method: decentralized architecture.

[11] these paper propose a semi-heterarchical control architecture is expected to reduce myopic behavior according to current plant conditions (i.e., adaptability) while favoring reactivity and low complexity. Besides the autors explain the control problem in FMS based on the flexibilities (i.e. sequencing, machine routing, and material handling flexibility) and present contributions relevant to pure static heterarchical architectures control, then static semi-heterarchical architectures, and last, they describe contributions which take the dynamic switching between these two architectures into consideration.

This is based in a flexible decision-making technique so as to reduce myopic behavior of local decisional entities. This architecture is a reference for research development in the area of material handling system in FMS, especially the AGV. Method: semi-heterarchical architecture.

6 Discussion

The requirement for adaptable, configurable and reactive control systems for manufacturing has emphasized the inadequacies of traditional centralised control approaches. For this reason, control systems, including FMS control, must incorporate architectures that effectively use the flexibility for reaching better decision-making processes, either before production begins (i.e., predictive phase) and during production (i.e., reactive phase) to deal with real manufacturing needs.

Another feature that current control paradigms prescribe is the autonomy of each component of the system. An acceptable control architecture allows to ensure the quality of the resulting control system and reducing the usually costly and time-consuming development process.

The architecture of a production system has been the most important paradigm for the reduction of complexity in control systems. To meet the objectives of efficient manufacturing, it is necessary to replace the rigid centralized control architecture with heterarchical architectures that have advantages of modularity, extensibility, reconfigurability and fault tolerance.

In these circumstances, the challenge is to develop manufacturing control architecture for FMS and AGV with reactivity to the environment changes, scalability, robustness against the occurrence of disturbances, easier integration of manufacturing resources, and intelligence capabilities. Although specific research in this topic has achieved a number of great successes, the general framework for the development on architectural level has not been defined by the community.

7 Conclusion

This paper presents a literature review of FMS and AGV control architectures. This provides a perspective of the challenges, research needs, and future directions for manufacturing control. In general, they are rigid architectures (i.e., centralized or decentralized hierarchical) and are not simple to modify or configure, being vulnerable to disturbances. In this traditional structures there is limited intelligence for the recognition of stopping points, inflection points or obstruction and reporting of internal conditions or having the limited ability to manage the functions of scheduling, routing and dispatching.

While the literature on AGV design, navigation, routing, scheduling production and transport is extensive, few works has been focused on the adopted architecture for controlling AGVs in FMS in order to overcome the limitations of classical hierarchical architectures, especially the inflexibility for adaptations, low autonomy, and reduced reactivity under perturbations. Despite the importance of the FMS control architecture FMS including AGV, research has often overlooked this topic as the related literature is scarce and does not provide developers with a comprehensive framework to reach more effective FMS control.

From the state of arts reviewed, the authors propose the research question it can be: How to achieve a FMS control framework that includes AGV focused on AGVs autonomy, flexibility, and reactivity under dynamic environments in order to improve the efficiency of the FMS?

Acknowledgements. This research was supported by Centro Tecnologico de Automatizacion Industrial (CTAI)- Pontificia Universidad Javeriana PUJ, Bogota Colombia and the fundation Centro de Estudios Interdisciplinarios Basicos y Aplicados (CEIBA).

References

1. Jimenez, J.-F., Bekrar, A., Trentesaux, D., Rey, G.Z., Leito, P.: Governance mechanism in control architectures for flexible manufacturing systems. IFAC-PapersOnLine **48**(3), 1093–1098 (1999, 2015)
2. Jimenez, J.F., Bekrar, A., Trentesaux, D., Montoya-Torres, J.R., Leito, P.: State of the art and future trends of optimality and adaptability articulated mechanisms for manufacturing control systems. In: IEEE International Conference on Systems, Man, and Cybernetics, pp. 1265–1270 (2013).
3. Kustak, A.: Flexible manufacturing systems: a structural approach. Int. J. Prod. Res. **23**(6), 1057–1073 (1985)
4. Rey, G.Z.: Reducing Myopic Behavior in FMS Control: A Semi-Heterarchical Simulation Optimization Approach, Universitate de Valenciennes et du Hainaut-Cambresis (2014).
5. Groover, M.P., de la Pea Gomez, C.M., Sarmiento, M.M.: Fundamentos de Manufactura Moderna: Materiales, Procesos Y Sistemas. Pearson Educación (1997).
6. Vukovi, N., Miljkovi, Z.: New hybrid control architecture for intelligent mobile robot navigation in a manufacturing environment. FME Trans. 37(1), 9–18 (2009).
7. Brennan, R.W., Norrie, D.H.: Evaluating the performance of reactive control architectures for manufacturing production control. Comput. Ind. **46**(3), 235–245 (2001)
8. Pach, C., Berger, T., Bonte, T., Trentesaux, D.: ORCA-FMS: a dynamic architecture for the optimized and reactive control of flexible manufacturing scheduling. Comput. Ind. **65**(4), 706–720 (2014)
9. Vieira, G.E., Veiga, C.P.: Hierarchical and centralized architectures for distributed production planning, scheduling and control activities. In: IEEE/INFORMS International Conference on Service Operations, Logistics and Informatics, SOLI 2009, pp. 341–346 (2009)
10. Salvendy, G.: Handbook of Industrial Engineering: Technology and Operations Management. Wiley, Hoboken (2001)
11. Rey, G.Z., Bonte, T., Prabhu, V., Trentesaux, D.: Reducing myopic behavior in FMS control: a semi-heterarchical simulation-optimization approach. Simul. Model. Pract. Theory **46**, 53–75 (2014)
12. Baykasoğlu, A., Özbakir, L.: Analysing the effect of flexibility on manufacturing systems performance. J. Manuf. Technol. Manag. **19**(2), 172–193 (2008)
13. Merdan, M., Valle, M., Moser, T., Biffl, S.: A layered manufacturing system architecture supported with semantic agent capabilities. In: Eli, A., Kon, M.T., Orgun, M.A. (eds.) Semantic Agent Systems. SCI, vol. 344, pp. 215–242. Springer, Heidelberg (2011)
14. Erol, R., Sahin, C., Baykasoglu, A., Kaplanoglu, V.: A multi-agent based approach to dynamic scheduling of machines and automated guided vehicles in manufacturing systems. Appl. Soft Comput. **12**(6), 1720–1732 (2012)
15. Trentesaux, D.: Distributed control of production systems. Eng. Appl. Artif. Intell. **22**(7), 971–978 (2009)
16. Dilts, D.M., Boyd, N.P., Whorms, H.H.: The evolution of control architectures for automated manufacturing systems. J. Manuf. Syst. **10**(1), 79–93 (1991)
17. Pach, C., Berger, T., Sallez, Y., Bonte, T., Adam, E., Trentesaux, D.: Reactive and energy-aware scheduling of flexible manufacturing systems using potential fields. Comp. Ind. **65**(3), 434–448 (2014)
18. Dai, Q.: On relation of manufacturing system, manufacturing mode and manufacturing technology. In: International Technology and Innovation Conference, ITIC, pp. 889–893 (2006)

19. Van Brussel, H., Wyns, J., Valckenaers, P., Bongaerts, L., Peeters, P.: Reference architecture for holonic manufacturing systems: PROSA. Comput. Ind. **37**(3), 255–274 (1998)
20. Berger, T., Sallez, Y., Valli, B., Gibaud, A., Trentesaux, D.: Semi-heterarchical allocation and routing processes in FMS control: a stigmergic approach. J. Intell. Robot. Syst. **58**(1), 17–45 (2009)
21. Sallez, Y., Berger, T., Raileanu, S., Chaabane, S., Trentesaux, D.: Semi-heterarchical control of FMS: from theory to application. Eng. Appl. Artif. Intell. **23**(8), 1314–1326 (2010)
22. Aized, T.: Materials handling in flexible manufacturing systems. Department of Mechanical, Mechatronics and Manufacturing Engineering, KSK, Lahore (2006)
23. Hernandez-Martinez, E.G., Foyo-Valdes, S.A., Puga-Velazquez, E.S., Meda-Campaa, J.A.: Hybrid architecture for coordination of AGVs in FMS. Int. J. Adv. Robot. Syst. **11** (2014)
24. Butler, L.J.: Autonomous Materials Handling Robot for Reconfigurable Manufacturing Systems. University of KwaZulu Natal, Durban (2011).
25. Badr, I.: Agent-based dynamic scheduling for flexible manufacturing systems (2011)
26. Vis, I.F.A.: Survey of research in the design and control of automated guided vehicle systems. Eur. J. Oper. Res. **170**(3), 677–709 (2006)
27. Berman, S., Schechtman, E., Edan, Y.: Evaluation of automatic guided vehicle systems. Robot. Comput.-Integr. Manuf. **25**(3), 522–528 (2009)
28. Antsaklis, P.J., Passino, K.M., Wang, S.J.: An introduction to autonomous control systems. IEEE Control Syst. **11**(4), 5–13 (1991)
29. Torres, I.R., Dessens, L.F.R., Flores, J.L.M., Bentez, E.O.: Review of comprehensive approaches in optimizing AGV systems. In: Operational Excellence and Supply Chains, p. 203
30. Herrero-Perez, D., Martinez-Barbera, H.: Modeling distributed transportation systems composed of flexible automated guided vehicles in flexible manufacturing systems. IEEE Trans. Ind. Inf. **6**(2), 166–180 (2010)
31. Bian, X., Qin, Z., Yan, Z.: Design and evaluation of a hierarchical control architecture for an autonomous underwater vehicle. J. Mar. Sci. Appl. **7**, 53–58 (2008)
32. Lau, H.Y.K., Wong, V.W.K., Lee, I.S.K.: Immunity-based autonomous guided vehicles control. Appl. Soft Comput. **7**(1), 41–57 (2007)
33. Arkin, R.C., Murphy, R.R.: Autonomous navigation in a manufacturing environment. IEEE Trans. Rob. Autom. **6**(4), 445–454 (1990)
34. Mantel, R.J., Landeweerd, H.R.A.: Design and operational control of an AGV system. Int. J. Prod. Econ. **41**(1), 257–266 (1995)
35. Laengle, T., Lueth, T.C., Rembold, U.: A distributed control architecture for autonomous robot systems. Ser. Mach. Percept. Artif. Intell. **21**, 384–402 (1995)
36. Miura, J., Ito, M., Shirai, Y.: A three-level control architecture for autonomous vehicle driving in a dynamic and uncertain traffic environment. In: IEEE Conference on Intelligent Transportation System, ITSC 1997, pp. 706–711 (1997)
37. Medeiros, A.A.D.: A survey of control architectures for autonomous mobile robots. J. Braz. Comput. Soc. **4**, 35–43 (1998)
38. Alami, R., Chatila, R., Fleury, S., Ghallab, M., Ingrand, F.: An architecture for autonomy. Int. J. Robot. Res. **17**(4), 315–337 (1998)
39. Amigoni, F., Luperto, M., Schiaffonati, V.: Toward generalization of experimental results for autonomous robots. Rob. Auton. Syst. (2016)

40. Estlin, T., et al.: Decision-making in a robotic architecture for autonomy. In: Proceedings of the International Symposium on Artificial Intelligence, Robotics, and Automation in Space (2001)
41. Heikkil, T., Kollingbaum, M., Valckenaers, P., Bluemink, G.-J.: An agent architecture for manufacturing control: manAge. Comput. Ind. **46**(3), 315–331 (2001)
42. Farahvash, P., Boucher, T.O.: A multi-agent architecture for control of AGV systems. Robot. Comput.-Integr. Manuf. **20**(6), 473–483 (2004)
43. Weyns, D., Holvoet, T., Schelfthout, K., Wielemans, J.: Decentralized control of automatic guided vehicles: Applying multi-agent systems in practice. In: Companion to the 23rd ACM SIGPLAN Conference on Object-oriented programming systems languages and applications, pp. 663–674 (2008)
44. Nakhaeinia, D., Tang, S.H., Noor, S.M., Motlagh, O.: A review of control architectures for autonomous navigation of mobile robots. Int. J. Phys. Sci. **6**(2), 169–174 (2011)
45. Ye, J., Li, D., Li, S., Liu, X., Hu, Y.: A Hybrid System Framework to Behavior Control of Nonholonomic AGV. In: Proceedings of the World Congress on Engineering and Computer Science, vol. 1 (2011)
46. Quintero Barrios, E., García Olaya, A., Borrajo Millán, D., Fernández Rebollo, F.: Control of autonomous mobile robots with automated planning (2011)
47. Zhao, P., Chen, J., Song, Y., Tao, X., Xu, T., Mei, T.: Design of a control system for an autonomous vehicle based on Adaptive-PID (2012)
48. Sharma, M.: Control classification of automated guided vehicle systems. IJEAT **2**, 191–196 (2012)
49. Digani, V., Sabattini, L., Secchi, C., Fantuzzi, C.: Hierarchical traffic control for partially decentralized coordination of multi AGV systems in industrial environments. In: IEEE International Conference on Robotics and Automation (ICRA), pp. 6144–6149 (2014)
50. Ambroszkiewicz, S., Bartyna, W., Skarżyński, K., Stpniak, M., Szymczakowski, M.: Architecture of an autonomous robot at the it level. J. Autom. Mob. Robot. Intell. Syst. **9**, 34–40 (2015)
51. Behere, S., Tröngren, M.: A functional reference architecture for autonomous driving. Inf. Softw. Technol. **73**, 136–150 (2016)
52. Sinreich, D., Shnits, B.: A robust FMS control architecture with an embedded adaptive scheduling mechanism. J. Manuf. Syst. **25**(4), 301–312 (2006)
53. Li, Q., Udding, J.T., Pogromsky, A.Y.: Modeling and control of the AGV system in an automated container terminal. In: Proceedings of the AsiaMIC (2010)

Design and Implementation of a Low Cost RFID ISO 11784/11785 Reader for the Automatic Livestock Management in Nicaragua

Joseling Sanchez, Mario Garcia(✉), and Maria Virginia Moncada

Faculty of Electronic and Computational Engineering,
National University of Engineering, Managua, Nicaragua
vanessasanchez2408@gmail.com, rodolfomario4444@gmail.com,
mariavirginia1965@gmail.com

Abstract. This paper presents the design and implementation of a low cost radio frequency identification (RFID) stick reader for livestock management in Nicaragua. A stick reader for livestock is a low frequency RFID base station that is compliant to the international standards of electronic animal identification described in ISO11784 and ISO11785. This device will allow the farmers in Nicaragua to implement a traceability system, so they can prevent theft of cattle and automatically maintain livestock health control.

1 Introduction

According to the National Agricultural Census of Nicaragua on 2011, Nicaragua is the country with the greater potential for cattle breeding in Central America [1], as shown in the Table 1. The livestock business in Nicaragua represents 10% of the national gross domestic product (GDP) and 50% of the agricultural GDP. However, the country has many difficulties to register the information related to the animal identification and health control [2]. Despite the existence of two traceability databases known as TRAZAR-NIC and SNITB, the farmers do not have technical knowledge about international RFID standards ISO 11784/11785 and the products that comply with these standards [3]. That is why the whole livestock traceability system is accomplished by manually collecting the data and eventually registering the data into the corresponding SNITB or TRAZAR-NIC database.

Manually writing all the data produces a lot of errors, and it is inefficient too, due to the large amount of animals that need to be constantly monitored by the farmers. Also, the lack of control in the corrals results in cattle robbery, which means a huge loss every year for the farmers. Thus, it is urgent to change the way farmers monitor their animals and how they insert their traceability data into the databases. To solve those problems we propose the design and implementation of a low cost RFID ISO 11784/11785 reader to: identify the animals, write their breeding data and automatically insert this information in the corresponding database. Figure 1(a) illustrates the main objective.

I. Chang et al. (eds.), *Advances in Automation and Robotics Research in Latin America*, Lecture Notes in Networks and Systems 13, DOI 10.1007/978-3-319-54377-2_15

Table 1. Potential of central america in the livestock entry

Country	Livestock heads	Exported meat in metric tons
Costa Rica	1,080,900	12,430
El Salvador	1,259,210	1,800
Guatemala	2,540,000	2,270
Honduras	2,500,020	2,160
Nicaragua	3,500,000	51,537
Panamá	1,600,000	4,329

2 Standard ISO 11784/11785

The international electronic animal identification is defined by the standards: ISO 11784, which describes the code structure for the RFID chips implanted in animals [4], ISO 11785, on the other hand, establish the technical concepts concerning to the communication between transponder and reader [5]. Both standards have been updated under the ISO14223, which includes details about the so called advanced transponders. Advanced transponders are the same transponders defined in ISO 11784/11785 but they have enhanced features as extra memory blocks for storage or integrated sensors [6].

2.1 Transponders

There are different animal transponders for different applications, as shown in Fig. 1(b), but the internal structure of the chip should respect the ISO 11784/85 standard [7]. The ID code for animals in the standard ISO 11784/11785 is given by a 15 digit decimal number, which is represented by 64 bits of binary data inside the chip memory blocks. The structure details for these 64 bits of data is show in Table 2. Usually the ID code for each transponder is printed on the outside of package [8].

Within the standard ISO 11784/85 the transponder topology may vary in two types, known as FDX-B and HDX. FDX-B stands for Full Duplex-B and HDX stands for Half Duplex. The FDX-B protocol is associated with the ASK modulation scheme and the load modulation method to transmit the information. On the other hand, the HDX protocol is related to the FSK modulation scheme and uses the backscatter method to convey the information [5]. When the reader queries any ISO 11784/85 tag, the total information received continuously from a FDX-B transponder equals 128 bits, and the total information received continuously from a HDX transponder equals 112 bits, Table 3 shows an example of the complete code structure for FDX-B transponders and Table 4 shows a complete code structure example for HDX transponders. Also, the operating frequency for both topologies is on the LF range (134.2 KHz) [6]. The ISO 11784/85 compliant transponders are usually passive transponders, this means

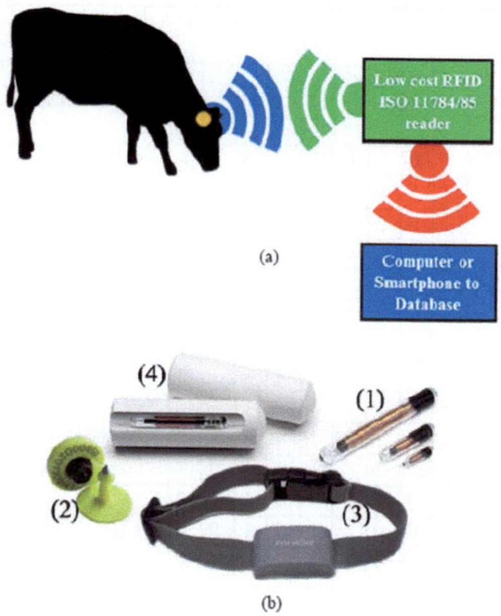

Fig. 1. (a) Reader communication with the livestock transponders and database. (b) Sample ISO 11784/11785 transponders. (1) Subcutaneous tags for pets, (2) ear tags for livestock, (3) collar for cattle and pets, (4) bolus for ruminants.

Table 2. ID code structure for ISO 11784/85

#bit	Information	Description
1	Animal application (1)/ Not animal application (0)	Specify if the transponder is used to tag an animal or not
2-15	Reserved	For future applications
16	Follows an extra memory block (1), do not follow an extra memory block (0)	If the transponder is advanced there will be an extra memory block
17-26	Country code according to ISO 3166.	Specify the country of the animal, 999 indicates that is a test transponder.
27-64	National ID code	Unique country code

that they do not have any own power source to operate, so they extract energy from the reader electromagnetic field when they are in its presence.

The reader electromagnetic field that activates the transponders is called activation field. The read distance will depend on several factors, such as the: power coming from the antenna drivers, antenna size, and tag position. Typically the read range can vary from some few centimeters up to one or one and a half meters. The short read range, compared with other RFID standards, is due to the behavior of the electromagnetic waves in the LF range.

Table 3. Complete code structure for the FDX-B protocol (Example Data)

MSB										LSB	Description
1	0	0	0	0	0	0	0	0	0	0	Header pattern(11 bits)
A control bit appears every 8bits to avoid sending the header pattern again		1	1	1	1	1	0	0	0	0	National code (38 bits equivalent to 12 decimal digits)
		1	0	0	0	0	0	0	1	1	
		1	0	0	0	0	0	0	0	0	
		1	0	0	0	0	0	0	0	0	
		1	1	1	1	0	0	0	0	0	
		1	1	1	1	1	1	0	0	1	Country code (10 bits)
		1	-	-	-	-	-	-	-	1	Extra indicator (1 bit)
		1	1	-	-	-	-	-	-	-	App indicator (1 bit)
		1	1	1	0	1	0	1	1	0	CRC checksum (16 bits)
		1	0	1	0	1	1	1	0	1	
		1	0	1	0	1	0	1	1	0	Extra data (24 bits, if present)
		1	0	0	1	1	0	1	0	0	
		1	0	0	0	1	0	0	1	0	

Table 4. Complete code structure for the HDX protocol (Example Data)

MSB							LSB	Description
0	1	1	1	1	1	1	0	Header pattern (8 bits)
1	1	1	0	0	0	0	0	National code (38 bits equivalent to 12 decimal digits)
0	0	0	0	0	0	1	1	
0	0	0	0	0	0	0	0	
0	0	0	0	0	0	0	0	
1	0	0	0	0	0	0	0	
1	1	1	1	1	0	0	1	Country code (10 bits)
-	-	-	-	-	-	-	1	Extra block indicator (1 bit)
1	-	-	-	-	-	-	-	App indicator (1 bit)
1	1	0	1	0	1	1	0	CRC checksum (16 bits)
0	1	0	1	1	1	0	1	
0	1	0	1	0	1	1	0	Extra data (24 bits, if present)
0	0	1	1	0	1	0	0	
0	0	0	1	0	0	1	0	

2.2 Reader

The reader will switch on the activation field periodically for 50 ms period of time, and then it will switch off the activation field for 3 ms. While it is switched on it will wait for the response from a FDX-B transponder [6]. A FDX-B transponder does not require any charging time or pause from the activation field, that is why it can answer while it is switched on, but a HDX transponder

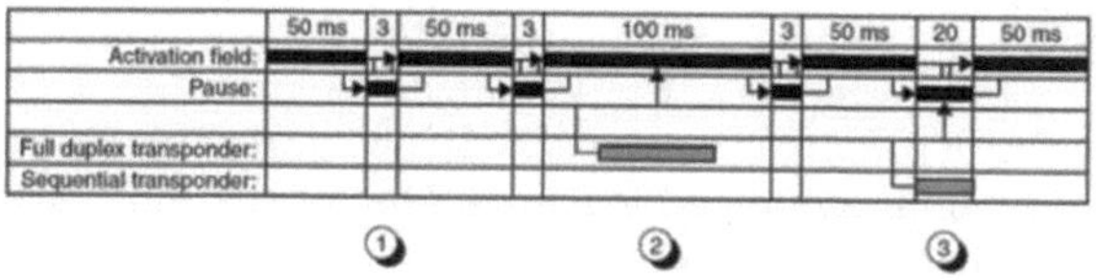

Fig. 2. RPID ISO 1L78D4/85 time diagram. (1) No transponder in the interrogation zone, (2) FDX-B transponder in the activation field, (3) HDX transponder in the activation field.

requires the activation field to charge up its charging capacitor and then answer in the 3 ms pause, when the activation field is switched off. For the FDX-B transponders, while data is being received, the operating interval of the field could be extended to 100 ms if the data transfer is not completed within 50 ms. For the HDX transponders, if the data transmission is not complete within the 3 ms pause, then the pause will be extended to a maximum of 20 ms. Figure 2 shows details about the reader synchronization to communicate to both types of transponders [8].

3 Market Research

In our investigation we classified the products for livestock management that already exist in the market into four categories. The classification is related to the functionalities that these products offer, and it was helpful to define the requirements for our own design. Some of the features that we take into account for our reader classification are: traceability data collection capabilities, user interface, compatibility with other devices, memory storage, read range, portability (rechargeable batteries), rugged design, and operating system.

3.1 Category 1

The category 1 devices are RFID readers ISO 11784/85 compliant, but their unique functionality is to retrieve the ID code from every ISO 11784/85 transponder. So they are useful to identify pets, however, the lack of any user interface to write traceability data makes them unsuitable to operate in the livestock management field.

3.2 Category 2

The category 2 devices are RFID readers ISO 11784/85 compliant that can retrieve the ID code from every ISO 11784/5 transponder. Also, they have an interface composed of a display and keypad, rechargeable batteries, and are compatible with several devices such as smartphones or computers by USB, Bluetooth or RS232 communication. This category is the most abundant in the market, because it is an affordable solution to implement a traceability system.

3.3 Category 3

Devices in the category 3 are very similar to devices into category 2, however, category 3 has additional enhanced features. One of the most attractive features of these devices, is that they collect the traceability data directly in the reader through a software without the need of a smartphone or computer. They also have Wi-Fi and GPRS capabilities, which allows the traceability data to inserted into the database automatically.

3.4 Category 4

Readers in category 4 are RFID ISO 11784/85 readers, but they work with operating systems like Windows or Linux. Category 4 devices are the evolution of category 3 devices.

3.5 Prices

We estimated the prices of the RFID ISO 11784/85 readers according to their category and brand, Fig. 3 shows details about the international costs to acquire a portable reader.

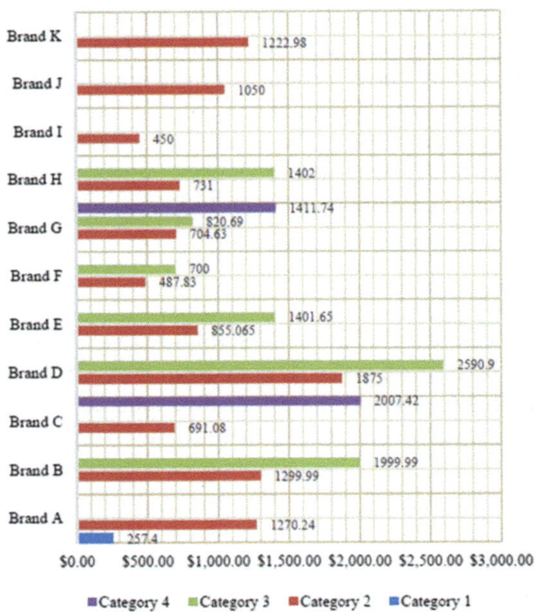

Fig. 3. Different RFID ISO 11784/85 reader brands and their average price according to their category in US dollars.

4 Related Work

There are several brands in the market of the RFID ISO 11784/85 technology, however, there are few public works that explain the development of the RFID ISO 11784/85 readers. In our research, we found out a document related to the ISO 11784/85 standard. This paper is called: Full ISO11784/11785 compliant RFID reader in a programmable analog-digital, integrated circuit. This works is about a programmable-digital Cypress CY8C27443 integrated circuit utilized to implement a fully compliant ISO11784/11785 RFID reader module, working at 134.2 kHz [9]. One of the advantages of this work compared to ours, is the increased the read range up to 33 cm because of the digital signal processing (DSP) capabilities of the chip CY8C27443. This work helped us to define in our design some of the technical parameters that are not stated in the ISO 11784/85 document.

5 Stick Reader Design

Our reader design belongs to category 2. It has a simple user interface and is compatible with multiple devices by means of USB or Bluetooth communication. Also, it is rechargeable and incorporates solar panel capabilities. Figure 4 shows the reader functional block diagram. The microcontroller selected is an ATMEL ATmega2560 chip due to its number of UART ports [10] that allow communication with every single block in the diagram. Likewise the, the ATmega2560 microcontroller works over 5 V logic levels which is useful since the rest of the system works with 5 V logic levels. The RFID ISO 11784/85 transceiver is accomplished by synchronizing two RFID front end chips. The first integrated circuit is an EM4095 which is dedicated to perform the RFID reader functionality for the FDX-B protocol, and the TMS3705 is the second chip dedicated to perform the RFID reader functionality for the HDX protocol.

5.1 RFID Base Station for FDX-B Protocol

The EM4095 integrated circuit, previously called P4095, is an analog front end chip manufactured by EM Microelectronic and is intended for its use in RFID base stations designs. This device can provide the following functionalities: antenna driving with carrier frequency, AM modulation of the field for writable transponders, AM demodulation of the antenna signal modulation induced by the transponder, and communication to any microcontroller via a serial interface [11]. The basic equation to calculate the EM4095 resonant frequency is given by (1). Figure 5(a) shows details about our calculated configuration

$$f_0 = \frac{1}{2\pi\sqrt{L_A C_0}} \tag{1}$$

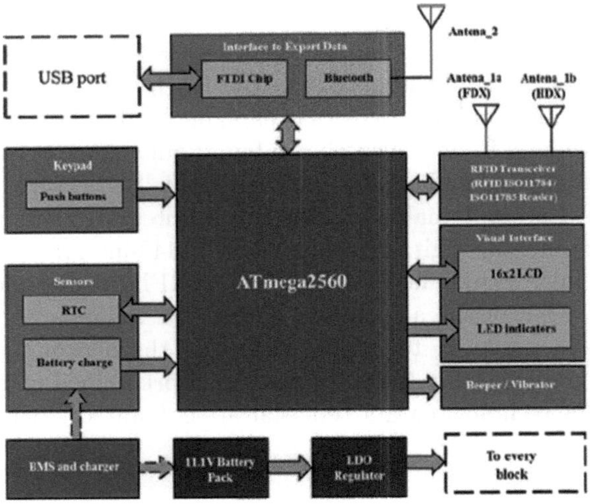

Fig. 4. RFID ISO 1178D4/85 reader functional blocks diagram.

5.2 RFID Base Station for HDX Protocol

The TMS3705 base station integrated circuit, manufactured by Texas Instruments, is used to drive a RFID antenna, and is the recommended solution for the HDX protocol. It is important to mention that the HDX protocol patent belongs to Texas Instruments. The TMS3705 functionalities are similar to EM4095 functionalities, but the transponder codification and modulation scheme are very different. Some of this chip features are: drive an antenna, send modulated data to antenna, signal demodulation from transponders response in FSK, and serial communication with any microcontroller. Figure 5(b) shows the TMS3705 configuration scheme designed for the HDX protocol [12]. Once we configured the base station chips EM4095 and TMS3705 according to the datasheets and applications notes given by the manufacturers, the next step is to synchronize the signals generated by both base stations. The program in the microcontroller must be able to switch on and off each chip in the specific time according to ISO 11784/85, so the information received can be captured without interference between both base stations. It is necessary to synchronize the base stations since they operate in the same frequency 134.2 kHz.

5.3 Antenna Design

To form a big inductance coil in a limited space, it is more efficient to use multilayer coils. For this reason, a typical RFID antenna coil is formed in a planar multi turn structure [13]. The inductance for this kind of antenna is calculated in (2) and the number of turns is calculated in (3).

$$L = \frac{0.31(aN)^2}{6a + 9h + 10b}(\mu H) \tag{2}$$

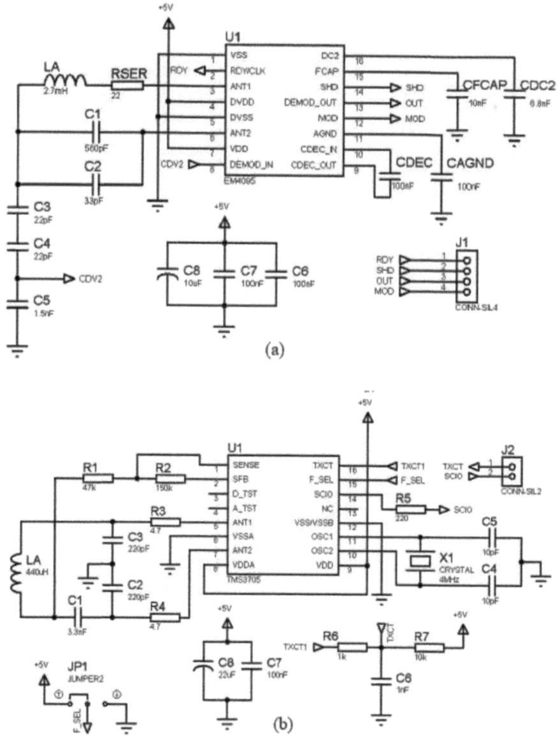

Fig. 5. (a) EM4095 schematic solution to read the FDX-B protocol, (b) TMS3705 schematic solution to read the HDX protocol.

$$N = \sqrt{\frac{L_{\mu H}(6a + 9h + 10b)}{(0.31)a^2}} \tag{3}$$

The DC resistance for a conductor with a uniform cross-sectional area is found by (4). Where "l" is the total length of the wire, σ is the conductivity and "S" is the cross section area.

$$R_{DC} = \frac{l}{\sigma S}(\Omega) \tag{4}$$

The wire resistance increases with the frequency due the "skin effect". An approximated equation for the AC resistance is given by (5). Where δ is the skin depth and the variable "a" is the coil radius.

$$R_{AC} = \frac{1}{2\sigma\pi\delta}(R_{DC})\frac{a}{2\delta} \tag{5}$$

5.4 Measurements

For the read range, we tuned the base stations from Fig. 6 at the resonance frequency by varying the resonance capacitors from both circuits until we reached

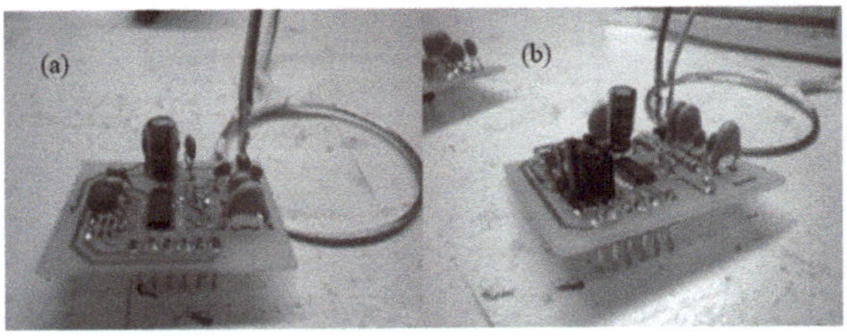

Fig. 6. Loop antenna coil and base stations for the RFID-ISO 11784/85 reader, (a) EM4095, (b) TMS3705.

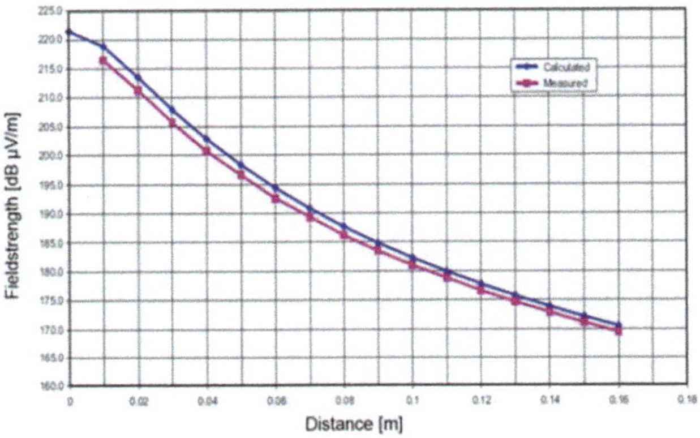

Fig. 7. Fields strength of the signal versus distance.

the desire frequency value according to ISO 11784/85. By means of an oscilloscope, and a Allflex livestock transponder parallel to a spy coil we measured the antenna signal at the proximity of the reader, then we move away the spy coil and the transponder until was not possible to get any coded signal at the output of the reader. The maximum read range is up 5 cm for FDX-B protocol and 8 cm for HDX protocol. Figure 7 shows the behavior of the electromagnetic field over the distance

5.5 USB and BLuetooth

The USB interface is performed by the FTDI FT232RL chip that converts the serial TTL output coming from the ATmega2560 into a virtual connection which is compatible with multiple operating systems like Windows, Linux or Mac [14]. The Bluetooth function is accomplished by an HC-05 Bluetooth module which is very cheap and versatile.

5.6 Powe Source

The system designed to power the RFID ISO 11784/85 reader is given in Fig. 8. We use a common 12 V AC/DC adapter @ 1 A, alongside a mini solar panel to

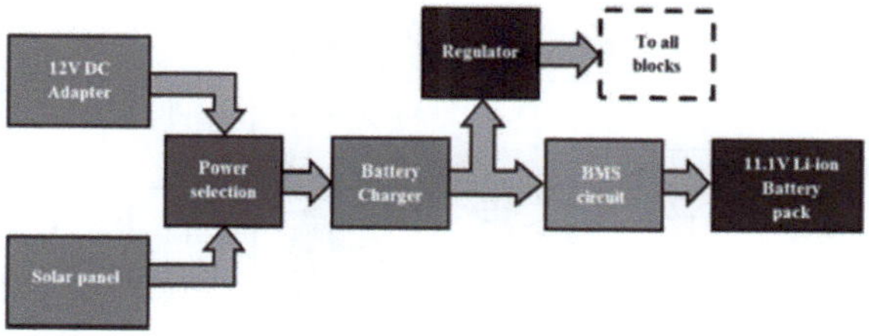

Fig. 8. Blocks diagram for the power source of the RFID ISO 11784/85 reader.

charge the device. The battery charger is based on a single multi-cell charging chip SY6912A and the battery pack is composed by three 18650 Li-ion cells arranged in series to achieve 11.1 V [15]. The battery management system (BMS) for the battery pack is performed by the module HH-P3-10.8. The regulator that powers all the blocks is a LM7805 at 1.5 A.

5.7 Package Details

The reader is being assembled in PVC plastic material. Figure 9 shows details about the package designed for the RFID ISO11784/85 reader.

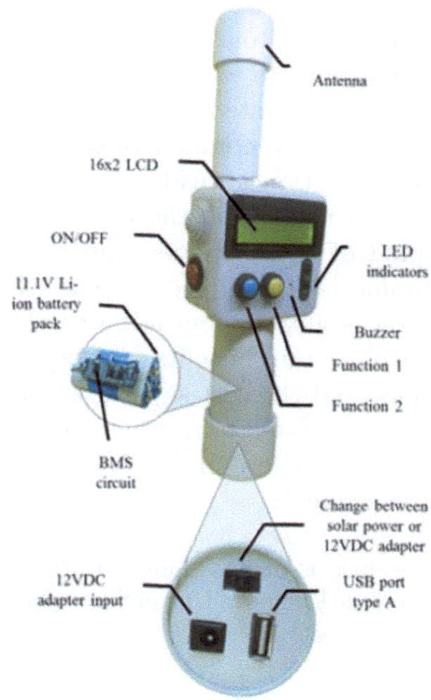

Fig. 9. Case and details about RFID ISO 11784/85 reader.

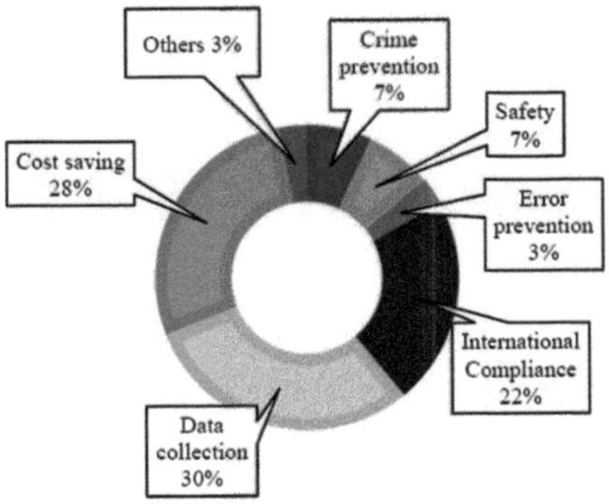

Fig. 10. Benefits of the ISO 11784/85 RFID implementation.

6 Conclusion

Implementing the electronic traceability is really important to Nicaragua because it will help the farmers to optimize their cattle breeding methods. Also, the products that include the traceability information leave better revenue to the country than the products without it. This is because international consumers nowadays are demanding information about the food they buy. The price of one metric ton of meat can increase about 100% of price when it includes the traceability data. On the other hand, our goal is to reduce the cost of the RFID reader up to USD 400.00 to compete with Brand I as shown in Fig. 3. It will be possible since we are using integrated products that allow the reduction of the electronic parts needed in a conventional reader, but also, we keep the same functionalities and features like a typical reader. Furthermore, we are developing the product by ourselves, and usually all the brands in the market outsource the research and developing of their readers, and that is one of the reasons to the high cost. The graphic in Fig. 10 shows some of the benefits leaded by the implementation of a traceability system using the RFID technology, as the graphic indicates, the most important benefits are the automatic data collection, the international compliance and the cost savings which contribute to increase the profit. Our design is being updated to create the definitive software that will manage the base stations shown before in Fig. 5(a) and (b). After that we will be able to manufacture the product.

Acknowledgment. Authors in this papers thank to: Ph.D. Pritpal Singh, Caroline Connolly and Matthew Gregory at Villanova University in the United States of America for their collaboration in this project.

References

1. Sol, R.: Ganadera, in Estructura Productiva de Nicaragua, Managua, Nicaragua (2007). Chap. 8
2. Sanchez, M.L.T., Orozco, W.J., Orozco, V.L.: UNAN-FAREM, Nivel de conocimiento de productores ganadores sobre el programa, trazabilidad bovina en tres comunidades de San Ramón-Matagalpa (2011)
3. Vergara, J.P.O.: "Trazabilidad Investigación", presented at "Camera di Industria e Comercio Italo-Nicaragüense-CCIN", pp. 5–6, 18th August, Managua, Nicaragua (2011)
4. Banks, J., Pachano, M., Thomson, L., Hanny, D.: RFID Applied
5. Zheng, F., Kaiser, T.: Digital signal processing for RFID
6. Finkenzeller, K.: RFID Handbook: fundamentals and applications in contactless smart cards, radio frequency identification and near field communication
7. Radio frequency identification for animals, code structure, ISO 11784 (1996)
8. Radio frequency identification for animals, technical concepts, ISO 11785 (1996).
9. Arnaud, A., Bellini, B.: Full ISO11784/11785 compliant RFID reader in a programmable analog-digital, integrated circuit
10. ATMEL, ATmega2560 datasheet
11. EM Microelectronic, EM4095 datasheet
12. Texas Instruments, TMS3705 datasheet
13. Lee, Y.: Microchip Technology Inc., AN678, RFID Coil Design
14. FTDI chip, FT232R USB UART IC datasheet
15. SILERGY, SY6912 datasheet

Setup of the Yaskawa SDA10F Robot for Industrial Applications, Using ROS-Industrial

Carol Martinez(✉), Nicolas Barrero, Wilson Hernandez, Cesar Montaño, and Iván Mondragón

Department of Industrial Engineering, School of Engineering, Pontificia Universidad Javeriana, Cr. 7 No 40-69, Bogotá, Colombia
carolmartinez@javeriana.edu.co

Abstract. The ROS-Industrial open source project was intended to expand the applicability of industrial robots by reducing the gap that existed between researchers and manufacturers, in order to start developing state of the art applications for the industry. This is part of the scope of the PIR (Perception for Industrial Robots) project from the Pontificia Universidad Javeriana Bogotá (PUJ). PIR project looks for expanding robot's capabilities for working in dynamic industrial environments. This paper presents the starting point of this project: the setup of the Motoman SDA_10 dual-arm robot with ROS-Industrial. In this paper, the different steps and modifications carried out to setup the system are presented. Simulation experiments and real tests were conducted in order to analyse and check the behaviour of the system. Results show some of the advantages of using ROS-Industrial, and the wide range of tasks that can be explored with these kind of robots from the research and application point of views (in Latin America only a few robots with this type of configuration is available for research purposes).

1 Introduction

In industrial processes automation systems are an essential factor in reaching high levels of productivity. In the last years, the application of robotics in industrial automation has taken an important role. With more efficient processes and demanding productivity indicators required, the implementation of automated systems, based on robotics, becomes an invaluable improvement opportunity for enterprises. Those systems would be more useful if they could be adapted to work in different tasks, and also if they could recognize changes in the environment where these activities are developed.

A huge challenge when using robotics in automated processes is configuration. Each process has specific requirements that must be arranged in the robot previously. Industrial robots have been limited to work in the same kind of tasks, such as handling and dispensing objects because they are cost-effective for repetitive and high volume tasks, but are not cost-effective for high-mix/low-volume

I. Chang et al. (eds.), *Advances in Automation and Robotics Research in Latin America*, Lecture Notes in Networks and Systems 13, DOI 10.1007/978-3-319-54377-2_16

production [1]. A disadvantage of developing applications based on robotics in industrial environments, is the complexity of writing the software, since the proficiency required is beyond the skills of only one researcher. Trying to solve this situation, robotics researchers have developed different frameworks, that make it easier to manage the complexity of writing software. These frameworks also help simulating the different applications without requiring to use the real hardware [2], making it possible to test different feasible sceneries and choosing the optimal one without spending money in unnecessary tests.

An important framework for researchers is ROS (Robot Operating System) [3]. It promises to make designing robotic software easier and less expensive. This framework provides software for different robots, especially the ones most used in industry.

It is important to remark that ROS is not just an operating system, it also provides software modules for performing typical robot activities, for example object recognition. ROS makes possible to simplify the task of writing complex software, so the design of the robotic application for a company could be done in shorter time.

Currently, the biggest challenge for ROS is to become the dominant robotics platform [4]. The software is distributed under BSD license and is free for anyone. It is used with Linux operating systems, so researchers can use ROS for commercial or non-commercial purposes. In fact, the ROS community has developed ROS-I [5], which stands for ROS-Industrial, as an open-source project that is focused in industrial robotics applications, making the programming of robots in the industry even easier, in addition to bringing support and documentation for industrial robots.

Based on ROS-Industrial, the Centro Tecnológico de Automatización Industrial (CTAI) [6] at Pontificia Universidad Javeriana Bogotá launched the PIR project: Perception for Industrial Robots. This project aims to incorporate different sensors to allow industrial robots to adapt easily to different tasks (adding avoidance capabilities, which are commonly found when working in dynamic environments).

PIR project will start developing research on the Motoman SDA10F Yaskawa dual-arm robot [7], and will focus on finding industrial applications of this robot, where the robot's adaptability to a complex environment is required for improving productivity of companies.

This paper presents the different steps and modifications carried out to setup the Motoman SDA10F robot with ROS-Industrial. Simulation experiments and real tests are conducted in order to analyze the behavior of the system highlighting the advantages of using ROS-Industrial.

The paper is organized as follows. First, an analysis of state of the art of industrial robots is presented. Then, the hardware and software architectures of the Motoman SDA10F robot is presented. Section 4 presents tests and results. Finally, Sect. 5 presents the conclusions and the direction of future work.

2 State of the Art

During the last few years, interest in anthropomorphic or dual-arm manipulators had increased. In the industry, it is expected that anthropomorphic robots can replace human workers in the future. These kind of robots are dual arm manipulators, whose main characteristic is the way of manipulation. It makes distinction between non-coordinated (the two arms are performing two different tasks) and coordinated manipulation (where the arms perform different parts of the same task) [8].

The use of a dual arm robot system for performing operations is currently being studied. For example, [9] presented a case study of the final assembly area of an automotive plant. The study presented the mechanical and programming aspects of using a dual arm robot in activities that are typically performed by humans. The implemented system is based on a COMAU Smart Dual arm robot platform. They have found that by using dual arm robots the cost of setting up a cell decreased, avoiding the cost of investing in expensive grippers which are required when conducting the same task with single arm robots. Additionally, the programming tasks are simplified due to the synchronized movements that already come with dual arm robots.

In [10] a computationally efficient and robust kinematic calibration algorithm for industrial robots is presented. The HP20D manipulator provided by Yaskawa (Yaskawa Motoman Robotics, Inc. manipulators are known to be fairly accurate) with the DX100 controller, were used to test the algorithm. They tested 135 points through defined trajectories. Future work is focused on improving the processing speed, because the current prototype is in MATLAB without hardware acceleration.

A method based on human-likeness to solve the manipulation planning problem for articulated robots, was presented in [11]. To test the algorithm, the Motoman SDA10D industrial robot was used. All the manipulation planning experiments were carried out for the right arm, and the MATLAB Robotics Tool box was used to carry out basic robotics calculations, such as robot forward kinematics.

Programming industrial robots is a very demanding task. Each manufacturer has its own communication protocols and programming interfaces which make the development of industrial applications a difficult task. For this reason, the Robot Operating System (ROS) is being widely used for dealing with different robots. ROS is a flexible framework for writing robot software. It is a collection of tools, libraries, and conventions that aim to simplify the task of creating complex and robust robot behavior across a wide variety of robotic platforms [12]. ROS simplifies the task of programming robots by providing a robust framework where the designers are provided with direct control for the robot. Typically, a ROS implementation has the components shown in Fig. 1, [13]:

- Nodes: basic processes that perform computation.
- Topics: a method for exchanging messages.
- Services: provides a stricter communication model where there is an established request and response message.

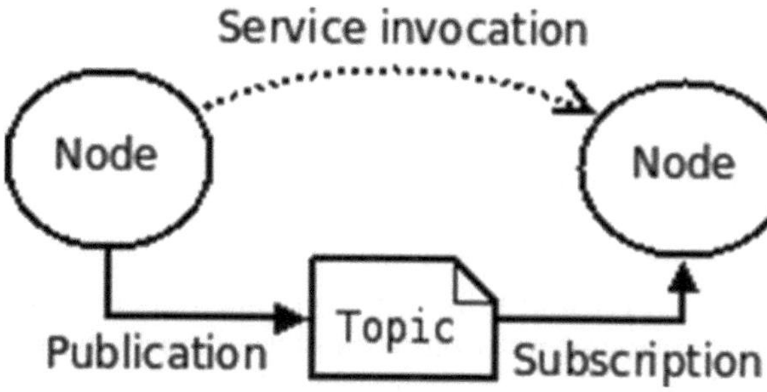

Fig. 1. Basic components for a ROS implementation [13]

Using ROS framework, in [13], a reconfigurable control system was developed. They present results with two robotic arms which require reconfiguration of the underlying control system in order to to carry out a task.

In [14] presents a simulation environment of an industrial manipulator based on ROS. The system structure consists of ROS as a middleware, visualization system, and simulation system. The robotic visualization simulates a five-joint augmented SCARA robot. The links and joint of the robot are defined as the Unified Robot Description Format (URDF) which is used by Gazebo to simulate the robot. A ROS based architecture is used in [15] to the coordination of human robot cooperative assembly tasks. The paper presents simulation results where an operator and a dual arm robot share assembly task and workspace. The assembly sequence is modeled in XML format, generated off-line. The COMAU dual arm robot is simulated and using ROS messages, the database communicates with the model retrieving, visualizing, and sending information to the robot controller through the TCP/IP communication protocol.

Yaskawa Motoman Robotics announced in 2004, his sponsorship in the Amazon Picking Challenge. Yaskawa provided robots, software (including MotoRos Driver), and technical support at the event [16]. On 2016, a team composed by collaboration between TU Delft Robotics Institute and the company Delft Robotics, won the Amazon Picking Challenge. The team built a flexible robot system based on industry standards. The system is equipped with a single arm Yaskawa Motoman robot with seven degrees of freedom, high-quality 3D cameras, and an in-house developed gripper. To control the robot, the team integrated advanced software components based on state of the art artificial intelligent techniques and robotics. The components are developed under the Robot Operating System for industry framework (ROS-Industrial) [17]. The ROS package used by the team is an open source system [18].

The collaboration between industrial robot manufactures, research institutions, and the use of open source software, allows to a rapid penetration of intelligent robot systems on modern industries. Several logistics companies (Amazon included) have announced new challenges and competitions that involve the use of industrial robots in applications like picking, packing, transportation, among others [16]. These competitions will improve the development of open robot control architectures, suitable for real time object recognition, tracking, and manipulation.

3 The SDA10F Motoman Robot

The industrial robot presented in this paper is the dual-arm robot SDA10F from Yaskawa [7] (see Fig. 2). Each arm has seven (7) degrees of freedom (DOF), and there is an additional axis of rotation in the base (torso), for a total of fifteen (15) DOF. The robot at Pontificia Universidad Javeriana (PUJ) features an additional axis (not used in this paper), which allows the robot to have linear displacement (forwards-backwards or from left to right).

Each arm of the SDA10F robot can be operated independently, or in a simultaneous way along with the torso. This robot is commonly used for material handling, machine tending, picking/packing, assembly, load and unload, and other applications, where the manipulation of parts is required.

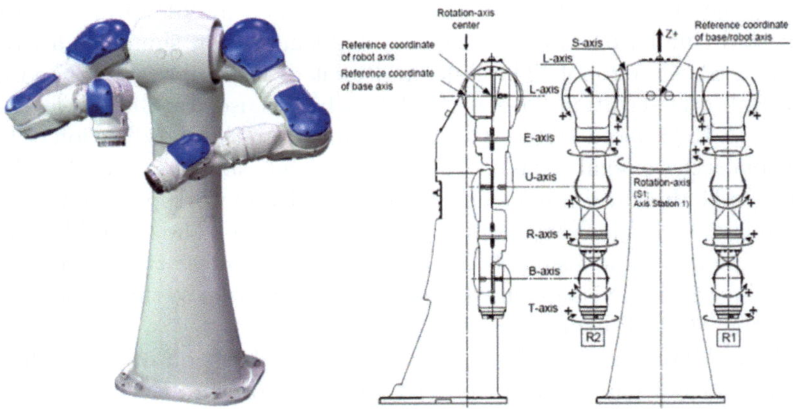

Fig. 2. Robot SDA10F. Image taken from [7]

Figure 2 shows the 15 axes of a typical configuration of the SDA10F robot (axes are represented by capital letters). The joints have the following ranges of motion: S 170°, L 180°, E 110°, U 135°, R 180°, B 110°, T 180°, and the torso (S1) 180°. Also the maximum speed of each joints is, S 170°/s, L 170°/s, E 170°/s, U 170°/s, R 200°/s, B 200°/s, T 200°/s and the torso 180°/s. The repeatability is ≈0.1 mm, its approximate weight is 220 kg, and the payload is 10 Kg per arm.

Figure 3 shows the work space of the robot. From the figure, it can be seen that the maximum range in the horizontal axis is 1970 mm and for the vertical one is 1440 mm. The way the robot is positioned is with an absolute encoder.

The controller of this robot is the FS100 from yaskawa [7]. This controller provides high performance in applications such as picking, packing, and parts

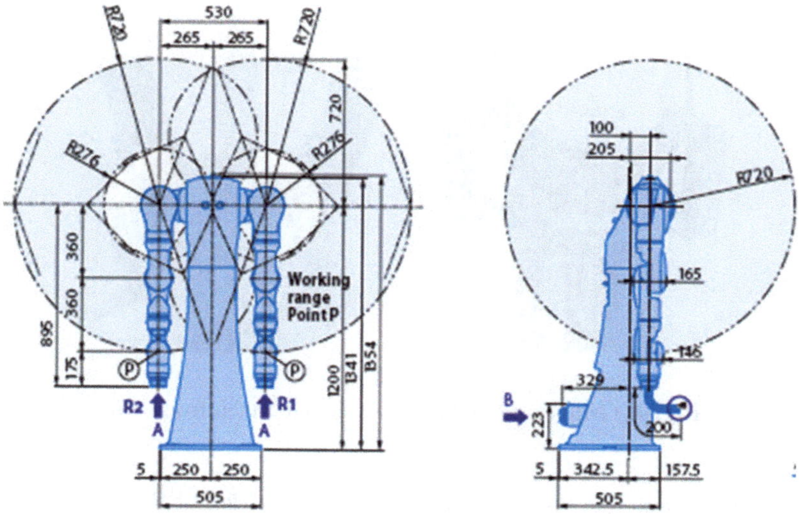

Fig. 3. Workspace of Robot SDA10F. Image taken from [7]

handling tasks. This controller has the advantage of offering an open software architecture, which allows the creation of customized industrial applications using C/C++/C#/and .NET. The controller supports MotoSync, MotoPlus, MotoCom, and ROS-Industrial software environments, and allows communication protocols such as Fieldbus and Ethernet. Additionally, the controller has 16 digital inputs and 16 digital outputs, along with a safety module where it is possible to connect extra push-buttons for an emergency stop and indicator lights.

The PUJ robot has two-different end-effectors from the company ROBOTIQ [19], one in each arm. A two-fingers gripper, as the one shown in the left side of Fig. 4, which has a load force of 5 kg, and 85 mm stroke. The grip force range is from 5 N up to 220 N. The second gripper is a three-fingers gripper, shown in the right side of Fig. 4. This gripper has a wider aperture range (155 mm), the force range is from 15 N up to 60 N, and its load force is 10 kg. Both grippers (two and three fingers) are equipped with different sensors: position sensor, force, and speed.

3.1 Software Architecture

The PIR (Perception for Industrial Robots) project at PUJ aims to augment the capabilities of industrial robots by incorporating different sensors in the robot's workspace. PIR project is based on the ROS-Industrial project. ROS-Industrial (ROS-I) [5] is an open source project which is used in this paper to interact with the robot. ROS-I comes with a simulation environment where it is possible to test the behaviors of different robots. Additionally, ROS-I features perception

Fig. 4. Gripper robot SDA10F. Image taken from [19]

sensors such as cameras, laser, radar, etc. The main idea of ROS-I is to be able to develop software that can run in different robot platforms.

ROS-Industrial extends ROS (Robot Operating System) [3] capabilities to the industrial area. Therefore, the development that has been reached so far in the research robotic area with ROS, can be extended to different applications, including industrial robotics in order to extend the capabilities of industrial robots. The Motoman robot SDA10F used in PIR project is one of the robots included in ROS-I project. This allows the integration of different functionalities that come with ROS and ROS-I facilitating the developing and testing of different industrial applications.

Many functionalities of ROS can be applied in ROS-I. Figure 5 shows an example of how the information flows between ROS and ROS-I. ROS can be used to define the path for the robot, for example using MoveIt! (this occurs in the PC-side). ROS-I is in charge of decomposing the path into smaller points, and then those points are passed to the robot controller. In this part, ROS-I will be in both sides, in the PC side; and robot side, where the server (based on ROS-I) is installed.

Figure 6 shows the ROS-based architecture of the Motoman package. ROS-Industrial consists of 4 layers: some are generic to ROS-I, others depend on the robot manufacturer, and others still under development.

ROS-Industrial packages used in this paper includes:

- Simple Messages: defines the communication protocol to an industrial robot.
- Industrial Robot Client: library for communicating with an industrial robot.
- Vendor specific package (Motoman): this is a ROS package that allows communication with industrial robot controllers. It contains URDF models for different robots, and MoveIt! navigation packages with default configurations.

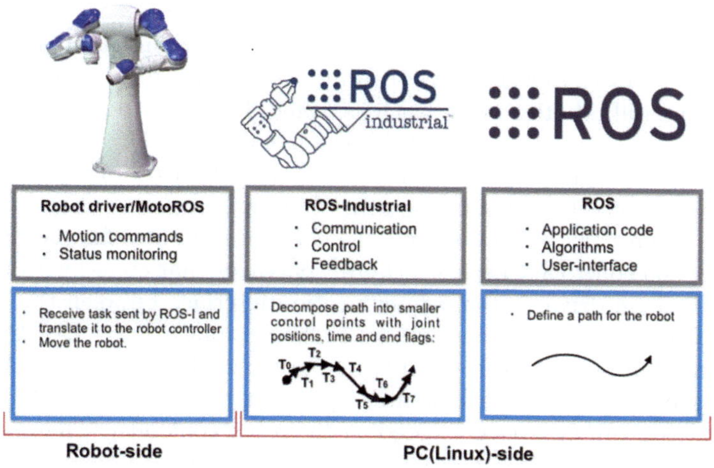

Fig. 5. ROS framework with industrial robots. Different functionalities of ROS can be used with ROS-I. ROS-I framework runs in the PC-side, as well as in the robot controller

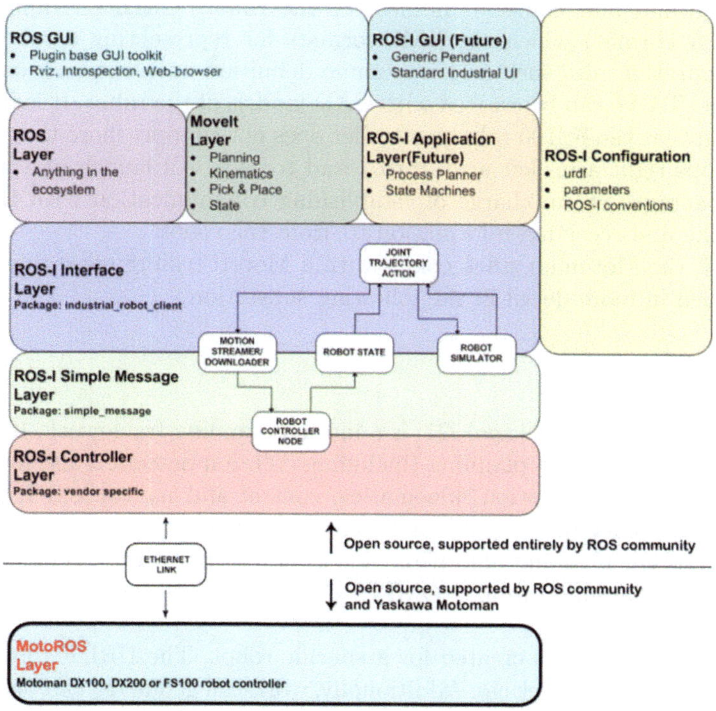

Fig. 6. Motoman package hierarchy. Image taken from [3]

3.2 Motoman Stack

Figure 6 shows the high level architecture of ROS-Industrial (ROS-I) when using the Motoman robot. ROS-I allows to bring well-known functionalities of ROS into the industrial robotics world, such as using MoveIt! for dealing with motion planning and kinematics, or using Rviz as a simulation tool.

The Motoman stack can be downloaded from [20]. It contains libraries, configuration files, and ROS nodes for controlling a Motoman robot using ROS-Industrial [3]. In our case, it contains the configuration files for the SDA robot with the FS100 controller, i.e. the SDA10F packages that comes with the stack.

The Motoman_driver is the communication interface between the controller and the ROS computer. It contains the ROS-I Interface Layer (see Fig. 6), the ROS-I Simple Message Layer, the ROS-I Controller Layer, and the MotoROS Layer. As shown in Fig. 6, the first three layers are located in the ROS computer, whereas the MotoROS layer is located on the robot controller. The communication is based on the simple_message package and some Motoman-specific messages. The MotoROS layer (MotoPlus folder inside the motoman_driver package) should be ordered from Motoman who will be the one in charge of configuring the robot controller for allowing ROS-Industrial integration with the robot.

On the other hand, the Motoman stack contains support packages (for the PUJ robot is the motoman_sda10f_support package). These packages contain meshes (normal and collision meshes) of the robot, URDFs (Unified Robot Description Format) which are XML formats for representing a robot model, and configuration files such as joints name definition and motion group definitions. The URDFs can be created with CAD models of the robot. It is important to mention that the FS100 robot controller does not support more than 4 groups (only groups right arm, left arm, torso1 and torso2). All launch files inside the support packages are in charge of establishing communication with the robot, and sending and receiving information to/from the robot.

Finally, the Motoman stack comes with a MoveIt! config package which will be described in more detail in the following subsection.

3.3 MoveIt!

MoveIt! (not a ROS-I package) [21] is a motion planning framework. It encapsulates software for motion planning (including collision detection and avoidance), manipulation, 3D perception, kinematics, control and navigation. It has been designed as an "easy-to-use platform for creating advanced robotics applications" [21]. The motion planning algorithms incorporated by MoveIt!, come from the Open Motion Planning Library (OMPL) [22], which contains implementation of different planners.

A MoveIt! package is created for a specific robot. The URDF robot data is required to create the package. Additionally, work-cell geometry can be included. Both URDF and work-cell geometry are used by MoveIt!, for visualization purposes and collision checking, among others. When creating a MoveIt! package (which is done following a set-up assistant GUI), it creates automatically all the required nodes, and generic config and launch files for the robot and/or sensors. In the set-up assistant GUI, it is possible to define motion groups which will be used for planning (very useful when working with dual arms robots, e.g. group right_arm, group left_arm), to define end-effectors, the kinematics solvers, and specific poses for a specific group (e.g. home position for left_arm group).

As mentioned before, the Motoman stack comes with a pre-configured MoveIt! packages for different Motoman robots (for the PUJ's robot, it is the motoman_ sda10f_moveit_config). This package contains config files automatically created by the MoveIt! setup assistant. These files contain configuration information about the robot - .sdrf file- (links, predefined poses, end-effectors, motion groups, collision information, etc.), information about the controllers, the joint limits, the defined kinematic solvers for each group, and the motion planning algorithm.

There are two launch files in the MoveIt! config package that are used to test the package created and allows you to start testing motion planning in your robot in a straight forward fashion. The demo.launch lets you test if your moveIt! configuration is correct (setting up of the links, end-effectors, poses, motion groups, etc.). This files does not create any connection to a real robot. It has a dummy joint-state-publisher which emulates the robot connection. When demo.launch is launched the Rviz simulator comes up with the robot in the scene. With this launch it is possible to move the end-effector of the robot manually, or creating random positions in order to plan and execute different motions with the robot.

The files generated by MoveIt! can be modified in order to provide information about a specific robot (controllers, etc.) to allow communication between MoveIt! and the robot nodes.

The second launch file: the moveit_planning_execution.launch lets you test in simulation and with the real robot, the config files of your package. The first test should be offline in order to test if the MoveIt! package is well-configured. By default, that launch file lets you test everything with the industrial robot simulator. Once sure about the configuration this launch file can be used to interact directly with the real robot by passing as parameter the robot ip address and the robot controller type.

Both previously mentioned launch files open the planning environment (Rviz + Motion Planning plugin). It is possible to see the robot and if connected to the real robot, the simulated robot should have the same pose as the real one (see Fig. 10).

For planning, it is possible to do it directly in the planning environment by moving the end-effector markers to the goal state and then pressing plan and execute commands from the Motion planning plugin. On the other hand the same command can be send through a customize node using C++, which can be created to make the robot follow specific trajectories or perceive the environment and based on that move the robot towards a specific object. MoveIt! API for sending the motion request to the planners and for executing the commands in the simulated or real robot. Planning with both the GUI and by C++ let you plan with obstacle avoidance capabilities.

4 Tests and Results

The following section presents some of the capabilities of the Motoman stack for controlling dual arm robots. In this section the PUJ's robot setup is presented. Additionally, an analysis of the strengths and drawbacks of the driver are discussed.

4.1 Experimental Setup

- MotoRos setup: as can be seen in Fig. 6 MotoROS should be configured in the controller (ROS Server). A pre-built binary is available in the Motoman Stack, and should be selected based on the robot controller. A guide of how to install the binary and the INFORM Code in the robot controller is found in [20]. Make sure you have loaded the .DAT, .JOB, .out files. Additionally, some controllers require modification of the ALL.PRM file from the controller. In the PUJ's case, this file was set by the support people from YASKAWA. It is important to make sure that none .out file is in the controller. We had to remove MotoSight2D.out MPLM.out. These two files were part of the camera that came with the robot. However, MotoROS requires that motoros.out has to be the only application running on the robot. Additionally, our MotoROS software had to be modified, this is due to the particular configuration of our robot (5 groups, we have and additional motor) and MotoROS allows to configure only four groups. Therefore, YASKAWA support team modified the MotoROS part to ignore our fifth group (the additional axis of the PUJ's robot, see Sect. 3 for more details), this is because the FS100 controller will never support more than 4 groups.

- Universal outputs checks: the universal outputs of the controller are used by MotoROS to indicate its state. In order to make sure the controller can receive remote commands, check the universal output signals in the pendant: OUT#0889 ROS_READY, OUT#0890 ROS_DONE, OUT#0891 INIT_DONE, OUT#0892 CONNECTION_SRV, OUT#0893 MOTION_SERVER, OUT#0894 STATE_SERVER OUT#0896 FAILURE. Output #899 should be ON, if the controller is well configured to receive motion from ROS. Output #891 confirms that the initialization was completed. Output #892, that the server threads are running properly, and Output #893 and Output #894 will only be ON when at least one client is connected, and should be OFF when the clients disconnect. if Output #896 is ON, a failure occurred and MotoROS should be reset (reboot the controller). For more details check document M2092-EDS [23].

- ROS computer setup: the computer used was a DELL Intel Core I5 2.6 GHz, 8 GB RAM memory, running Ubuntu 14.04 with ROS-indigo installed with the following packages: motoman driver indigo-devel, ROS-Industrial indigo-devel, and MoveIt.

- Motoman Stack configuration: In order to be able to echoing information from the robot, as mentioned in [20], it was required to modify the file sda_10f_motion_ interface.yaml. Instead of having groups IDs from 1–4, we had to modify them to 0–3. By doing this, it was possible to have communication with the real robot and to be able to read joints state information. This change is required only because of the non-standard configuration of the SDA10F robot the PUJ has.
- MoveIt! configuration: Motoman stack comes with a pre-configured MoveIt! configuration package (motoman_sda10f_moveit_config). By launching the de-mo.launch file of this package it is possible to check if the package is working properly. If so, you should be able to plan and execute commands, using the Motion Planning plugging from Rviz. Rviz should show the Robot, the goal positions you are defining and the motion executed in the simulated robot. In order to test the real robot, launch the motoman_sd10f_motion_streaming_interface.launch file passing the robot ip and the controller type, as arguments. The pendant of the robot must be in REMOTE mode. After launching this file, the current position of the robot should be shown in the Rviz simulated robot. The SDRF file that comes with the MoveIt! configuration package was modified using the MoveIt! setup assistant in order to define some poses for each motion groups. The five motion groups that are defined in the package are: arm_left, arm_right, torso, sda10f, and arms. Group sda10f allows to control all the joints of the robot, and the arms group allows to control both arms simultaneously. Some of the poses defined for the different groups were: “HOME_ARMS”, “LEFT_FRONT”, “RIGHT_ELBOW”, “RIGHT_PICK”, etc. See Fig. 7 that illustrates the different pre-configured poses.

4.2 Testing the ROS-I Motoman Stack

The MoveIt! API (Application Programming Interface) has different functions that let you define goal states, plan and execute them. In these tests we will show some of this functionalities using both the simulated and the real robot.

4.2.1 Pre-defined Poses

When configuring MoveIt! with the MoveIt! setup assistant, it is possible to configure specific robot poses, for the different motion groups that has been defined. Figure 7 shows 4 of the multiple poses that were defined for the different motion groups of the Motoman robot. Figure 7 shows poses for the motion groups arm_left and arm_right: “LEFT_FRONT and RIGHT_FRONT”, “LEFT_ELBOW and RIGHT_ELBOW”, “RIGHT_PICK”

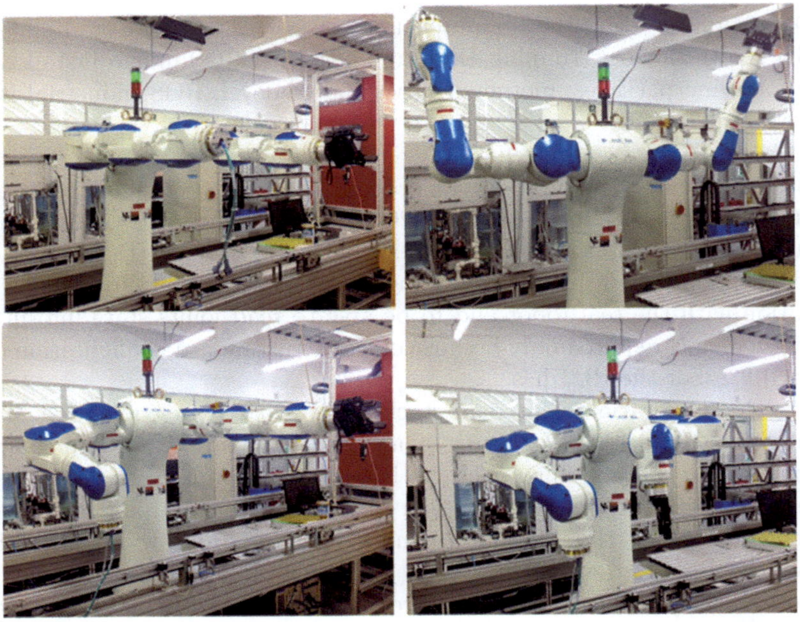

Fig. 7. Poses defined in the SDRF file from Motoman MoveIt—config file

and "LEFT_FRONT", and "RIGHT_PICK"-"LEFT_PICK". These poses can be called from a ROS node using the MoveIt! API or by using the Rviz graphical user interface.

4.2.2 Joint Commands

Another way to interact with the robot instead of using predefined poses is by moving specific joints. Figure 8 shows how the position of the robot changes, when the value of specific joints are modified. The robot starts with its elbows bended, as can be seen in Fig. 8, left image. Then, the position of joints L and U (see Fig. 2 for clarifying the names of the joints) is modified. Figure 8, right image shows the position of the robot, after modifying joints L and U.

4.3 Obstacle Avoidance

One of the advantages of using ROS-Industrial is that many ROS functionalities can be used. MoveIt! is one of those. It is a motion planning framework [21], and here in this test we show one of its great attributes: motion planning with obstacle avoidance capabilities. By creating the MoveIt! config package of the robot, it is straightforward to plan avoiding obstacles.

In the following tests, two ROS nodes where created: one for publishing objects in the scene, and the second one is in charge of planning and executing a specific trajectory. The trajectory was programmed using functions from

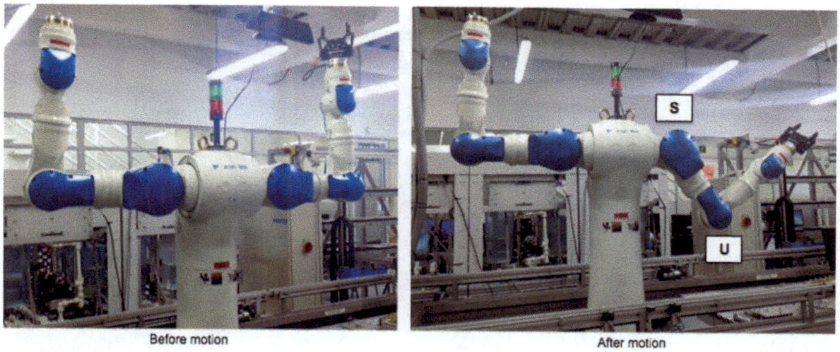

Fig. 8. Joints motion test. The MoveIt—'s API allows to send joint commands to the robot. The figure shows the robot when moving joints L and U. Left image shows the starting position, and right image the final position.

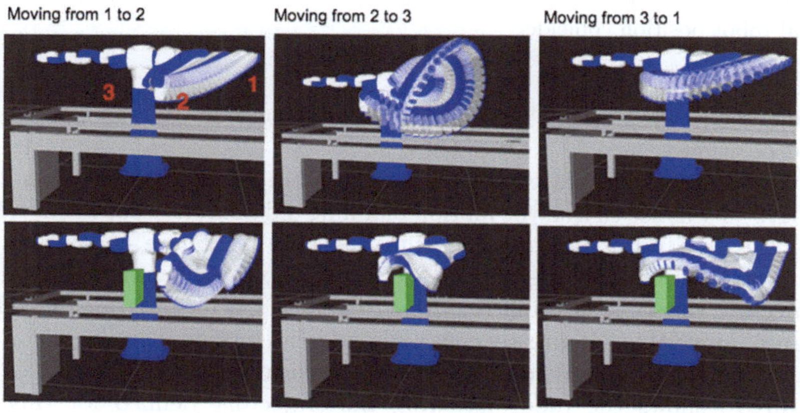

Fig. 9. Collision avoidance capabilities with MoveIt!. First row shows the motion of the robot without obstacles. Second row shows the motion of the robot when an obstacle was added to the scene. Check the different trajectories of the robot.

the MoveIt! API. The programmed trajectory moves the robot from position 1 (extended arm, see Fig. 9, number 1) to position 2 (see Fig. 9, number 2), and from position 2 to position 3 (moving towards the right arm of the robot, see Fig. 9, number 3).

MoveIt! comes with a simulator based on Rviz that lets you plan the robot's movements without requiring the real hardware. All the motions that occur in the simulator are the ones the real hardware will follow.

Figure 9 shows the path followed by the simulated robot when moving towards the three previously mentioned positions. The first row shows the motions when no object was in the environment. The second row, shows the simulated path, followed by the robot, when a green box is placed on the conveyor. Comparing figures shown in rows 1 and 2, it is possible to see how the

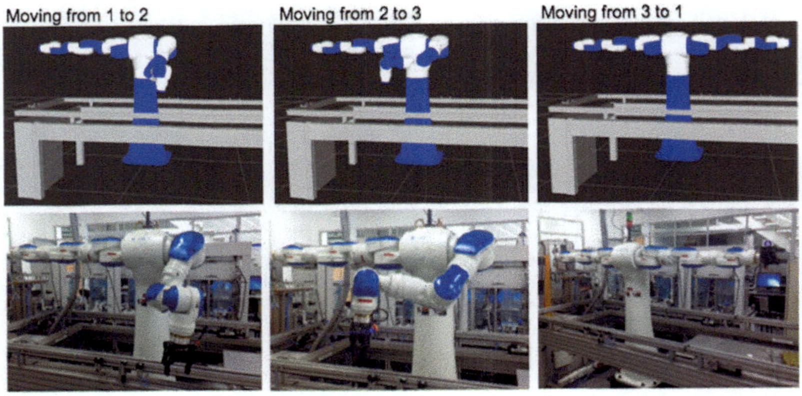

Fig. 10. Collision avoidance capabilities, real test. First row shows the motion of the simulated robot. Second row shows the motion of the real robot. Check that simulated and real robot position coincide.

planning algorithm gets in charge of finding collision-free path for reaching the desired positions.

A closer look at the trajectories when the robot moves from position 2 to 3 (see Fig. 9), let us see that, if the motion planning did not have obstacle avoidance capabilities, the robot had crashed with the box. Therefore comparing both trajectories from Fig. 9 shows that indeed, MoveIt! plans different path based on the objects in the scene, without requiring to write a specific program for collision avoidance. MoveIt! has it already encapsulated in its motion planning algorithms which represents an big advantage for rapid applications development.

Figure 10 shows the robot following the trajectory defined for this tests (reaching the 3 previously explained positions). Comparing the images from the first and second row, it can be seen that the same positions reached by the simulated robot, are the ones followed by the real one.

4.3.1 Trajectories

In this test, we show how to program the robot to follow a specific trajectory. MoveIt!'s API allows you to stack different waypoints, and then execute them. These poses can be created using the Rviz GUI, by moving the simulated robot with the interactive markers and saving the different coordinates. Additionally, it is possible to generate the trajectory by specific increments of a previous pose.

For this test, the left and right arms were programmed to move following a square trajectory in the YZ plane. An initial pose was given to each arm, and then different way points were programmed by increments in the Y (left/right) and Z (up/down) axis of the robot. A ROS node was created to plan and execute the trajectory. Figure 11 shows the trajectory followed by the real robot, which is represented in real-time in Rviz using the simulated robot, when the real robot moves. The green line in the figure represents the trajectory of the end effector.

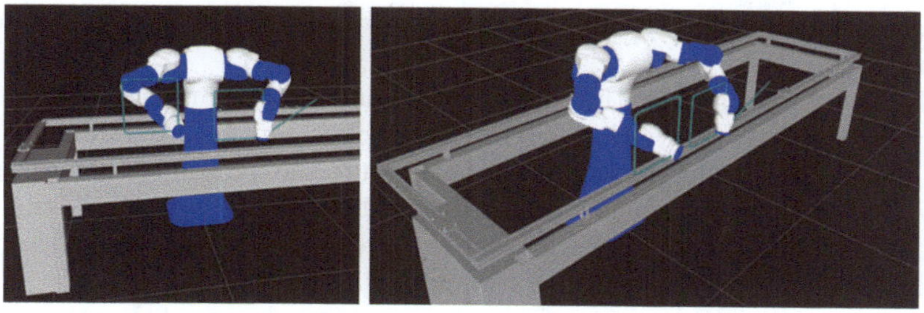

Fig. 11. Executing trajectories using MoveIt!: simulated robot. Right and left arms of the robot were commanded to follow a square trajectory. The image shows the simulation environment with the trajectory (green line) followed by the robot.

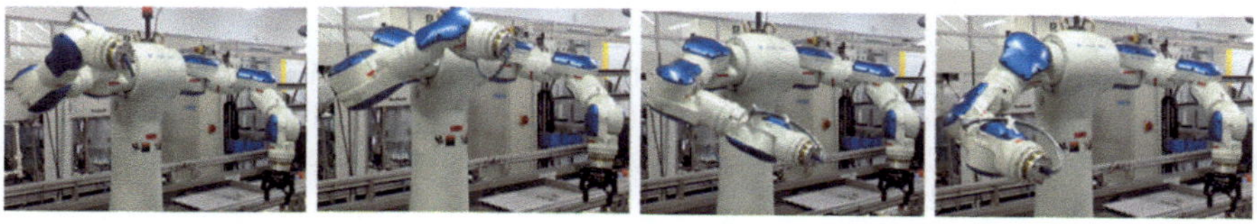

Fig. 12. Executing trajectories using MoveIt!: real robot. Right and left arms of the robot were commanded to follow a square trajectory. The image shows the real robot when reaching the four corners of a square trajectory.

Both arms performed movements that generated a square trajectory in the YZ plane.

Figure 12 shows representative images of the real robot when moving its right arm. The first image (from left to right) shows the arm positioning in the upper left corner of the square trajectory. In the second image, the arm is in the upper right corner of the square. The third and fourth images show the arm in the lower right corner and lower left corner of the square, respectively. A video of the test can be seen in [24].

5 Conclusions and Future Work

This paper presented the starting point of PIR (Perception for Industrial Robots) project at Pontificia Universidad Javeriana Bogotá. This project aims to augment the capabilities of industrial robots by including different sensors, that allows them to adapt dynamically to the environment. Therefore, expanding the robot's capabilities to different tasks.

In this paper, the configuration set up of the robot with ROS-Industrial was presented. Additionally, different capabilities of ROS and ROS-I ecosystems were described, providing a general view of the functionalities that can be used for programming an industrial robot.

During the tests, it was possible to experience the advantages of MoveIt!, not only for solving the planning problem, but also for providing a powerful simulation environment, that let conduct tests without requiring to be connected with the real robot.

The tests presented in this paper were conducted using the arm_left and arm_right groups of MoveIt!. Currently, the robot can be programmed, using ROS and ROS-I, for developing tasks that require the motion of the arms separately. However, simultaneous motion of the robot (arms + torso) or simultaneous motion of the arms (both arms at the same time) can not be achieved yet. This is due to the particular configuration of the robot at PUJ.

Future work in PIR project is focused on enabling the use of the other motion groups, and on incorporating cameras, that will be used to locate in real-time the position of objects; and a laser, that will provide 3D information of the objects present in the scene. By incorporating these sensors, it is expected to allow the robot to conduct different tasks, based on what it perceives.

Acknowledgements. We would like to thank the Pontificia Universidad Javeriana for supporting this research. We also want to thank the Centro Tecnológico de Automatización Industrial (CTAI) for letting us use the installations and equipments for developing this research.

References

1. The challenge: http://rosindustrial.org/the-challenge/
2. Quigley, M., Conley, K., Gerkey, B., Faust, J., Foote, T., Leibs, J., Wheeler, R., Ng, A.Y.: Ros: an open-source robot operating system. In: ICRA Workshop on Open Source Software, vol. 3, no. 3.2, p. 5 (2009)
3. ROS, Robot operating system (2016). http://www.ros.org
4. Garber, L.: Robot os: a new day for robot design. Computer **46**(12), 16–20 (2013)
5. ROS-I, Robot operating system-industrial (2016). http://rosindustrial.org
6. Centro Tecnógico de Automatización Industrial, Pontificia Universidad Javeriana, Bogot (2016). http://www.javeriana.edu.co/blogs/ctai
7. YASKAWA, Motoman SDA10F robot (2016). https://www.motoman.com/industrial-robots/sda10f
8. Smith, C., Karayiannidis, Y., Nalpantidis, L., Gratal, X., Qi, P., Dimarogonas, D.V., Kragic, D.: Dual arm manipulation-a survey. Robot. Auton. Syst. **60**(10), 1340–1353 (2012)
9. Tsarouchi, P., Makris, S., Michalos, G., Stefos, M., Fourtakas, K., Kaltsoukalas, K., Kontrovrakis, D., Chryssolouris, G.: Robotized assembly process using dual arm robot. Procedia CIRP **23**, 47–52 (2014)
10. Messay, T., Ordóñez, R., Marcil, E.: Computationally efficient and robust kinematic calibration methodologies and their application to industrial robots. Robot. Comput. Integr. Manuf. **37**, 33–48 (2016)
11. Gan, Y., Dai, X.: Human-like manipulation planning for articulated manipulator. J. Bionic Eng. **9**(4), 434–445 (2012)
12. Martinez, A., Fernndez, E. (eds.): Learning ROS for Robotics Programming. Packt Publishing. http://www.packtpub.com/learning-ros-for-robotics-programming/book
13. Aitken, J.M., Veres, S.M., Judge, M.: Adaptation of system configuration under the robot operating system. In: 19th IFAC World Congress. IFAC Proceedings Volumes, no. 3, pp. 4484–4492

14. Ergur, S., Ozkan, M.: Trajectory planning of industrial robots for 3-d visualization a ros-based simulation framework. In: IEEE International Symposium on Robotics and Manufacturing Automation (ROMA 2014), pp. 206–211, December 2014
15. Tsarouchi, P., Makris, S., Michalos, G., Matthaiakis, A.-S., Chatzigeorgiou, X., Athanasatos, A., Stefos, M., Aivaliotis, P., Chryssolouris, G.: Ros based coordination of human robot cooperative assembly tasks-an industrial case study. Procedia CIRP **37**, 254–259 (2015)
16. Amazon, Amazon robotics challenge. https://www.amazonrobotics.com/pickingchallenge
17. RobotValley TuDelft, Team delft wins the amazon picking challenge (2016)
18. Bharatheesha, M., Burger, R., Vries, M.D., Ko, W., Tan, J.: Motion module for the amazon picking challenge - team delft, August 2016
19. Robotiq, Industrial grippers (2016). http://robotiq.com
20. Motoman_Driver, ROS-I driver for Mottoman robots (2016). http://wiki.ros.org/motoman_driver
21. MoveIt, Motion planning framework (2016). http://moveit.ros.org
22. OMPL, The Open Motion Planning Library (2016). http://ompl.kavrakilab.org
23. Motoman_Driver-Especifications, Motoman driver M2092-EDS document. motoplus-ros especifications (2016). http://wiki.ros.org/motoman_driver
24. Video, Planning and executing trajectories with ROS-I (2016). https://youtu.be/mS05tVPRCq0

Fault Detection for Sinusoidal PWM Inverter

Jesús A. Esquivel(✉), J. Aarón Rios, Pedro Galván, and Isela Carrera

Facultad de Ingeniería Mecánica Y Eléctrica, UAdeC, Monclova, Mexico
{jesus.esquivel,pedrogalvan,iselacarrera}@uadec.edu.mx,
rios-galindo1@hotmail.com

Abstract. This paper applies basic concepts of fault detection for a sinusoidal power inverter when a fault is induced in the electronic control using the internal model principle for a sinusoidal reference waveform.

1 Introduction

As the advances of control engineering are more complex the reliability is more hard to achieve, [3]. When the number of components involved in the automatic control systems is incremented, the chances to fail of each one rises, doing likewise the odds of failing of the entire system. A way to increment the reliability of the system is to ensure a good behavior of their components. The quality of these requires in turn a mayor development of science and technology related with the specific area. Some examples are difficult theoretical approaches such as the geometric concepts, see for example [1,6]. Another way to improve the reliability is through the redundancy of their sensors and/or actuators, which is an expensive choice due the number of elements involved in the control process is greater than desired or needed in normal conditions. Duplicating elements is expensive due the costs involved in the additional items and the implications for installing them into the system, however might be an alternative when the reliability is imperative for the safety of the process. When doing so the fault of a sensor or actuator can be highly tolerated keeping the good performance of the global system. Redundancy of elements in a control system installed for support and provide plenty of extra information has the advantage of incrementing the reliability and the good performance of the control law.

When the system fails the component may be detected and replaced automatically for the system, this process is called *accomodation* or *self-reorganization*. Although this may be the easiest part of the *failure detection and isolation* (FDI) the way to change the control law to accommodate the system behavior is no that easy. The fault has to be detected on time to avoid the collapse of the entire process. Some types of failures are detected and located almost immediately, *e.g.*, loss of pressure of an hydraulic system.

Redundancy of the sensors and/or actuators generally is not the most effective way to solve these kind of problems. Sometimes the fault detection might give false alarm due another situations, *e.g.*, perturbations, slow parameter variations, etc. For these cases the tools needed may be of different nature: adaptive control, robust control, etc. Some approaches in power electronics are in

I. Chang et al. (eds.), *Advances in Automation and Robotics Research in Latin America*, Lecture Notes in Networks and Systems 13, DOI 10.1007/978-3-319-54377-2_17

the literature, for example, in [7] is presented a model-based fault detection and identification (FDI) method for switching power converters where they construct a library of fault signatures for possible component and sensor faults with responses for fault detection within the 400 μs and its identification in under 10 ms. The approach shown in this work is focused only in the inputs and is based mainly in a basic decoupling between inputs and outputs.

2 Fault Detection

As pointed out in the foregoing section the fault of an element in a control process may be a hard problem to solve, [2,5], so the tunning of the instruments is critical for implementing the approach for detecting the fault whenever a deviation appears according to the mathematical models and in this way estimate their presence.

Assume a linear time invariant system expressed by

$$\dot{x} = Ax + Bu \tag{1}$$

$$y = Cx \tag{2}$$

where $x \in \mathbb{R}^n$ is the state, $y \in \mathbb{R}^m$ is the output, $u \in \mathbb{R}^r$ is the input and the matrices have the corresponding dimensions, *i.e.*, $A \in \mathbb{R}^{n \times n}$, $B \in \mathbb{R}^{n \times r}$ and $C \in \mathbb{R}^{m \times n}$. An important feature of the systems must be their controllability which is the capacity of bring the state from an initial point $x(t_0) = x_0$ to origin in a finite time given some input $u(t)$. If the systems has this characteristic then is said to be full controllable which means that the columns of the matrix $W \equiv [B, AB, \cdots, A^{n-1}B]$ have complete rank, *i.e.*, n, then it is said that (A, B) is controllable.

There are three places where a fault can appear: actuators, sensors and within the system. The last may be interpreted as a perturbation or an unmodeled dynamics. This work assume faults only in actuators.

Equation (1) is rewritten as follows,

$$\dot{x} = Ax + b_1 u_1 + \cdots + b_r u_r \tag{3}$$

where b_i are vectors of B and u_i are the elements of u.

Note that the subspace generated by the columns of W, namely W_m, is invariant under A, *i.e.*, $AW_m \subset W_m$ and contains the image of B, *i.e.*, Im $B \subset W_m$. When the system is completely controllable $W_m \equiv \mathbb{R}^n$, and when this is not the case there exists a subspace complement of W_m which it is denoted by W_m^c. This contains the subspace no controllable with vectors x that satisfy $W^T x = 0$ where T means the transpose of W. Having said this the uncontrollable subspace is made up the kernel or null space of the matrix W^T.

The above is taken as a basis to define the function

$$v_h(t) = h^T x(t) \tag{4}$$

where $h \in \mathbb{R}^n$ is constant. If h is in the uncontrollable subspace then there might be actuators with null effect in the behavior of the system. To simplify exposure it is assumed only one active actuator, namely the i, others are set to zero. The solution of state space is simplified and has the shape

$$x(t) = \Phi(t, t_0)x_0 + \int_{t_0}^{t} \Phi(t, \tau)b_i u_i(\tau)d\tau \tag{5}$$

where $\Phi(t, t_0)$ is the transition matrix which has the following properties

$$\frac{d}{dt}\Phi(t, t_0) = A\Phi(t, t_0) \tag{6}$$

$$\Phi(t_0, t_0) = I \tag{7}$$

where I is the identity matrix.

Below, it is defined W_i as the matrix obtained from W, where the vectores are canceled except b_i, *i.e.*, $B \equiv b_i$. In this case the Cayley-Hamilton theorem and the power expansion series of the transition matrix must satisfy

$$\int_{t_0}^{t} \Phi(t, \tau)b_i u_i(\tau)d\tau = W_i g(t) \tag{8}$$

for a given $g(t)$. It is important to take into account the rank of W_i which defines in a unique way $g(t)$ if it is n. Otherwise there exists an infinite number of representations for the vector that satisfies Eq. (8). The function $v_h(t)$ can be rewritten as

$$v_h(t) = h^T\Phi(t, t_0)x_0 + h^T W_i g(t) \tag{9}$$

if h is in the uncontrollable subspace this equation is reduced to

$$v_h(t) = h^T\Phi(t, t_0)x_0 \tag{10}$$

where it is clearly seen that the control $u_i(t)$ does not affect the function $v_h(t)$.

The analysis exposed until this point shows that the rank of action of the actuators is established by the subspace of controllability from which can be taken advantage for $v_h(t)$.

2.1 Supplements

An important problem is determine the minimal number of actuators that needs the control system to work properly. In other words, which is the minimal value of r for keeping the pair (A, B) controllable? The answer is obtained from the invariant polynomials of the matrix A represented by $\{i_1(s), \ldots, i_n(s)\}$ of order $k_1, \ldots, k_n$ respectively. The polynomials are monic and their product yield the characteristic polynomial of A. The quantity of non trivial polynomials is equal to the minimal quantity of inputs needed for keeping the controllability condition of (A, B).

As previously seen, a linear function $v_h(t) = h^T x(t)$ allows to determine the degree of dependence of a control action with respect the others inputs. For that, the supplementary controls are defined as $\{b_{i_1}, \ldots, b_{i_j}\}$ which control some linear function $v_h(t)$, that is, each $i-$th input of this set satisfies the condition $h^T W_i \neq 0$. The number j indicates the degree of vulnerability of the function $v_h(t)$ to the fault of actuators.

2.2 Filters

Given the following detection filter

$$\dot{z}(t) = Gz(t) + Dy(t) + Bu(t) \tag{11}$$

and assume the same number of inputs, state variables, and outputs, *i.e.*, $n = m = r$. The inputs $u(t)$ and $y(t)$ are the input and output of the system, respectively. The error is defined as $e(t) = x(t) - z(t)$, and its dynamics is

$$\dot{e}(t) = (A - DC)x(t) - Gz(t) \tag{12}$$

doing $G = A - DC$ the equation is reduced to

$$\dot{e}(t) = (A - DC)e(t) \tag{13}$$

which means that (1) is observable. The selection of D must be so the eigenvalues of $A - DC$ have negative real part to guarantee $\lim_{t\to\infty} e(t) = 0$, which means that in the absence of perturbations the state of the filter approaches asymptotically to the system state. In the special case of $m = n$ and if C has inverse then $D = (A - G)C^{-1}$ and if $m < n$ the solution is not unique and is given by $D = (A - G)C^T(CC^T)^{-1}$.

2.3 Information of the Fault

Assume the fault take place in the $i-$th actuator, this can be modelled by

$$u(t) = u_d(t) + e_i n(t) \tag{14}$$

where e_i is the vector of n zeros except one in the $i-$th position and $n(t)$ is an arbitrary scalar function of time and whose value different of zero is the deviation of the desired control law $u_d(t)$. The dynamics of the error including the fault is given by

$$\dot{e}(t) = (A - DC)e(t) + b_i n(t) \tag{15}$$

If all eigenvalues of the characteristic equation are placed in $-\sigma_f$ with $\sigma_f > 0$ the solution of the error equation is

$$e(t) = e^{-\sigma_f(t-t_0)}e_0 + b_i \int_{t_0}^{t} e^{-\sigma_f(t-\tau)} n(\tau)d\tau \tag{16}$$

when the transient response of the fault appears, *i.e.*, when the first term of the right side of the equation disappears, then the error signal set up a unique direction with vector b_i and magnitud, of fault, given by

$$\int_{t_0}^{t} e^{-\sigma_f(t-\tau)} n(\tau) d\tau \tag{17}$$

The deviation of the system and filter due to a fault is reflected directly in a scalar quantity that changes with time and is located in the position of the actuator in trouble.

3 Example in Simulink

The foregoing concepts are applied next in an abstract example with the following conditions. The system is of second order, has two inputs, two outputs, observable and controllable, and it is represented by

$$\dot{x} = \begin{bmatrix} 0 & 1 \\ -100 & -2 \end{bmatrix} x + \begin{bmatrix} 5 \\ 0 \end{bmatrix} u_1 + \begin{bmatrix} 0 \\ 100 \end{bmatrix} u_2 \tag{18}$$

$$y = \begin{bmatrix} 1 & -1 \\ 1 & 1 \end{bmatrix} x \tag{19}$$

The eigenvalues of A are conjugate complexes with time constant equal to a second and the matrix C is invertible. The time constant of the filter is set smaller in such a way the error is detected long before finishes the system's transient response; in this way the time constant of both modes are fixed in half a second, *i.e.*, $G = A - DC = -2I$, matrix D is obtained directly,

$$D = (A + 2I)C^{-1} \tag{20}$$

$$= \begin{bmatrix} 2 & 1 \\ -100 & 0 \end{bmatrix} \begin{bmatrix} 1 & -1 \\ 1 & 1 \end{bmatrix}^{-1} = \begin{bmatrix} \frac{1}{2} & \frac{3}{2} \\ -50 & -50 \end{bmatrix} \tag{21}$$

the filter is rewritten as

$$\dot{z} = \begin{bmatrix} -2 & 0 \\ 0 & -2 \end{bmatrix} z + \begin{bmatrix} \frac{1}{2} & \frac{3}{2} \\ -50 & -50 \end{bmatrix} y + \begin{bmatrix} 5 & 0 \\ 0 & 100 \end{bmatrix} \begin{bmatrix} u_1 & u_2 \end{bmatrix} \tag{22}$$

or

$$\dot{z} = \begin{bmatrix} -2 & 0 \\ 0 & -2 \end{bmatrix} z + \begin{bmatrix} \frac{1}{2} & \frac{3}{2} & 5 & 0 \\ -50 & -50 & 0 & 100 \end{bmatrix} \begin{bmatrix} y_1 \\ y_2 \\ u_1 \\ u_2 \end{bmatrix} \tag{23}$$

where y_1 and y_2 are the entries of output y. It is assumed zero initial conditions for the system and the filter and the input u_2 is a unit step at the half second of simulation, then at $t = 6\,\mathrm{s}$ appears a second step at u_1. A fault is simulated in

u_2, at $t = 3\,\mathrm{s}$ and displace the amplitude of 1 to 1.2, see Fig. 1. The simulation was performed with Simulink of Matlab. (Figs. 2 and 3)

In Fig. 4 is shown clearly the beginning of the fault in u_2 in $t = 3\,\mathrm{s}$ and how this is projected on the second state variable. In the transient response of the error can be seen the time constant of the filter which is 0.5 s. Note that the amplitude depends on the amplitude of the simulated fault at u_2. Then the question, when is viable to sound the alarm that a fault has appeared? In another situation an input can have a more complex behavior as shown in Fig. 5 where is simulated a random fault beginning at 3 s. Its effect in the error, shown in Fig. 6, is random too, so the decision to determine a fault or not may be very subjective or taken according the design of the control system. In Fig. 7, for example, the squared error is shown to see the trend and thus facilitate the detection of the fault.

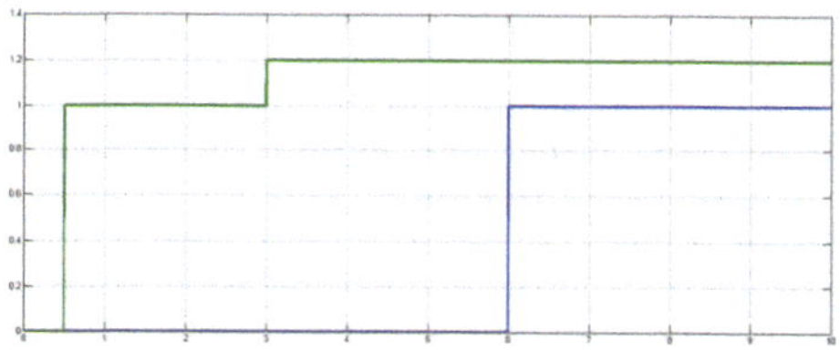

Fig. 1. Fault as a displacement in the input u_2

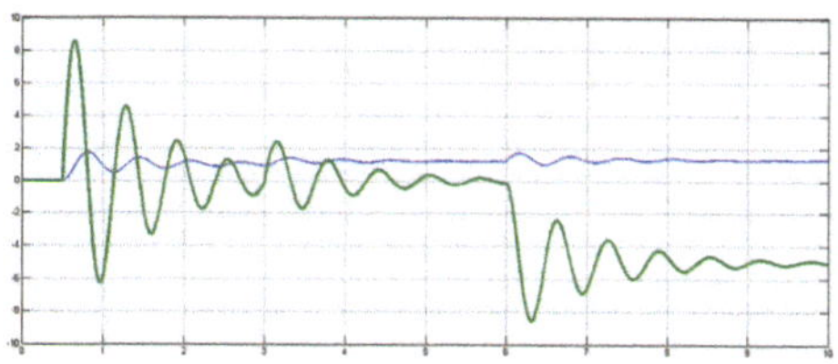

Fig. 2. State variables x_1 and x_2

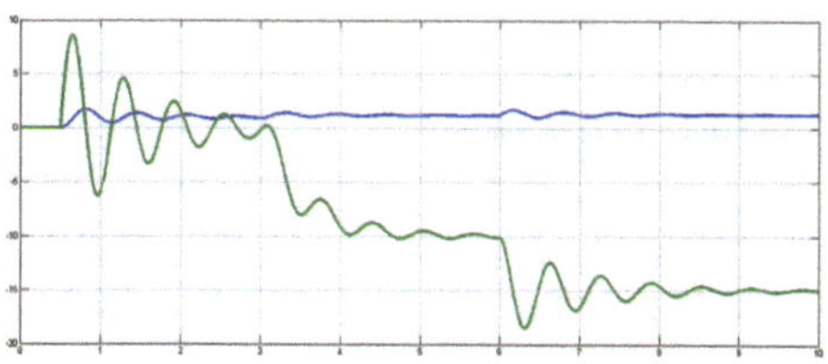

Fig. 3. State variables of the filter, z_1 and z_2

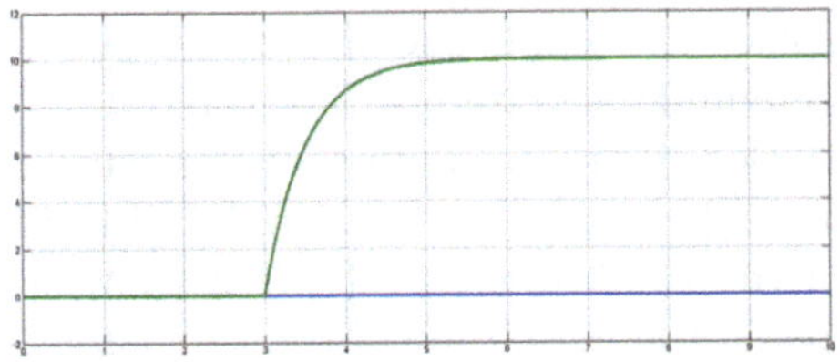

Fig. 4. Error between x_2 and z_2

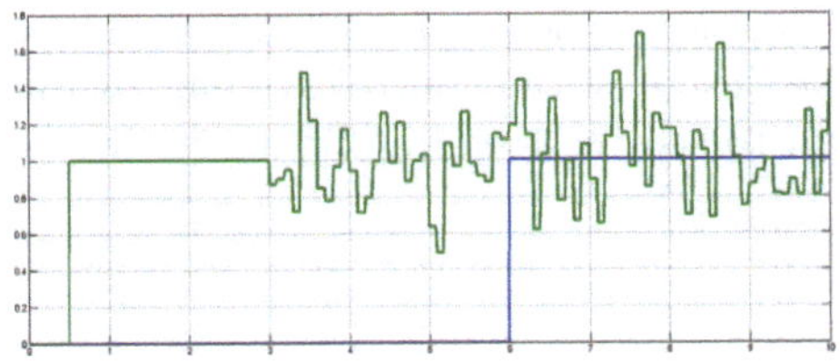

Fig. 5. Fault in u_2 as a random input after 3 s

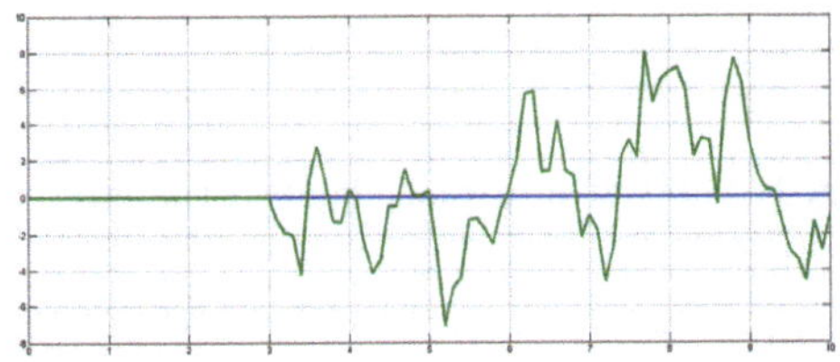

Fig. 6. Effects of random fault in error of x and z

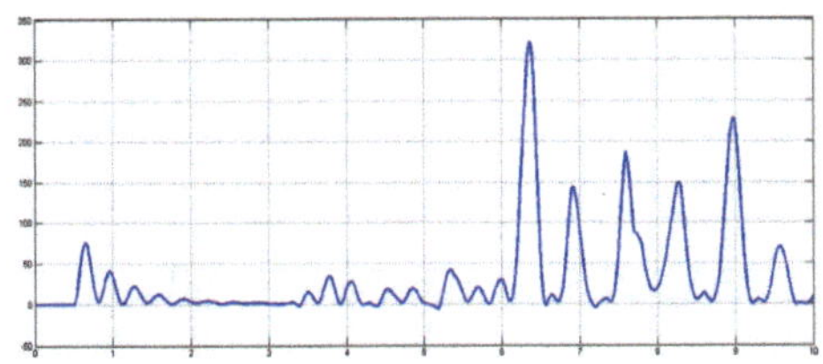

Fig. 7. Square response of Fig. 6 to show trend of error

4 Sinusoidal PWM Inverter

This section is based on [8] where a robust controller for a *sinusoidal pulse width modulation inverter*, see Fig. 8, and [4] too, is designed using LMI approach. An inverter is an electrical device that converts power from direct current to alternate current. The design in this reference uses the so-called *internal model principle* where a compensator consisting of a sine wave generator, or transfer function, $ks/(s^2+\omega^2)$ is implemented. This controller can be put in parallel with

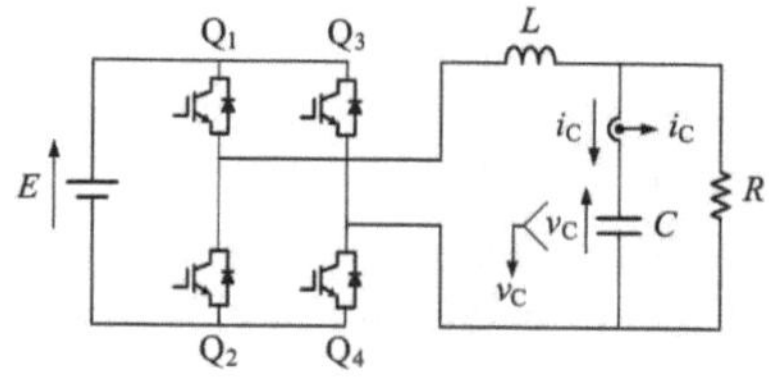

Fig. 8. Sinusoidal PWM inverter circuit

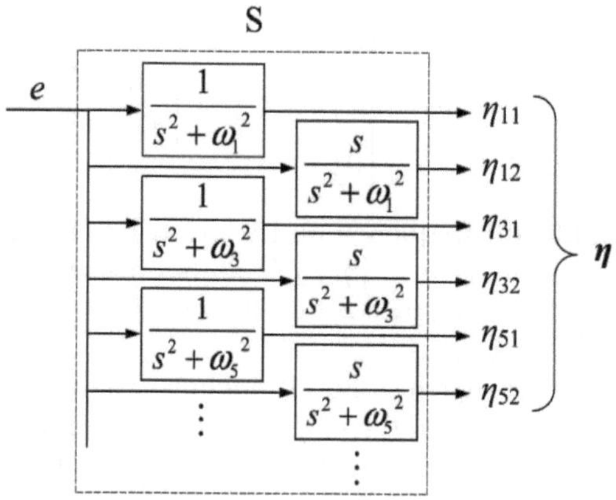

Fig. 9. Structure of controller S

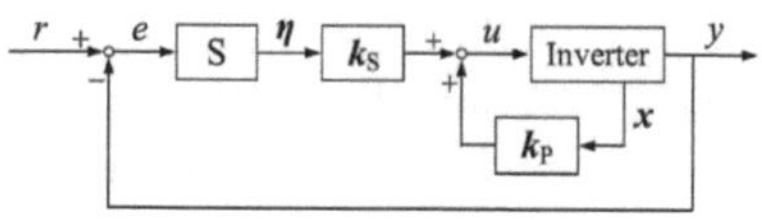

Fig. 10. Closed-loop system for controller design

proportional and integral compensators for which they receive the name of PIS compensators. The optimization done is on the pole-placement constraints and control performance indices based on H^2 and H^∞.

The closed-loop system is given by the Fig. 10 where the signal u is limited to $[-1, 1]$ and the output y is the capacitor voltage. In PWM control the input is compared with a triangular carrier signal to generate the pulses in the gates. In the Fig. 10 the error signal is passed through a controller given by S which stands for the internal model.

The notations E,R,C and L are the source voltage, resistance, capacitance and inductance respectively. The state vector of the system is given by $x = [v_C, i_C]^T$, so the circuit is modeled by

$$\dot{x} = A_P x + b_P u \tag{24}$$

$$y = c_P x \tag{25}$$

where

$$A_P = \begin{bmatrix} 0 & \frac{1}{C} \\ -\frac{1}{L} & -\frac{1}{RC} \end{bmatrix}, b_P = \begin{bmatrix} 0 \\ \frac{E}{L} \end{bmatrix}, c_P = [1, 0] \tag{26}$$

and the controller of Fig. 9 is implemented only for the fundamental frequency. A reduced form from that given in [8], and is given by

$$\dot{\eta}_S = A_S \eta_S + b_S e \tag{27}$$

where $\eta_S = [\eta_{S1}\, \eta_{S2}\, \eta_{S3}\, \eta_{S4}]^T$ and

$$A_S = \begin{bmatrix} 0, & 1 & 0 & 0 \\ -\omega^2 & 0 & 0 & 0 \\ 0 & 0 & 0 & 1 \\ 0 & 0 & -\omega^2 & 0 \end{bmatrix}, \quad b_S = \begin{bmatrix} 0 \\ 1 \\ 1 \\ 0 \end{bmatrix}, \quad C_S = \begin{bmatrix} 1\,0\,0\,0 \\ 0\,0\,1\,0 \end{bmatrix}, \quad D_S = \begin{bmatrix} 0 \\ 0 \end{bmatrix} \tag{28}$$

where ω is the fundamental frequency and is given by $\omega = 2\pi\, 50$. The input is the error e and the output is $\eta = [\eta_1\, \eta_2]^T = [\eta_{S1}\, \eta_{S3}]$ which are $[\eta_{11}\, \eta_{12}]^T$ of Fig. 9. The product of the controller S and its gain $k_S = [k_{S1}, k_{S2}]$ are given by $(k_{S2}s + k_{S1})/(s^2 + \omega^2)$ or $k_S\eta$.

The closed-loop system shown in Fig. 10 includes the model which can generate the input reference as sinusoidal and the system can achieve zero steady-state error tracking according to the internal model principle so long as the closed-loop system is stable.

In Fig. 10 the input of the inverter is calculated as

$$u = k_P x + k_S \eta \tag{29}$$

Thus, with the inverter (24, 25), controller (27), input (29) and $\zeta = [x^T, \eta^T]^T$ as the global state, the closed-loop dynamic is given by

$$\dot{\zeta} = A\zeta + b_1 r + b_2 u \tag{30}$$

$$y = c_1 \zeta \tag{31}$$

$$u = [k_P\; k_S c_S]\, \zeta \tag{32}$$

where

$$A = \begin{bmatrix} A_P & 0_{2\times4} \\ -b_S c_P & A_S \end{bmatrix}, \quad b_1 = \begin{bmatrix} 0_{2\times1} \\ b_S \end{bmatrix}, \quad b_2 = \begin{bmatrix} b_P \\ 0_{4\times1} \end{bmatrix},$$

$$c_1 = [c_P, 0_{1\times4}], k = [k_{P_{1\times2}}, k_{S1}\; 0\; k_{S2}\; 0] \tag{33}$$

Also, the output of the filter can be defined as

$$y_\zeta = \begin{bmatrix} c_1 \\ [k_P\; k_S c_S] \end{bmatrix} \tag{34}$$

where the control parameters k_P and k_S are found, in [8], to be

$$k_P = [-0.13223, -1.2189], \; k_S = [4.8469 \times 10^4, 176] \tag{35}$$

To facilitate the simulation in Simulink of Matlab the system was changed to its discrete form with sampling period of 50 μs. The controller, see [8], is given by

$$G_c(z) = \frac{k_2 z + k_1}{z^2 - 2z\cos(\omega_1 T_S) + 1} \tag{36}$$

where T_S is the sampling period and k_1 and k_2 are given by

$$k_1 = -8.7391 \times 10^{-3} \tag{37}$$

$$k_2 = 8.8603 \times 10^{-3} \tag{38}$$

the reference was $r = 10\sin(\omega t)$. In the Fig. 11 is shown the reference input and the capacitor voltage v_C, which is different to that shown in [8].

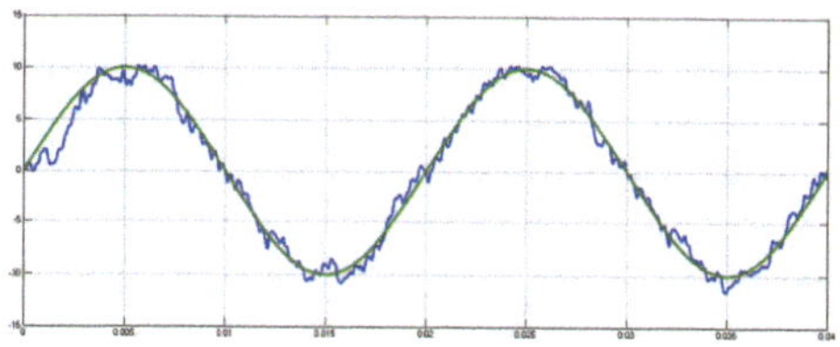

Fig. 11. Output of inverter

5 Fault Detection in Inverter

The inputs of the inverter are the reference r and control u. A fault is incited on this last. The dynamics (Fig. 14) taken into account was that of Eq. (30), (31) and (32) so the fault detection was done for this system which has two inputs, *i.e.*, r and u in this order, and it has two outputs too, *i.e.*, y and u. The parameter values are $E = 50\,\mathrm{V}$, $L = 2.78\,\mathrm{mH}$, $C = 60\,\mu\mathrm{F}$ and $R = 50\,\Omega$. The closed-loop system is of order 6 and the filter for the detection is given by the matrices $G = 1e - 4I_6$ where I_6 is the identity matrix in $\mathbb{R}^6$. Matrix B in the filter (11) is defined from (30) as $B = [b1b2]$. Matrix D in (11) is found with the pseudoinverse due the output dimension is minor than that of the state; with Matlab is found to be, $D = [1e - 4, -359.7122, 0, -1, -1, 0]^T$ So, appending the input vectors b_1 and b_2 to matrix D the filter (11) has an input matrix for the three inputs, that is, the output of the system *i.e.*, v_C (named x_1 or y) and its two inputs, *i.e.*, r and u. Hence, the filter has the input vector $[x_1\, r\, u]$. The state variables of the filter which are taken as output $[z_1\, z_3\, z_5]$ which corresponds with $[y\, \eta_{S_1}\, \eta_{S_3}]$. A fault is simulated in the closed-loop of the inverter in the actuator changing with an offset of two from its actual value. The result is shown in Fig. 12. The error of the state of the inverter against the first two state variables of the fault detection filter is shown in the Fig. 13. It is worthy to remark the shape of the error signals for detecting easily the fault, which might not be possible, or hard to find, in the signal of Fig. 12.

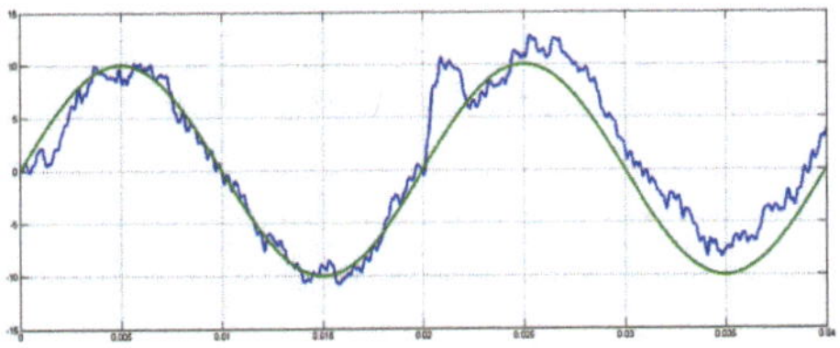

Fig. 12. Fault on $t = 0.02$s

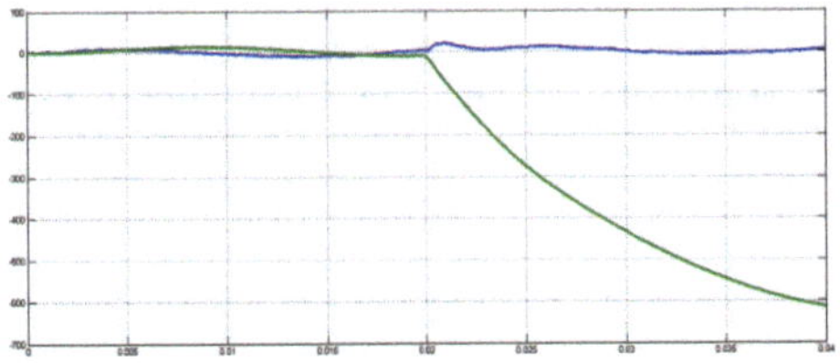

Fig. 13. Errors $x_1 - z_1$ and $x_2 - z_2$

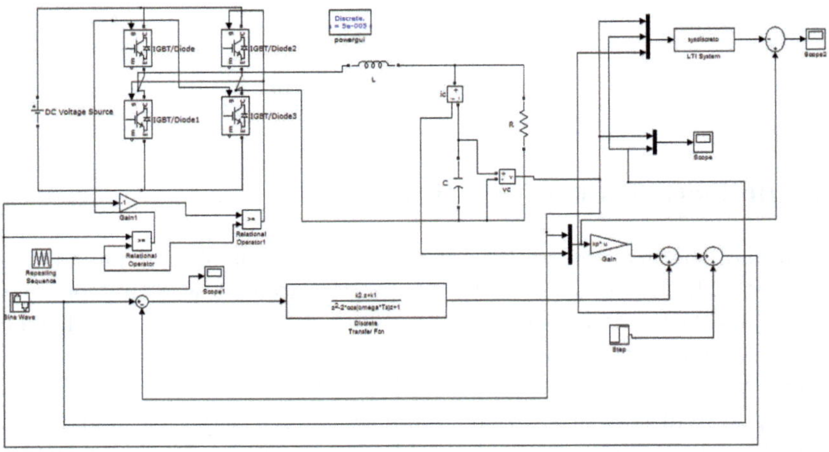

Fig. 14. Inverter on Matlab

6 Conclusion

This paper presents a small survey about of the theory of fault detection in linear invariant systems under the approach given by [2,5] and applies successfully these concepts for detecting a fault in a power inverter with Proportional-Integral-Sinusoidal (PIS) controller designed in base of the internal model principle shown in [8].

References

1. De Persis, C., Isidori, A.: A geometric approach to nonlinear fault detection and isolation. IEEE Trans. Autom. Control **46**(6), 853–865 (2001)
2. Beard, R.V.: Failure accomodation in linear systems through self-reorganization, Ph.D. dissertation, Massachusetts Inst. Technol., Cambridge, MA (1971)
3. Hwang, I., Kim, S., Kim, Y., Seah, C.E.: A survey of fault detection, isolation, and reconfiguration methods. IEEE Trans. Control Syst. Technol. **18**(3), 636–653 (2010)
4. Kato, T., Inoue, K., Kita, T., Kuroda, S.: Sinusoidal waveform following method for optimum digital control of PWM inverter. In: The 4th International Power Electronics and Motion Control Conference IPEMC, vol. 3 (2004)
5. Jones, H.L.: Failure detection in linear systems, Ph.D. dissertation, Massachusetts Inst. Technol., Cambridge, MA (1973)
6. Massoumnia, M.A.: A geometric approach to the synthesis of failure detection filters. IEEE Trans. Automat. Contr. 31, 839–846 (1986)
7. Poon, J., Jain, P., Konstantakopoulos, I.C., Spanos, C., Panda, S.K., Sanders, S.R.: Model-based fault detection and identification for switching power converters. IEEE Trans. Power Electron. **32**(2), 1419–1430 (2017)
8. Wakasa, Y., Baba, Y.H., Tanaka, T., Tanaka, K.: Robust controller design for sinusoidal PWM inverters. Electr. Eng. Japan **187**(2), 44–52 (2014)

High-Accuracy Localization in Wireless Sensor Neworks Based on Ray-Tracing and Fingerprinting Techniques

Antonio del Corte-Valiente[1](✉), Oscar Gutierrez-Blanco[2], and Jose Manuel Gomez-Pulido[2]

[1] Department of Computer Engineering, University of Alcala, Alcalá de Henares, Spain
antonio.delcorte@uah.es

[2] Department of Computer Science, University of Alcala, Alcalá de Henares, Spain
{oscar.gutierrez,jose.gomez}@uah.es

Abstract. This paper aims to apply a novel high-accuracy localization technique in wireless sensor networks to increase the security in airports when coverage and resolution of the surveillance systems deployed is degraded. We propose a new localization method based on fingerprinting techniques using the multipath ray-profile due to effects of the ray-tracing as cost function instead of power levels measurements over the received signal strength. The accuracy obtained with this technique is very high, while the main contribution to the localization accuracy is provided by the dominant rays of the ray-tracing.

1 Introduction

The objective of airports is to provide a cost-effective and efficient operation to deliver high-quality service to customers while also ensuring their safety [1]. Several technologies are deployed to achieve this objective [2]. To detect approaching airplanes in flight, primary radar is used, while for ground surveillance, cameras, surface radar and trilateration systems are deployed [3]. The ultimate goal of these systems is to improve the efficiency and the security of airport operations. However, in the apron and stands areas in the airports, where aircraft are parked, loaded, refueled and boarded, the risk of accidents is high due to the large number of vehicles and people involved in the handling activity [4]. In this paper, the problem of localization based on the signals available from the wireless devices [5] that compose the WLAN standards is presented. The localization process is performed using the fingerprinting technique [6,7] that operate on the relationship between the signal information by multipath reflections. In comparison with other techniques, such as the angle of arrival (AOA) or time of arrival (TOA) that present several challenges due to multipath effects and non-line-of-sight (N-LOS), the fingerprinting technique is relatively easy to implement. We propose a novel algorithm based on the dominant rays of ray-tracing in wireless sensor

I. Chang et al. (eds.), *Advances in Automation and Robotics Research in Latin America*, Lecture Notes in Networks and Systems 13, DOI 10.1007/978-3-319-54377-2_18

networks to improve the localization accuracy in critical areas where GPS or radar systems does not provide good resolution.

2 Surveillance Technologies in Airports

There are currently three normalized types of radar technologies involved in airports. The basic radar is the primary surveillance radar (PSR). It is a passive system because it indicates that the aircraft simply 'is there' and does not convey any useful information for identifying the aircraft. The secondary surveillance radar (SSR) is a collaborative system because the detection does not depend on the size of the object reflected, but on the response from a transponder on the aircraft. This transponder operates in a combination of transmitter-receiver mode and it is capable of responding to the ground interrogation system to identify the plane. Finally, the passive radar is the surface movement radar (SMR). It is used to provide aircraft identification to the controller in low visibility conditions. More recently, airports have been deploying other surveillance information sources based on MLAT (multilateration) and ADS-B (Automatic Dependent Surveillance Broadcast). The multilateration technology [8], also known as hyperbolic positioning, is the process of locating airplanes and vehicles by calculating the time difference of arrival (TDOA) of a signal emitted by the source to three or more receivers (see Fig. 1). An aircraft equipped with ADS-B uses a conventional GNSS (Global Navigation Satellite System) to obtain precise positioning within the satellite constellation [9]. However, GPS and ADS-B systems are degraded due to lack of satellite coverage near the terminal building (see Fig. 2), while radar systems are affected by the low-resolution surface radar in the parking areas. MLAT system is affected by shadows, multipath reflections, garbling and de-activation of aircraft transponder that inhibits multilateration detection and cause holes in coverage and degraded accuracy [10].

3 Ray-Tracing

Localization algorithms have been performed using ray-tracing techniques that have been implemented within the newFASANT [11] simulation tool. The newFASANT is able to make a 3D propagation analysis by means of deterministic methods and it is based on geometric optics (GOs) and the uniform theory of diffraction (UTD) [12,13]. The electric field can be obtained by means of rays directed, reflected, transmitted, diffracted or by any combinations of these effects. An advantage of using the ray-tracing techniques is that information about multipath effects such as the relative delays between rays (see Fig. 3) in a series of points can be obtained in addition to the power level. This information can be used as a fingerprint in the fingerprinting technique proposed with the purpose of improving the efficiency of the localization system.

4 Fingerprinting Technique

The fingerprinting technique can be divided in two phases [14]. In the first one, it obtains the radio map or fingerprints database. The radio-map of fingerprints

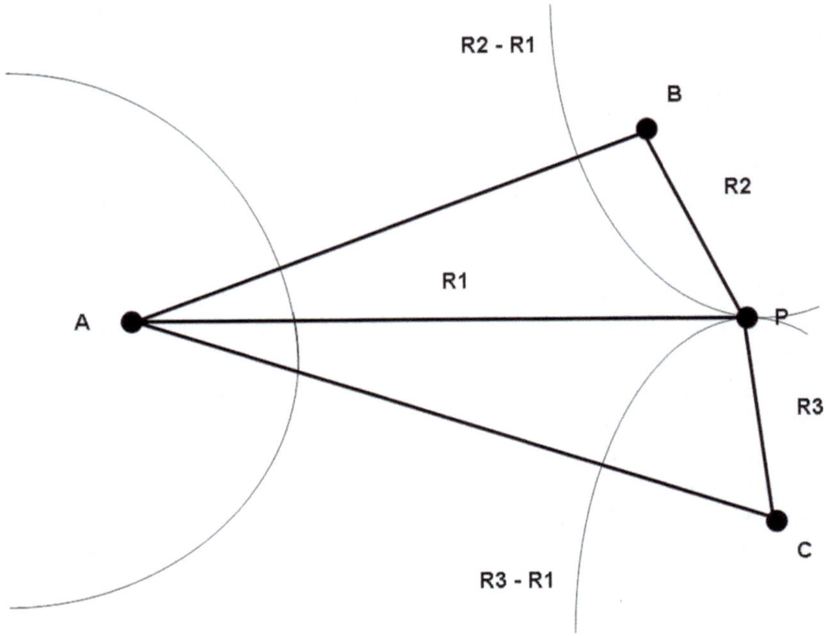

Fig. 1. Time difference of arrival.

Fig. 2. Air side of terminal area.

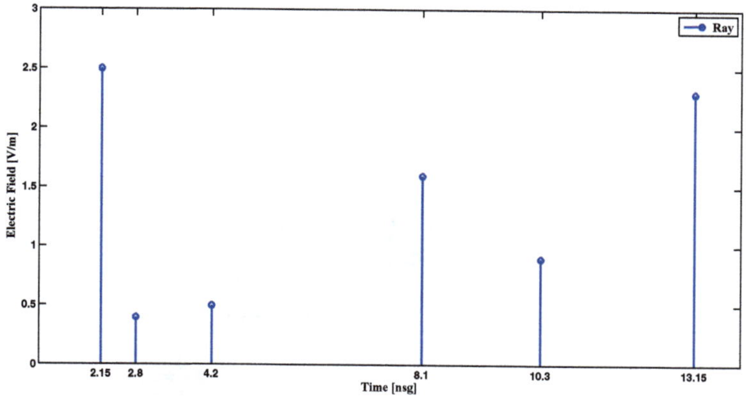

Fig. 3. Relative delays between rays.

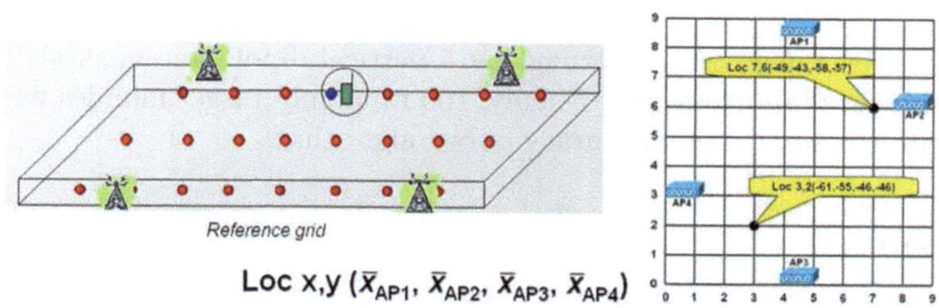

Fig. 4. Fingerprinting technique

is obtained by performing an analysis of the information available from devices and multiples access points over a defined grid. The vector of received signal of power and relative ray-delay at a position on the grid is called the location fingerprint of that point. In the second phase, it analyzes the accuracy obtained in the localization process. For this purpose, the developed technique places a significant number of mobile stations into the area covered by the radio map and it obtains the vector of received samples from different Access Points (APs) [15,16]. The location estimation is made by an algorithm that computes the distance between each measured mobile sample and all the fingerprints stored in the radio map. The coordinates associated with the fingerprint that results in the smallest distance are returned as the position for the mobile [17]. Vectors of signal power as well as relative ray-delays are available from all the access points to the mobiles (see Fig. 4).

5 Airport Simulations

The simulation area used to validate the technique was a section of 120×140 m of Adolfo-Suarez Madrid-Barajas Airport (see Fig. 5). In this case, 16800 fingerprints for a grid resolution of 1 m were simulated. As radiation pattern,

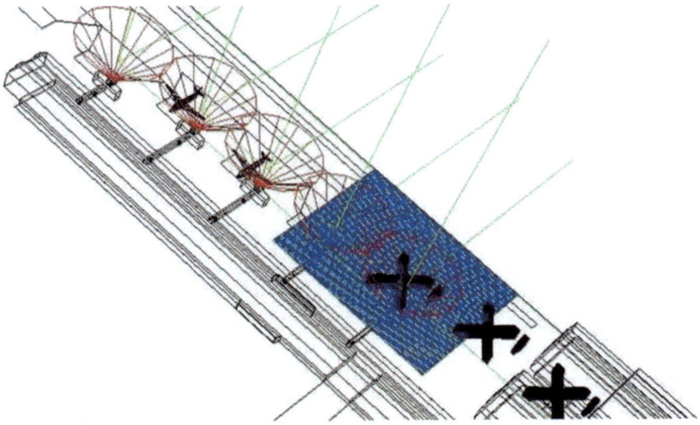

Fig. 5. Airport simulation area

a wireless sensor network conformed by 5 vertical-dipole antennas at 2.4 GHz of frequency were implemented. Finally, 100 randomly target mobiles were used to estimate the localization accuracy of our algorithm.

6 Results

Four distance measures [18] and two detection methods were compared (see Fig. 6). The relative ray-delay detection method provides better localization results than the power level detection method [9], and the Mahalanobis distance (1) better results than the Euclidean distance (2). By the other hand, Fig. 7 shows the impact of the dominant rays into the localization accuracy. It is clear that only the dominant rays ($E < 3$) reduce the localization error for all the metrics analyzed.

$$DMah(x,y) = \sqrt{\sum_{i=1}^{N}[\sum_{j=1}^{E}[(Rm_{ij} - Rf_{ij})\prime * Cov(Rm_{ij}) * (Rm_{ij} - Rf_{ij})]]} \quad (1)$$

$$DEuc(x,y) = \sqrt{\sum_{i=1}^{N}[\sum_{j=1}^{E}[(Rm_{ij} - Rf_{ij})^2]} \quad (2)$$

where:

- N = number of antennas
- E = number of rays
- Rm = number of mobiles
- Rf = number of fingerprints

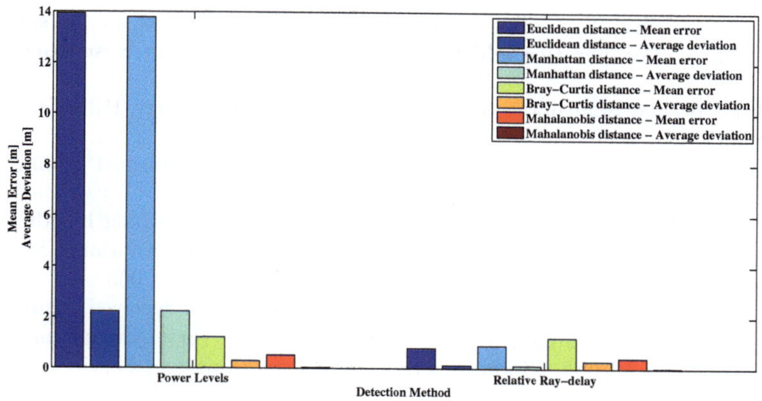

Fig. 6. Detection methods and metrics.

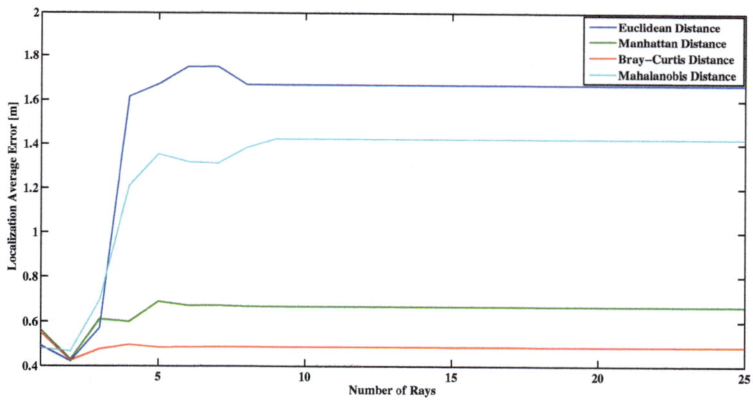

Fig. 7. Number of rays and localization error.

7 Conclusions

This paper presents a new localization technique over WLAN networks based on ray-tracing. This technique utilizes the fingerprinting method with the information about the relative ray-delay due to multipath effects instead of power levels. The exact localization of mobiles can now be performed with higher accuracy, and only by means of the dominant rays available in the relative-ray profile. Using this technique could help to increase the safety in critical areas of airports and to reduce the implementation effort in a real-time environment.

References

1. Bouisset, J.F.: Security technologies and techniques for airport security systems. J. Test. Eval. **22**(3), 247–250 (1994)

2. Casaca, A., Silva, T., Grillo, A.: The use of wireless networks for the surveillance and control of cooperative vehicles in an airport. Telecommun. Syst. **36**(1–3), 141–151 (2007)
3. Perl, E.: Review of airport surface movement radar technology. IEEE Radar Conf. **1–2**, 692–695 (2006)
4. Ribak, J., Cline, B., Froom, P.: Common accidents among airport ground personnel. Aviat. Space Environ. Med. **66**(12), 1188–1190 (1995)
5. Kaemarungsi, K.: Distribution of WLAN received signal strength indication for indoor location determination. In: Proceedings of First International Symposium on Wireless Pervasive Computing, p. 6, CD-ROM, January 2006
6. Kaemarungsi, K., Krishnamurthy, P.: Properties of indoor received signal strength for WLAN location fingerprinting. In: Proceedings of First Annual International Conference on Mobile, Ubiquitous Systems: Networking and Services, pp. 14–23, August 2004
7. de Adana, F.S., et al.: Propagation model based on ray tracing for the design of personal communication systems in indoor environments. IEEE Trans. Veh. Technol. **49**, 2105–2112 (2000)
8. Wan, X., Bischoff, O., Laur, R., Paul, S.: Localization in wireless ad-hoc sensor networks using multilateration with RSSI for logistic applications. Proc. Eurosensors Conf. **1**(1), 461–464 (2009)
9. Campbell, S., Grappel, R., Flavin, J.: Multi-sensor processing for aircraft surveillance in mixed radar ADS-B environments. In: Aviation Week Space Technology, pp. 27–32 (2008)
10. del Corte, A., Gmez, J.M., Gutirrez, O.: Increasing airport security with the development of high accuracy location algorithms based on ray-tracing techniques over WLAN networks. In: International Conference on Electromagnetic in Advanced Applications (ICEAA), Sydney, Australia, September 2010
11. newFASANT simulation tool: www.fasant.com
12. Ctedra, M.F., Prez-Arriaga, J.: Cell Planning for Wireless Communications. Artech House Publishers, Boston (1999)
13. del Corte-Valiente, A., Gmez-Pulido, J.M., Gutirrez-Blanco, O.: Analysis of fingerprinting distance measures for RF indoor location applications based on WLAN networks. In: Third International Workshop on User-Centric Technologies and Applications, (Madrinet), pp. 12–25, Salamanca, Spain, June 2009
14. Kjærgaard, M.B.: A taxonomy for radio location fingerprinting. In: Hightower, J., Schiele, B., Strang, T. (eds.) LoCA 2007. LNCS, vol. 4718, pp. 139–156. Springer, Heidelberg (2007). doi:10.1007/978-3-540-75160-1_9
15. Honkavirta, V., Perala, T., Ali-Loytty, S., Piche, R.: A comparative survey of WLAN location fingerprinting methods. In: 6th Workshop on Positioning, Navigation and Communication, WPNC, Proceedings, pp. 243–251 (2009)
16. Kaemarungsi, K.: Efficient design of indoor positioning systems based on location fingerprinting. In: International Conference on Wireless Networks, Communications and Mobile Computing, vol. (1–2), pp. 181–186 (2005)
17. Zhao, Y.X., Zhou, H.B., Li, M.F., Kong, R.S.: Implementation of indoor positioning system based on location fingerprinting in wireless networks. In: 4th International Conference on Wireless Communications, Networking and Mobile Computing, vol. (1–31), pp. 5246–5249 (2008)
18. del Corte-Valiente, A., Gmez-Pulido, J.M., Gutirrez-Blanco, O.: Efficient techniques for improving indoor localization precision on WLAN networks applications. In: 2009 IEEE International Symposium on AP and USNC/URSI, Charleston, South Caroline, USA, June 2009

Real-Time Monitoring Human Motion System for Ergonomic Posture

Thiago Beckert Otto(✉), Alexandre Campos, and Marcos Aurelio de Souza

Santa Catarina State University – UDESC, Florianópolis, Brazil
thiagobeckert@gmail.com, alexandre.campos@udesc.br,
as.marcoss@gmail.com

1 Introduction

Market competition searches continuous improvement to reduce production costs in order to profit increase. In this sense, there are different approaches to improve the production process. Specifically in shop floor, an expertise methodology must evaluates work stations considering every operation tasks and determines the difficult levels [1].

Some methodologies are developed to find better situations in so many targets whether for costs reduction, manpower improvements or even related to safety topics [1]. However, measures to quantify the risks with focus on the individual are required.

In this sense, the present paper explores ergonomic concepts at work, i.e. during a task execution. The main goal is to analyze some methodologies to create a new one, collecting and analyzing angles. General ergonomics context is presented on the second section followed by techniques and cases studies under technical papers and books. Regularly, a checklist composes such ergonomic methods in order to follow the worker postures.

Besides, a lack of real-time ergonomic system to avoid human health issues is observed [2]. Those ones available are usually expensive.

Therefore, this study focuses in a human vestment development for instantaneous ergonomics evaluation with low investment. Similar research projects is found under some case studies [3–5]. System for both capturing body positions and cross-sectional analysis, focusing an example of motion safety ranges for wrists according four range angles are developed [3]. Additionally, some automated approaches for posture estimation using a range camera to ergonomically evaluate a particular task are used [4]. This study consists on the analysis of: overhead work, squat or sit to lift load, bend to lift load and crawl. For each one, the body angles and heights at working are verified through a program.

Another case study explores a new input modality suited on the wrist as a joystick to perform on-handed movements on smartwatches [5]. The prototype performance and the results are demonstrate on the Fig. 1.

Owning this, in the present paper, the human body is divided by joints and the mainly angles are explored to evaluate risks for some tasks during execution. After that, dangerous postures are determined based on ergonomic methods explored at second section aiming at design a real time posture feedback system.

I. Chang et al. (eds.), *Advances in Automation and Robotics Research in Latin America*,
Lecture Notes in Networks and Systems 13, DOI 10.1007/978-3-319-54377-2_19

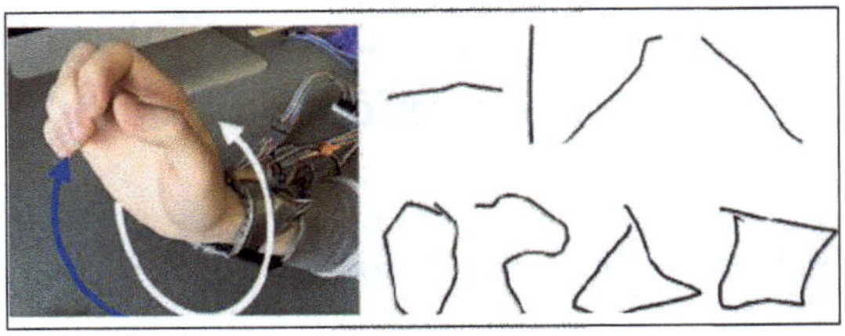

Fig. 1. Left: wrist whirling using a prototype. Right: example gestures drawn.

At third section, specific sensors, microcontrollers and other small devices, that are systems components, are presented. Finally, the prototype vestment is used to evaluate an office standard work routine and to warn the user immediate about risks.

2 Ergonomics and Ergonomic Methods

There are some studies regarding ergonomics to ensure better conditions and to avoid health issues at work. In this context, ergonomics is the science of worker pose adaptation [6–8]. The intention is to minimize the work fatigue, stress, errors and accidents, and to maximize the user security, comfort and health. Then, the resultant efficiency come as consequence.

Ergonomics science intention is to improve the human tasks and the operational conditions through man-machine-environment combination analysis [6, 7]. It occurs at the beginning of work stations design, known as project phase, but it may be necessary to adapt some existing operations based on human condition limitations.

Generally, this evaluation goes on deepening gradually, until it gets each work station level [8]. A work station consists in one part of the company structure where one or more workers act. The evaluation regards activities, postures and worker movements. According to the authors it is important to involve the operators in the evaluation process, noticing their opinion about the stations, their recommendations and possible solutions [8].

Another thought comes from the physiology team that distinguishes manpower effort between dynamic and static muscle effort [6]. The dynamic work is characterized by the muscles sequence in contraction and in extension during the activities. In this case, the muscle acts like a pump for blood circulation. The work, in this context, is expressed by the following formula that involves the muscles shorting and the dispensed effort [6].

Work = weight x raised height level

In another way, the static work refers to the status that the muscles are contracted for a long period time, resulting in a postural work. In this case, the blood circulation is pressed against muscular tissue generating less blood to the muscle. For these reasons, it is not recommended to execute any static work during a long time, otherwise, the dynamic work, followed by a good planning of the tasks, can be executed without resting.

Complementing, there are some cases study showing health problems due of static effort excess [6]. This effort increases some risks as:

- joints inflammation;
- tendon sheath inflammation;

- tendons ends inflammation;
- arthrosis joints chronic degenerate processes;
- invertebral disc ills;
- muscle cramps.

The muscle function is explained by following topics [8]:

- movements generation;
- maintaining corporal position and posture;
- articular stabilization;
- protection of viscera and internal organs;
- corporal cavity pressure control;
- maintaining corporal temperature;
- swallowing and defecation control.

Anyway, to human operation improvement, muscle fatigue and monotony should be investigated and reduced as much as possible, mainly by repetitive activities elimination [8]. In fact, the work upgrading includes, also, better environment conditions regarding good conditions of temperatures, noises, vibrations, toxics gas and illuminations [7, 9].

Another approach to achieve better ergonomic results is followed by five steps as described below [7]. Generally, it involves the evaluation phase, diagnostic phase and, finally, the correction step.

1. Evaluation of demand: describes a problem or a problematic real situation, trying to understand the current problems dimension.
2. Evaluation of the task: analysis the discrepancies between what is established to do and what really is being executed.
3. Evaluation of activity: evaluates the worker behavior through activities execution.
4. Diagnostic: finds out the reason that affects the problem described at the demand.
5. Recommendations: informs which providences should be executed to solve the problem explored before. It must be clearly specified, containing each needed step.

Besides, the ergonomic analysis consists in an real-time worker observation or by video recording with a description about the body postures codes [3]. This code is defined in different conducts by degree ranges in accordance of the used method.

The ergonomic methods are described and real implemented on several study cases [10–25]. Some authors suggest to develop a methodology for ergonomic evaluation involving an real-time observation, videos recording of tasks execution and body postures measurement data acquisitions.

The method OWAS (Ovako Working Analysis System) examines the whole body posture recording and analyzing activities based on two parts [10]. First step evaluates the work posture and the second classifies in a set of criteria to redesign the work procedures and places.

Another method is called by RULA (Rapid Upper Limb Assessment) which categorizes the body postures and expended forces. This tool requires no special equipment providing a quick assessment of the neck, trunk and upper limbs postures [11].

An ergonomic risks evaluation is developed by one-page checklist to features the assessment of legs, trunk and neck for repetitive tasks [12]. It is explored by persons with limited ergonomic training and by factors to assess postural risks on 335 jobs in four plants.

A national institute named by NIOSH (National Institute for Occupational Safety and Health) first developed an equation to assist safety and health professionals to analysis lifting demands in the sagittal plane [13]. This technique gradually evolves covering lifting index that can be used to identify hazardous lifting tasks [14].

Another method is an originally recommended by NIOSH in the concise exposure index called by OCRA (Occupational Repetitive Actions) [15]. It analysis the repetitive movements of the upper limb on the relationship between the daily number of actions performed by the upper limbs in repetitive tasks.

To identify the ergonomics hazards or risk factors can be use the method PLIBEL (Plan for Identifying Belastningsfaktorer). It contains a checklist with questions for different body regions to highlight muscle-skeletal risks in connection of workplace investigations [16].

Also, it is proposed a technique to assessment the jobs risk of upper extremity disorders based on knowledge and theory of the physiology, biomechanics and epidemiology [17]. The methodology measures six task variables, involving the duration of exertion per cycle, efforts per minute, wrist posture, exertion speed, duration of tasks per day and exertion intensity.

Due the same reason, ergonomic analysis, QEC (Quick Exposure Check) is a method created based on investigation of the user's needs [18]. It consists in a practical tool for the assessment of physical exposure to risks for work-related muscle-skeletal disorders.

Also, it is created a manual handling operation guidance in order to help manage, control and reduce the injury risks from manual tasks execution [19]. The main aim is to prevent injury on any body part, mainly the back area. Figure 2 presents an advisement of this guidance that reduces the injury risks to a single handler.

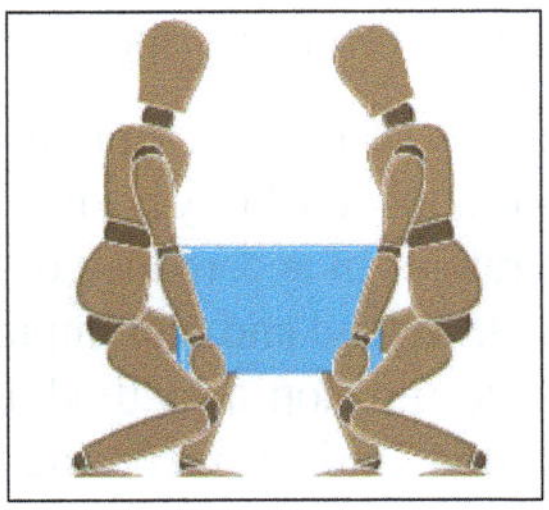

Fig. 2. Team handling: reduction of injury risks with two workers instead a single load handler. This sort of adaptation is implemented in operations that are classified as dangers work conditions.

An ergonomic assessment tool named as REBA (Rapid Entire Body Assessment) uses a systematic process to evaluate the whole body posture and risks associated with job tasks [20]. For this, a single page worksheet is used to revaluate each body section.

Another observation method results from observation of 127 work cycles at a food-processing plant and a paper mill [21]. At this time, six risk factors were included:

repetitive hand use, hand force use, pinch grip, non-neutral wrist posture, elevation of upper arms and local mechanical pressure.

An assessment technique for postural loading named by LUBA (Louvain University Body Assessment) includes the hand, arm, neck and back evaluation and the corresponding maximum holding times in static posture [22]. Twenty male subjects participated in the experiment designed to measure perceived joint discomforts.

A tool is created with manual assessment health and safety for inspectors in order to guide prioritization and intervention on main risk factors [23]. It is named by MAC (Manual Handling Assessment Charts) aiming at identify manual handling high risks due most common risk factors in lifting and lowering, carrying and team handling operations.

A paper grouped examples of ergonomic methods and classified them according what is evaluated as posture, load, movement, vibration and others (mechanical compression, glove use, environmental condition, visual demands, etc.), according Fig. 3, [24]. These techniques can be reach by several studies also.

Techinique	Posture	Load, Force	Movement frequency	Duration	Recovery	Vibration	Other
OWAS	x	x					
Checklist	x						
RULA	x	x	x				
NIOSH	x	x	x	x	x		x
PLIBEL	x	x					x
The Strain Index	x	x	x	x			x
OCRA	x	x	x	x	x	x	x
QEC	x	x	x	x		x	x
Manual Handling Guidance	x	x	x	x	x		x
REBA	x	x	x				x
FIOSH Risk Factor Checklist	x	x	x	x			x
ACGIH TLVs	x	x	x	x			
LUBA	x						
HSG60	x	x	x	x		x	x
MAC	x	x	x				x

Fig. 3. Some ergonomic methods for assessing exposure to risk factors for work-related muscle-skeletal disorders. It is classified the evaluation for such method.

Another method is created based on some analysis criteria [25]. According the author, the activities have to avoid any stress factors that could be harmful to health, both during the planning phase as well as for existing workplaces, considering 11 characteristics. The result will be indicated by traffic light from bad to good workstation condition:

1. Work height
2. Neck muscles stress
3. Above shoulder work

4. Trunk mobility
5. Arms mobility
6. Arms and shoulders stress
7. Wrists stress
8. Fingers stress
9. Knee joints mobility
10. Distribution of standing, walking, sitting
11. Loads handling

3 Vestment Prototype Design and Simulation

This section focuses the human vestment developed prototype to warn, in real-time, the user about the alert and dangers postures during tasks execution. Below it is described the considered angles, the script, the used devices to control the angles and a simulation.

Several ergonomic methods are considered according to the exploration of section two. Figure 4 maps the work conditions divided per body section, which contains angles ranges considered as low, medium and high risk for health conditions. As explained before, most muscle-skeletal disorders are assessed on upper regions of the body such as the back, neck, shoulder, arms and wrists [24].

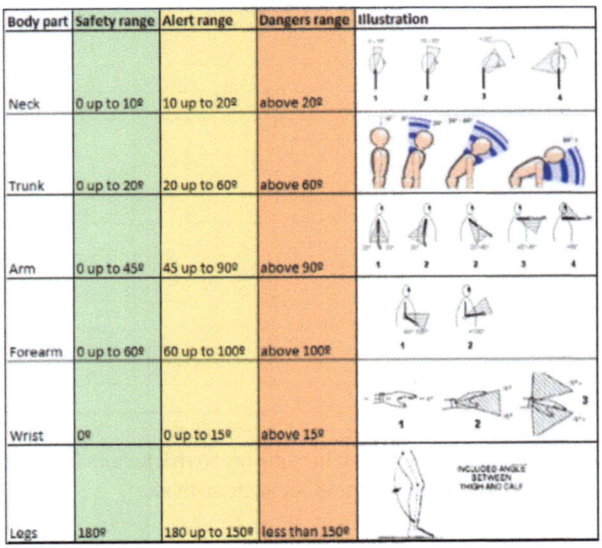

Body part	Safety range	Alert range	Dangers range	Illustration
Neck	0 up to 10º	10 up to 20º	above 20º	
Trunk	0 up to 20º	20 up to 60º	above 60º	
Arm	0 up to 45º	45 up to 90º	above 90º	
Forearm	0 up to 60º	60 up to 100º	above 100º	
Wrist	0º	0 up to 15º	above 15º	
Legs	180º	180 up to 150º	less than 150º	INCLUDED ANGLE BETWEEN THIGH AND CALF

Fig. 4. Ergonomic work condition classified for each body section: neck, trunk, arm, forearm, wrist and legs. This information is used in the script database of ergonomic vestment created in this paper.

It is developed a vestment prototype system to map these tasks ranges and to accuse, in real-time, unsafety positions. This system, below described, is designed to not disturb normal tasks execution.

The prototype is divided into two subsystems connected to one central controller which is connected to a computer for data acquisition. This computer has the script to command the leds turning on and off.

One subsystem takes care about arm, forearm and wrist. The second regards legs and trunk. The intention is to alert the user when green, yellow or red positions are achieved according Fig. 4.

Figure 5 presents both subsystem which basically are distinguished by devices numbered as six. In one subsystem there is a potentiometer, whereas in the other there is a flex sensor.

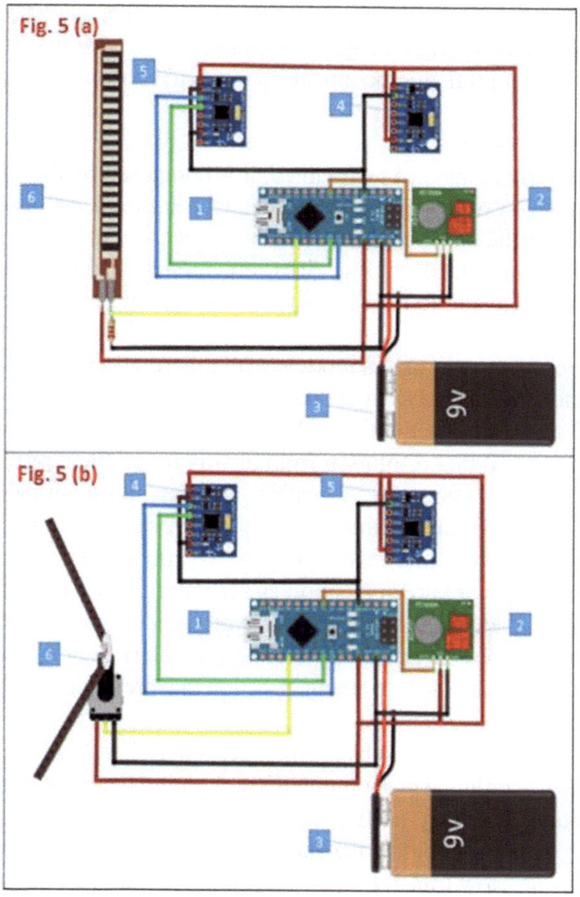

Fig. 5. Ergonomic vestment prototype diagrams. Composition: (1) Microcontroller Arduino Nano, (2) Transmissor 433 MHz, (3) 9 V battery, (4) Accelerometer MPU-6050 (Fig. 5a is linked at the Arm. Figure 5b is linked at the Calf), (5) Accelerometer MPU-6050 (Fig. 5a is linked at the Forearm. Figure 5b is linked at the Thigh), (6) Flex sensor 2.2 inches (Fig. 5a is linked at the wrist) or Potentiometer 10 kΩ (Fig. 5b is linked at the trunk). (a) Subsystem to accuse overhead condition plus wrist risks. (b) Subsystem to measure knee joints mobility plus trunk risks.

In this sense, to control the first subsystem devices are connected as shown in the Fig. 5(a). As mentioned, it links the arm, forearm and wrist together, and it is instrumented with the following numerated items:

(1) Microcontroller Arduino Nano;
(2) Transmissor 433 MHz;
(3) 9 V battery;
(4) Accelerometer MPU-6050 linked to the arm;
(5) Accelerometer MPU-6050 linked to the forearm;
(6) Flex sensor 2.2 inches linked to the wrist.

These sensors are designed to measure the wrist angular variations and overhead positions through arm and forearm angles measurements. The signal is sent by wireless communication to a central microcontroller connected to a computer which collect results.

Besides, a second subsystem is designed, see Fig. 5(b), in order to accuse the knee joints mobility and trunk angle. As explained, it is composed by the same devices 1 up to 5 but the sixth is controlled by a potentiometer 10 kΩ to measures the trunk angles. It is made by two acrylic bars positioned aiming at measure trunk inclination.

These devices, introduced on Fig. 5(b), are mounted together and designed to measure the trunk angle and to map the knee movement through accelerometer information.

There are several potentiometers models with this rotation angle [26]. Considering this study characteristics, it is used a 10 kΩ range potentiometer connected to a 5 VCC, that allows 260° as maximum rotation. This device is connected at a microcontroller analog input. Due the 10 bits resolution from analog input, the device possess a range of 1024 values [27]. Therefore, the calibration consider minimum parameter of 0 V to the value 0000 bits and for the maximum of 5 V to the value 1023 bits.

The accelerometers work in the same way for the arm and the leg subsystems. They are initially calibrated for a known position as zero mark. After that, the member positioning can change the angles in each axis for every accelerometer, then this generated data is received and processed, according to observed criteria. For the first analysis, each sensor is monitored through the movement direction that present major impact according observed criteria.

In addition, in this study the subsystems microcontrollers and the central receiver are Arduino boards. Aiming at programing these Arduino boards it is used a USB cable connected to a computer using an IDE available on the device [28]. With the central board connected to the computer, the data received on it, by wireless, through the 433 MHz modules are now transferred to the computer and other actions may be taken with this data available.

Led are used, illuminated by green, red and yellow colors in order to warn the user about alerts and dangers ergonomic conditions due the work angles. The user interface is through these led.

Finally, it is simulated a manual routine task using the presented vestment prototype. The complete monitor hardware system is fixed on the user. The tasks are executed by an office intern. The activity consists, basically, in transport paper sets from a copy machine to a workbench in order to deliver the papers to the costumer, normally student or teacher. The scenario is composed by one intern without any special equipment to handle the papers.

The following figures show the real system. Figure 6(a) represents the subsystem wrist, arm and forearm subsystem, Fig. 6(b) the legs system and Fig. 6(c) the trunk measurement.

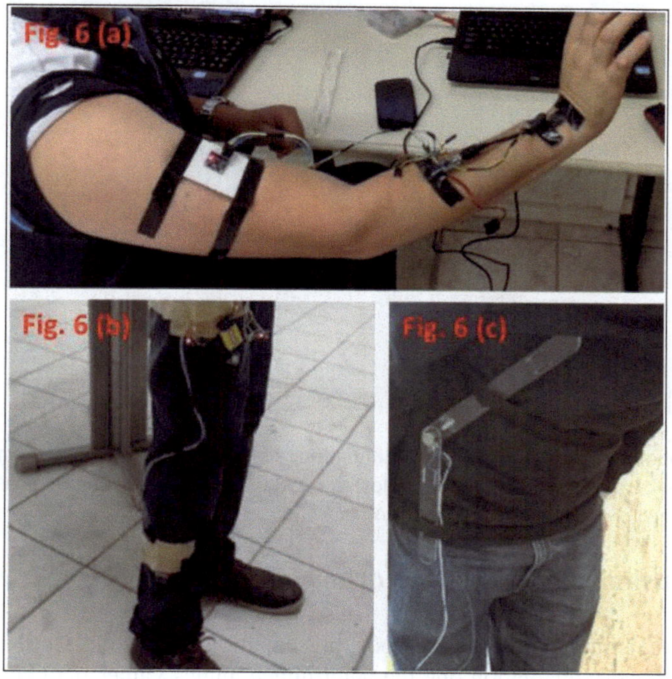

Fig. 6. Ergonomic vestment prototype real system. (a) Arm, forearm and wrist subsystem, related to the Fig. 5(a). (b) Legs subsystem, related to the Fig. 5(b). (c) Trunk subsystem, related to the Fig. 5(b).

The intern executed all the routine tasks using the prototype and the led yellow had turned on, which accused some risks to execute such tasks. It is related, specific, in the moment of books grip once the trunk twist more than 20°, but less than 60° otherwise it has accused the led red. Figure 7 presents this office operation, which can be noted safety position under Fig. 7(a) and the alert condition under Fig. 7(b).

Improvement actions from ergonomic evaluation should start in those tasks which contain critical ergonomics problems, analyzing the high numbers of mistakes, accidents, ills, absenteeism and turnover [7]. It is analyzed the posture and effort required in tasks execution, detailing the mainly sinews and muscles pains caused by tensions.

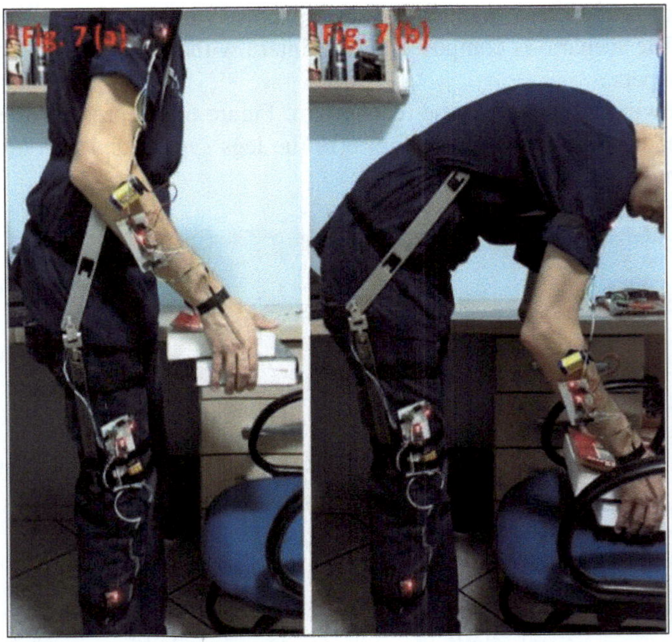

Fig. 7. Ergonomic vestment prototype simulation. Sensors devices mounted on the user to execute the tasks and to send the sign (angles) via Arduino to the computer database. (a) Safety position or green condition. (b) Alert position or yellow condition.

4 Conclusions and Future Works

Considering existing ergonomic methods, presented on the second section, it is noted that a simpler work evaluation due the determinate angles values may be established. Figure 4 presents the safe and unsafe work positions which are used on the database system for the presented ergonomic real-time vestment.

This paper presents a posture real-time feedback system developed as a prototype. It is divided into subsystems in order to be mounted in the worker body sections. Therefore legs, arms, forearms, trunk and wrists may be mapped. For the future, the presented prototype, Figs. 6 and 7, should progress aiming at professional vestment.

Finally, an office work is simulated using the developed prototype, warning the alert position though turning on yellow led according Fig. 7. This project attends the expectation, as the results accuse to the user, in real-time, a bad ergonomic condition. Nevertheless, an improvement point is to eliminate as much as possible the transmission cables once interferences during the simulation are observed.

Additionally, this kind of study is relevant for the industrial companies, as the ergonomic topics regard to avoid stress and consider some worker incentives [29, 30]. Therefore, for next step, it may include a time for each section body in evaluation, related to the ranges. Then, the industrial application can consider the cycle time about the operation to be analyzed.

References

1. Duell, N., Lindley, R., et al.: Ageing and employment: identification of good practice to increase job opportunities and maintain older workers in employment. Final report, Munich (2006)
2. Duarte, F.: Ergonomia e projeto na indústria de processo contínuo. Rio de Janeiro (2002)
3. Bartnicka, J., et al.: Ergonomic analysis of surgeries with the use of wireless body postures measurement system. Journal supported by statutory work (2015)
4. Ray, S.J., Teizer, J.: Real-time construction worker posture analysis for ergonomics training. Adv. Eng. Inform. **26**, 439–455 (2012)
5. Gong, J., et al.: WristWhril: one-handed continuous smartwatch input using wrist gestures, Dartmouth, 14 October 2016. http://xingdongyang.net/papers/Wristwhirl.pdf. Accessed 17 Oct. 2016
6. Grandjean, E.: Manual de Ergonomia: Adaptando o trabalho ao homem, 4th edn. Bookman, Porto Alegre (1998)
7. Iida, I.: Ergonomia: Projeto e Produção, 2nd edn. Edgard Blucher, São Paulo (2005)
8. Abrahao, J.: Introdução à ergonomia: da prática à teoria. Blucher, São Paulo (2009)
9. Mondelo, P.R., et al.: Ergonomía 4: El trabajo em oficinas. Universitat Politecnica de Catalunya (2001)
10. Karhu, O., et al.: Correcting working postures in industry: a practical method for analysis. Appl. Ergonomics **8**, 199–201 (1977)
11. McAtamney, L., Corlett, E.N.: RULA: a survey method for the investigation of work-related upper limp disorders. Appl. Ergonomics **24**, 91–99 (1993)
12. Keyserling, M., et al.: A checklist for evaluating ergonomic risk factors resulting from awkward postures of the legs, trunk, and neck. Int. Ind. Ergon. **9**, 283–301 (1992)
13. NIOSH. Work Practices Guide for Manual Lifting. NIOSH Technical report No. 81-122, US Department of Health and Human Services, Cincinnati, OH (1981)
14. Waters, T., et al.: Revised NIOSH equation for the design and evaluation of manual lifting tasks. Ergonomics **36**, 749–766 (1993)
15. Occhipinti, E.: OCRA: a concise index for the assessment of exposure to repetitive movements of the upper limb. Ergonomics **41**, 1290–1311 (1998)
16. Kemmlert, K.: A method assigned for identification of ergonomic hazards. In: Fundamentals and Assessment Tools for Occupational Ergonomics, vol. 40, pp. 1–14. CRC Press, Boca Raton (2006)
17. Moore, J., Garg, A.: The strain index: a proposed method to analyze jobs for risk of distal upper extremity disorders. Am. Ind. Hyg. Assoc. **56**, 443–458 (1995)
18. Li, G., Buckle, P.: Evaluating Change in Exposure to Risk for Musculoskeletal Disorders – A Practical Tool. HSE Books, Suffolk (1999)
19. Health and Safety Executive: Manual Handling Operations Regulations. Guidance on Regulations, 4th edn. HSE Books, Suffolk (2016)
20. Hignett, S., McAtamney, L.: Rapid entire body assessment (REBA). Appl. Ergonomics **31**, 201–205 (2000)
21. Ketola, R., et al.: Inter-observer repeatability and validity of an observation method to assess physical loads imposed on the upper extremities. Ergonomics **44**, 119–131 (2001)
22. Kee, D., Karwowski, W.: LUBA: an assessment technique for postural loading on the upper body based on joint motion discomfort and maximum holding time. Appl. Ergonomics **32**, 357–366 (2001)

23. Monnington, S., et al.: Development of Manual Handling Assessment Charts (MAC) for health and safety inspectors. In: McCabe, T. (ed.) Contemporary Ergonomics. Taylor & Francis, London (2003)
24. David, G.C.: Ergonomic methods for assessing exposure to risk factors for work-related musculoskeletal disorders. Occup. Med. **55**, 190–199 (2005)
25. Sanchez, A.B.: Adaptation of the ergonomics evaluation software ABATech to the needs of MAN Nutzfahrzenge AG. Technische Universität München, Institut für Produktionstechnik (2005)
26. Cuarelli, G.: Estimação de parâmetros da marcha humana a partir de medições feitas por acelerômetros e giroscópios e uso de mecanismo de referência, São Paulo (2013)
27. da Costa, E.M.G.: Goniômetro. Centro Universitário Positivo, Núcleo de Ciências Exatas e Tecnológicas, Engenharia da Computação. Curitiba (2007)
28. ARDUINO. http://www.arduino.cc/. Accessed 05 Mar. 2016
29. Hakola, T.: Economic Incentives and Labor Market Transitions of the Aged Finnish Workforce. VATT Government Institute for Economic Research, Helsink (2002)
30. Goedhard, W.J.A.: Ageing and Work. Pasmans, The Hague (1992)

Multipurpose Module Based on Infrared Light for Small Scale Swarm Robots

Jorge Gaspar Lira(✉), Edgar Alfonso Trujillo, and Jose Torres Millan

Instituto Tecnologico y de Estudios Superiores de Monterrey,
Tlalpan, Mexico City, Mexico
jorge.gaspar.lira@gmail.com, mosso.edgar@gmail.com, tor_ja0802@live.com

Abstract. There has been a rising interest in swarm robotics, nonetheless, being able to test algorithms in hardware is inaccessible to most people interested in the area due to the high price of the hardware required and the work area needed to test these robots. Our project aims to decrease this issue by providing a low cost and size efficient module that implements a variety of functions required in swarm robotics. The solution shown in this article is based on infrared (IR) transmitters and receivers arranged in a way that allows the robots to communicate, sense ambient light, triangulate the position of other robots and detect obstacles in a 360 radius.

1 Introduction

Swarm robotics is an approach to collective agents inspired by social insects which consists of a large number of robots that self-organize to do some tasks [1]. In order to call a robot a "swarm robot" it must meet three main characteristics. The first one is to be able to move in its surroundings, the second one is to be capable of sensing the environment and the last and most important is to be capable of communicating between them to coordinate their actions to satisfy their objectives as an organized group.

An important factor for robots to work as part of a swarm is the ability to communicate with each other. When trying to reduce the dimensions and price of swarm robots, it is always problematic trying to keep the same functionalities, which limits how much the system can be scaled down. The solution we suggest is implementing a module that serves multiple purposes as a communication system, distance measurement system, a programming interface and an ambient light intensity sensor. This helps to reduce the size and price of implementation of swarm robots.

Distance or proximity sensors are used to measure how close an object is from another one. There are many types of these sensors like capacitive sensors, Doppler-effect sensors, radars, ultrasonic, photoelectric, etc. The most frequently used are ultrasonic and photoelectric distance sensors [2,4,7,9].

Photoelectric sensors can also be used for optical wireless communication. Their greatest advantages are their low cost and fast response time.

I. Chang et al. (eds.), *Advances in Automation and Robotics Research in Latin America*,
Lecture Notes in Networks and Systems 13, DOI 10.1007/978-3-319-54377-2_20

Therefore, they are not only feasible for robotics for obstacle avoidance and mapping but also for wireless communications [5]. This technology utilizes different methods to measure distance. In this repertoire of methods, using the intensity of the light reflected from and object to approximate the distance is one of the simplest at the cost that it is vulnerable to variations due to color/reflectivity of the object and external light sources [8].

A key condition for communication between photoelectric devices to happen is to have line of sight (LOS) from the transmitter to the receiver and vice versa. In the event of lacking LOS, a reflecting surface can also be used. However, the presence of LOS between photoelectric devices does not ensure the absence of other problems like ambient light noise, interference and multipath dispersion which causes intersymbol interference (ISI) [3].

Despite most of the swarm robots use RF communication and non-IR technologies to measure distance (ultrasonic sensors, laser sensors, etc.) the ones that use IR as distance sensors and communication system have low reliability in performance. One of the most significant example is Kilobot Project [6] where they use an IR transmitter next to the receiver at the bottom of the robots body, so the radiation emitted is reflected on the ground and then received by other robots. This means the LOS is null and the communication and distance measurement depends just on the reflection.

Upon the need for a system with multiple functionalities that is a low-cost solution is that we proposed a design with IR as the optimal technology to achieve this. IR technology enables a system capable of measuring distances, ambient light sensing, positioning and communicating while freeing up implementation space by using only one circuit for all of the capabilities. Most swarm-robotic projects that use IR technology for communication or distance measurement have function reliably, so the idea is to have three receptors (IR phototransistors) and three emitters (IR LEDs) positioned in strategic points to have a 360° coverage.

The system is intended to be used in an application where LOS is possible for robots with dimensions below 3 cm with a maximum distance of 11 cm between them. The system must be capable of handling three reception and three transmission peripherals over two dedicated pins for serial communication. Over these two channels, the system will be able to calculate distance and triangulate a position between robots or between itself and an obstacle by measuring the level of light intensity on the IR receiver.

Along with this functionality, the system has fast serial communication to send data at different baud rates. These characteristics allow the system to fulfill two of the requirements for robots used in swarms, i.e., to be able to sense their environment and to communicate between them.

2 Methodology

The objective is to design, implement and test a solution to communicate robots with the added functionality of sensing the environment. This is why we considered several existing technologies that are used for these two purposes. Our solution can be used to communicate and measure distances between robots or obstacles, as well as, sense ambient light based on IR light.

It is important to remember that even though all design considerations were made for the robots described in the previous section, changes can be made to suit the operation for bigger designs or with slightly different requirements. This is why we will not go into too much detail when it comes to the calculations made since this article aims to present a general solution that doesnt depend heavily on our selection of components. However, we do explain the most important things to consider when choosing components to ensure that the communication and sensing are both accomplished.

2.1 Electronic Design

Our electronic design aims to use as little space with the lowest cost possible without compromising the functionality mentioned earlier. The general layout of the transmitter and receiver devices is shown in Fig. 1a and also the PCB layout is shown in Figure 1b. The operational amplifiers as well as some passive components are meant to be in the main board (where the microcontroller is placed).

2.1.1 Transmission

It is important to have omnidirectional communication on a horizontal plane so that robots can communicate between them without having to face in the same direction. This is accomplished by using 3 IR LEDs pointing 120° apart from each other, in that way they can transmit information in 360°, as shown in Fig. 1c.

This design does not modulate the signal due to several reasons. The most important one is to prevent the emitted signal from having a range that is too long compared to the size of the robots, which would translate to the impossibility of calculating distance accurately at a close range with an Analog to Digital Converter (ADC). Another important reason is that modulation and demodulation requires a carrier frequency greater than the communication baud rate. This becomes a problem when choosing the IR LED and IR receiver because the ones that have a rise time small enough to handle these frequencies tend to be more expensive. Finally, modulating would require more components both on the transmitting and the receiving end.

It is important to mention that this circuit has 3 enable pins, one for each transmitter to control in which directions the robot is sending information. This is done to reduce the interference caused by many robots communicating simultaneously in close proximity. The circuit was designed in a way such that any combination of LEDs can be turned on simultaneously without a reduction in the intensity of the emitted light, which is necessary to compute the distance between robots in an accurate way.

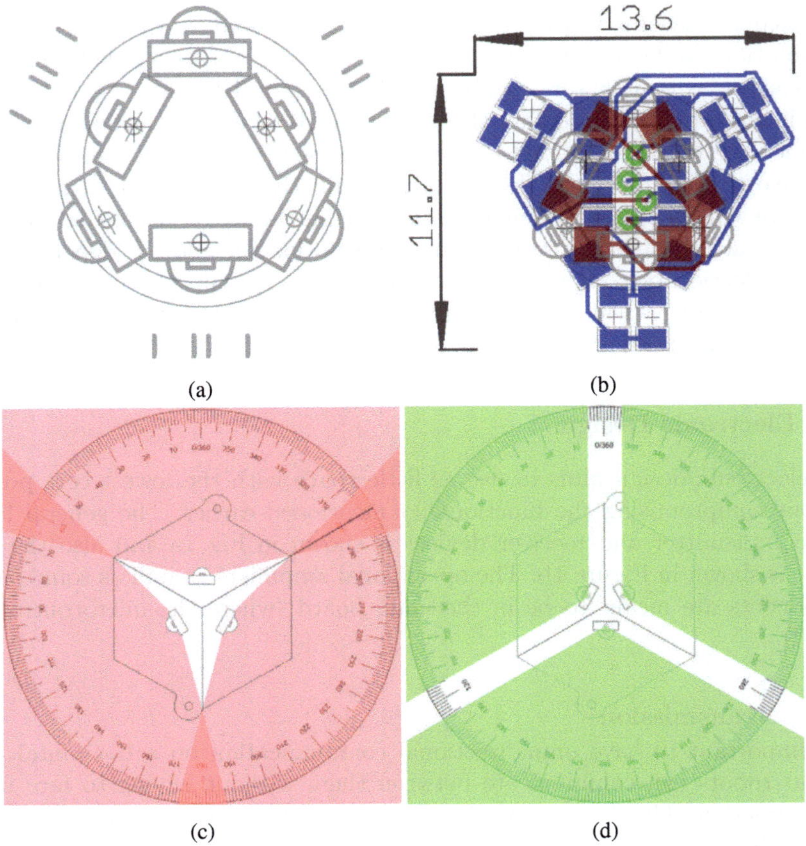

Fig. 1. (a) Transmitter and receiver layout, (b) PCB layout (dimensions in mm), (c) PCB Layout of the Transmitters and their transmission angle, (d) PCB Layout of the Receivers and their viewing angle (Fig. 6 shows that there are no blind spots in reception)

2.1.2 Reception

This is the most important part of the implementation since it obtains the analog voltage that depends on the intensity of the received light which can be used to compute the distance between robots, to triangulate positions, and to know the ambient lights intensity. Additionally, it manages to get the message reliably even with interference and with different intensities of the received signal. This is achieved with the circuit shown in Fig. 2b.

The phototransistor receives the signal and converts the intensity of the light into current. The voltage is filtered through a high pass filter in order to remove unwanted noise from ambient light as well as the DC component of the signal. Finally, this filtered signal is compared to ground to get TTL voltages at the output of the circuit

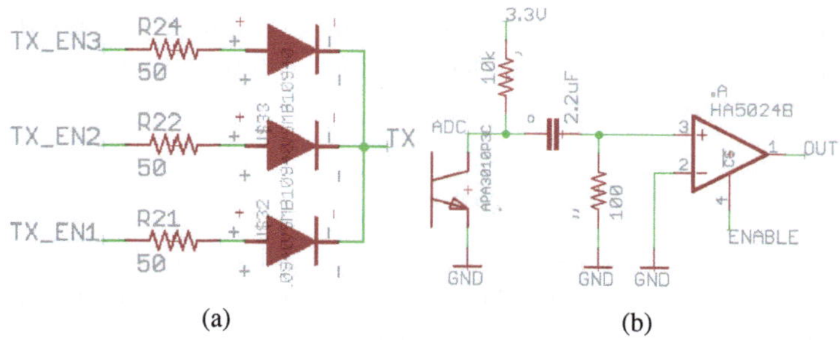

Fig. 2. (a) Transmission Circuit, (b) Single receiver circuit. The phototransistor used as a receptor is in the bottom left. The operational amplifiers have the advantage of having enable pins, which allow multiplexing the 3 signals reading only the signal we are interested in and disabling the others.

Just as with the emitters, the 3 receivers point 120° apart from each other to allow an omnidirectional horizontal reception. The photoresistor that is used for the reception can be selected with enable pins to avoid interference from signals that come from other directions. Additionally, even if the actual reception

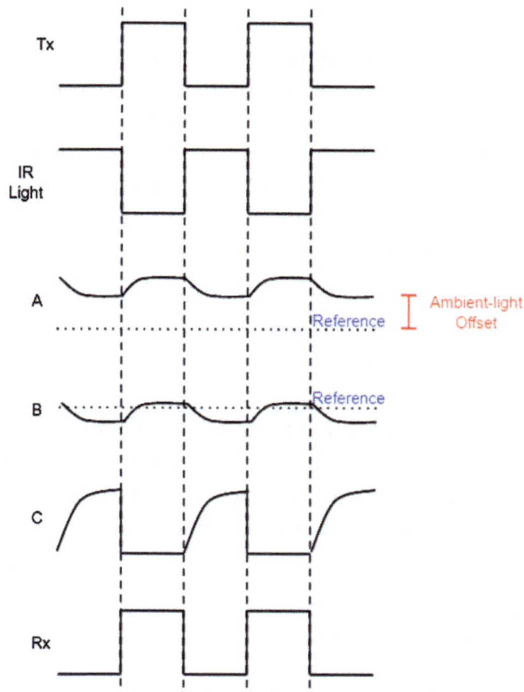

Fig. 3. Signal at different stages of analog processing. (Tx) The transmitted signal. (A) signal at point ADC of Figure 2b. (B) Filtered signal. (Rx) The received signal.

is disable, the analog values of each phototransistor can be measured independently. This allows the system to decide which receptor to use based on the intensity of the received signal.

2.2 Component Selection Parameters

After selecting the characteristics for our communication system, the components also needed a set of particular characteristics. i.e., both the emitter and receiver must also work in the same wavelength.

If an IR modulator and demodulator was being implemented, the rising and falling time would have been critical for the speed limit at which the communication could have been possible, however, in this design, the rising and falling time can be practically ignored because the system reacts to changes in the level of at the photoresistor resulting in a bigger baud rate.

After considering wavelength and response time, the angle of sensibility needs to be taken into account in order to have no blind spots in transmission or reception. Also, it is highly important to consider that the operational amplifier should have a fast-enough switching time for the baud rates that will be used while communicating.

2.3 Testing

In order to verify the performance of our implementation we test limits of velocity, and distance as well as, to get a value of Bit Error Rate (BER) of the system. Additionally, the intensity of the received message was measured with a microprocessors ADC unit. These measurements were done at varying angles and distances to determine how feasible it is to calculate the position of another robot based on the messages that are received. Finally, the intensity of the ambient light was also determined using this same principle of analog to digital conversion.

3 Results and Analysis

The results from the test are intended to highlight the advantages, as well, as the limitations of our module. This data is important to differentiate from various options when evaluating hardware for new projects. Hence, this performance results will support the functionalities of the module, i.e. distance measuring and 360° field of communication.

3.1 Transmission and Reception

To analyze this functionality, three signals were taken into account, the digital signal sent, the digital signal received and the analog signal received before being processed (Fig. 3). Each of these measurements were made at a baud rate of 9600 bauds per second at the distance marked in Table 1. Furthermore, ambient

Table 1. Bit Error Rate for different baud rate speeds and distances between transmitter and emitter.

Baud rate/Distance	4800	9600	19200	38400
4 cm	0.00%	0.00%	0.00%	0.00%
6 cm	0.00%	0.00%	0.00%	0.00%
8 cm	0.00%	0.00%	0.00%	30.78%
8.5 cm	0.00%	2.66%	0.02%	35.61%
9 cm	0.02%	5.00%	15.18%	37.11%
9.5 cm	2.38%	10.20%	16.89%	37.27%
10 cm	12.63%	13.52%	20.71%	28.97%

light under which the tests were conducted remain constant to avoid signal interference.

The next graphs in Figs. 4 and 5 show the tests made to the receiver (IR phototransistor) and transmitter (IR LED) circuit, also including the ADC of the receiver. The tests consist of measuring the signal received with an oscilloscope; the orange signal is the transmitted data, the green signal is the ADC lecture given by the collector pin of phototransistor, the blue signal is the received and rebuilt data by the receiver circuit.

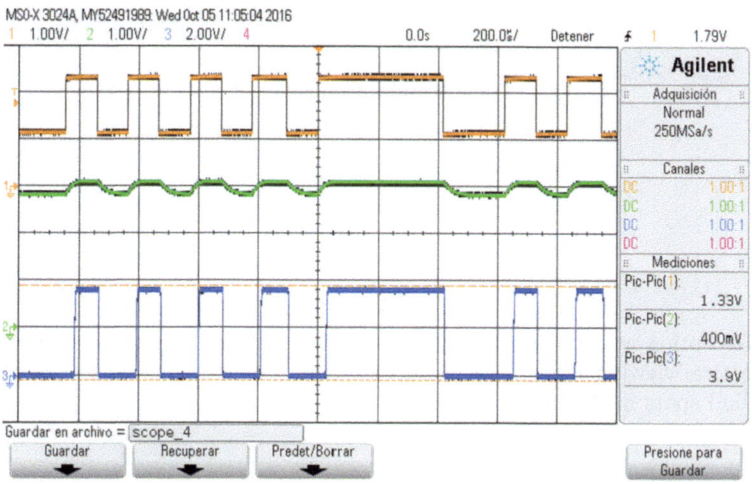

Fig. 4. 4 cm distance communication.

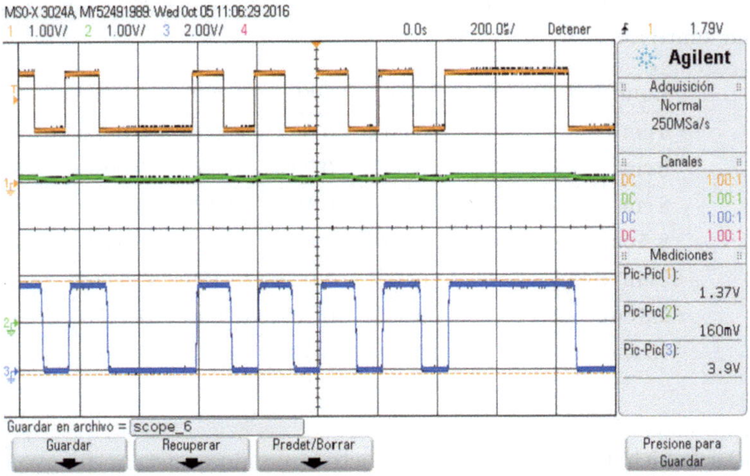

Fig. 5. 8 cm distance communication.

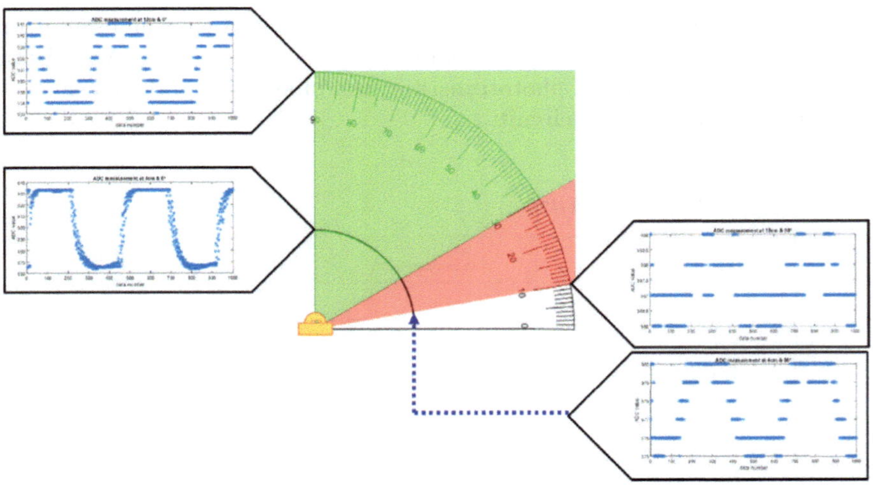

Fig. 6. Reception at different distances (4 and 12 cm) and angles. The green area is the theoretical viewing angle, tests in the red region were performed to show that there are no blind spots in the 360° view area

3.2 ADC Measurements

Using the ADC measurements, the precision is sufficient to calculate and triangulate the position of the emitter. Table 2 shows the values obtained with the microcontrollers ADC for different distances and angles.

Table 2. Comparative table of ADC readings at different angles and distances.

Distances	4 cm		8 cm		12 cm	
Degrees	Max	Min	Max	Min	Max	Min
0°	933	864	946	932	941	933
20°	948	941	951	934	948	941
40°	958	951	959	946	958	951
60°	964	943	968	955	969	964
80°	980	975	978	975	969	966

4 Conclusions

Swarm robotics is a field that has been in constant growth for the past decade. However, it is not an area of study that can be easily accessed due to high costs of the hardware and the space needed to deploy a swarm of robots. This is why there is a necessity of optimizing the cost and the size of the components used in this sort of project. Our project aims to address this issue by creating a module that implements both the communication and the sensing aspects. Our approach is based on three IR transmitters and three IR receivers per robotic unit to be able to communicate information and to triangulate the position of objects and other robots. Components have been carefully selected to advice this goal. This paper summarizes the parameters that are important to consider when implementing this module

As a proof of concept, several implementations of the module were created and tests were performed on them to demonstrate the feasibility of the system within a swarm robot. Given that the solution is meant for small size of robots (less than 3 cm in diameter), the communication system works properly on distances smaller than 8 cm while transmitting at a baud rate of 9600. The location of other robots and obstacles can be accurately calculated at distances smaller than 12 cm. In summary, the module has the functionalities needed in swarm robotics for communication and ambient sensing while reducing number of components and size.

Future work will research variations in the transmitted signal's intensity to regulate current consumption during communication.

Acknowledgments. We deeply thank Dr. Pedro Ponce Cruz for his guidance, suggestions and support while working on this project.

References

1. Sha, A.: Introduction to Swarm Robotics (N.A.). http://ewh.ieee.org/r6/scv/ras/intro_swarm.pdf
2. Baumer.: Operation and design of ultrasonic sensors (N.A.). http://www.baumer.com/cn-en/services/user-knowledge/ultrasonic-sensors/functionality/

3. Elgala, H., Mesleh, R., Haas, H.: Indoor optical wireless communication: potential and state-of-the-art, October 2011. https://www.researchgate.net/publication/224257503_Indoor_optical_wireless_communication_Potenti
4. Tanaka, K., Wada, K., Takayama, H.: Development of infrared communication device using RT middleware, August 2010. http://ieeexplore.ieee.org/stamp/stamp.jsp?arnumber=5603094
5. Mohammad, T.: Using ultrasonic and infrared sensors for distance measurement. http://waset.org/publications/6833/using-ultrasonic-and-infrared-sensors-for-distance-measurement
6. Self-Organizing Systems Research Group. The Kilobot Project, August 2014. https://www.eecs.harvard.edu/ssr/projects/progSA/kilobot.html
7. Shieh, W.-Y., Hsu, C.-C.J., Chen, H.-C., Wang, T.-H., Chen, C.-C.: Construction of infrared signal-direction discriminator for intervehicle communication, June 2015. http://ieeexplore.ieee.org/stamp/stamp.jsp?arnumber=6862912
8. Wikibooks. Robotics/Sensors/Ranging Sensors, November 2016. https://en.wikibooks.org/wiki/Robotics/Sensors/Ranging_Sensors#Ranging_Light-Based_Sensors
9. Zeng, M., Li, C.: Infrared communication link maintaining method for multiple mobile microrobots, December 2013. http://ieeexplore.ieee.org/stamp/stamp.jsp?arnumber=6739807

Identification of Aircraft in a non-Cooperative Surveillance System. The Case Sutudy of Aircraft Type Canadair Regional Jet

Antonio del Corte-Valiente[1(✉)] and Jose Manuel Gomez-Pulido[2]

[1] Department of Computer Engineering, University of Alcala, Alcalá de Henares, Spain
antonio.delcorte@uah.es
[2] Department of Computer Science, University of Alcala, Alcalá de Henares, Spain
jose.gomez@uah.es

Abstract. This paper presents the results obtained in the identification of the aircraft tail number by a non-cooperative surveillance system installed on the T4 apron of Adolfo-Suarez Madrid-Barajas Airport. The study focuses on the problems encountered by identifying the aircraft type Canadair Regional Jet through high resolution optical sensors whose response is affected by the shadow that, on sunny days, the engine produces over the tail number in this type of aircraft. The evaluation main objective is to quantify the probability of false identification produced by this phenomenon to determine the aircraft identification capacity of the Advanced Surface Movement Guidance and Control Systems deployed in the airport.

1 Introduction

In his document *ATM Airport Performance Framework* [1], the European Commission and Eurocontrol have established themselves as one of their goals for the program SESAR (Single European Sky ATM Research) that the functional design of the future ATM (Air Traffic Management) system provides a consistent management of all information to facilitate decision-making cooperative, providing air traffic services that allow continuous operations by integrating the *gate-to-gate* concept with all the processes at the airports. From this viewpoint, the airport must be completely integrated in the ATM network so that the scaling process acting as binding land and air segments. In this area, the efficient management of gates and parking areas will be particularly significant [2,3].

In the land segment, the airport aims increased timeliness, efficiency and optimized the use of available resources [4]. These objectives can be achieved through improvements in the management of operational activities in apron, both by automating some airport processes and through real-time knowledge of the events that occur in apron [5].

The airports are demanding tools that integrate surveillance information with data managed by the airport operational systems, so that they can develop the

I. Chang et al. (eds.), *Advances in Automation and Robotics Research in Latin America*, Lecture Notes in Networks and Systems 13, DOI 10.1007/978-3-319-54377-2_21

necessary procedures to improve the management and the monitoring of the operating activities in apron areas. In this vein, and in order to provide to the operations staff a tool that meets the demand mentioned before, it was proceeded to the deployment and evaluation of an aircraft surveillance system on the apron T4 of the airport [6].

2 System Architecture

The system is a non-co-operative and scalable Advanced Surface Movement Guidance and Control Systems (A-SMGCS) [7,8] solution for airports conformed by a distributed network of high definition compact millimeter-wave sensors [9] that determines target position and non-cooperative optical identification sensors that recognizes aircraft tail numbers as they pass in front of them. Both sensors determine the aircraft detection and the identification functionality (Fig. 1).

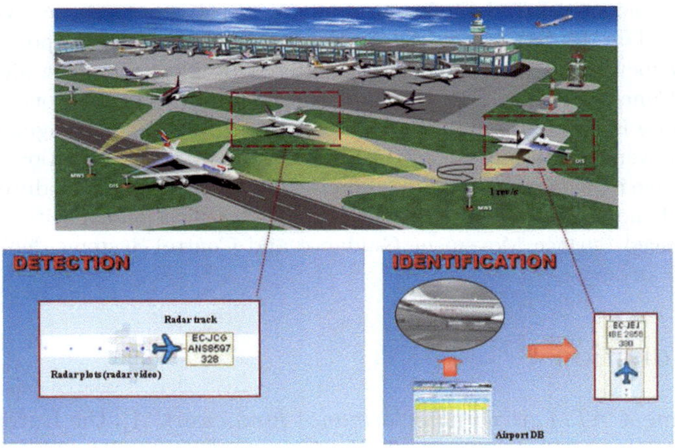

Fig. 1. Aircraft detection and identification.

On one hand, millimeter-wave sensors (MWS) are strategically located around the airfield to avoid any coverage holes that might be encountered by conventional single radar units. This surveillance data is supplemented by information from ground-based optical identification sensors (OIS) that read aircraft tail numbers and relay this data to the central processor. OIS sensors are placed at the apron entry points to recognize the aircraft tail number as soon as possible. On the other hand, MWS sensors are placed to guarantee the optimal coverage (Fig. 2).

The data is compared with flight plan information to provide a real-time display [10] of traffic on the airfield, using the tail number of the aircraft to correlate the information. This valuable information enhances sensor performances.

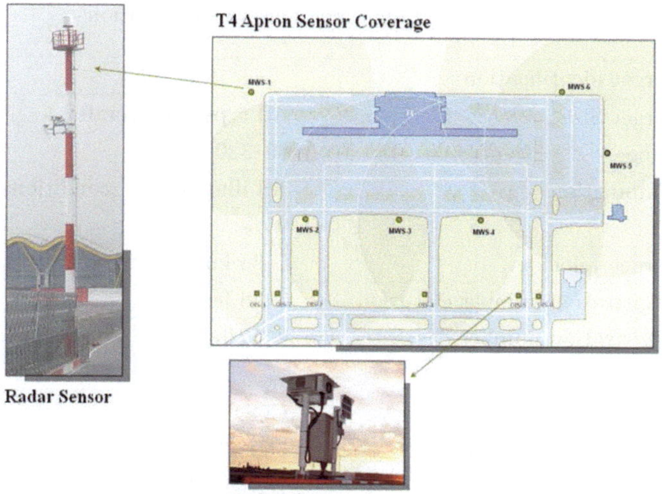

Fig. 2. SRR and OIS sensors.

3 Optical Identifiers Sensors

As mentioned previously, the surveillance data is supplemented by information from ground-based OIS (Fig. 3) that read aircraft tail numbers and relay this data to the central processor. See Table 1 for OIS specifications.

Fig. 3. OIS sensor in apron T4 of the airport.

Table 1. Optical identifier sensors specifications

Target identification	
Target Size	All types of aircraft
Range	50–330 feets
Illumination	All illumination conditions
Pd	95%
Target velocity	1–35 knots
Registration number elevation	3–28 feets
Registration number character size	2–12 in.
Sensor	
Imager	CCD
Horizontal resolution	550 TV lines
Picture elements	768 × 494 pixels
Pixel size	6.35–7.4 μ
Sensitivity	0.003 Lux
Scanning system	2:1 interlace
Video output	Composite video
S/N ratio	50 dB
Spectral sensitivity	Peak at 530 nm
Processing	Image processing and OCR

4 Aircraft Type CRJ-200

The Bombardier Canadair Regional Jet 200 (CRJ-200) (Fig. 4) was designed to provide superior performance and to operate efficiencies in the fast-growing regional airline industry. Against the closest competition, it flies faster and farther while burns less fuel and has lower operating costs. With over 1000 units in commercial service it has become the most successful regional airliner program the world has ever known. Designed specifically to provide superior operating efficiencies to regional and major airlines, the twin-turbofan Bombardier CRJ-200 has also established itself as the quietest and most environmentally friendly commercial jet aircraft of the world [11]. See Table 2 for CRJ-200 specifications.

Fig. 4. Aircraft type CRJ-200.

Table 2. Aircraft type CRJ-200 specifications

Performances	
Maximum range at LRC	1345 NM
High cruise speed	860 km/h
Normal cruise speed	785 km/h
Maximum operating altitude	12496 m
Noise level (Takeoff)	77.7 EPNdB
Noise level (Sideline)	82.5 EPNdB
External dimensions	
Length overall	26.77 m
Wingspan	21.21 m
Wing area (net)	48.35 m^2
Height overall	6.22 m
Fuselage maximum diameter	2.69 m
Turning circle	22.86 m
Weights	
Maximum ramp weight	23247 kg
Maximum takeoff weight	23133 kg
Maximum landing weight	21319 kg
Maximum zero fuel weight	19958 kg
Operating weight empty	14016 kg
Maximum fuel load	6489 kg
Maximum payload	5942 kg

5 Aircraft Identification

The indicators used to evaluate the system identification capacity are defined [12,13] in this section. The meaning of these indicators are described in detail.

1. Probability of Detection (P_D): The possibility of a real aircraft, presented on the apron, is represented as an aircraft target
2. Probability of Identification (P_ID): The possibility of a real aircraft, displayed as an aircraft target, has the right tail number associated
3. Probability of False Identification (P_FID): The possibility of a real aircraft, displayed as an aircraft target, has a wrong tail number associated
4. Probability of No Identification (P_no_ID): The possibility of a real aircraft displayed as an aircraft target has no tail number associated

$$P\_D = ((A_{ex} - A_{nd} - T_{un})/A_{ex}) * 100\% \quad (1)$$

$$P\_ID = ((A_{ex} - A_{nd} - T_{un} - A_{ci} - A_{ii})/(A_{ex} - A_{nd} - T_{un})) * 100\% \quad (2)$$

$$P\_FID = (A_{ii}/(A_{ex} - A_{nd} - T_{un})) * 100\% \quad (3)$$

$$P\_no\_ID = 100 - (P\_FID - P\_ID)\% \quad (4)$$

where A_{ex} *is the number of aircraft expected,* A_{nd} *is the number of aircraft non-displayed,* T_{un} *is the number of unknown targets,* A_{ci} *is the number of aircraft correctly presented in the system but that have no associated identification tag and* A_{ii} *is the number of aircraft correctly presented in the system with an incorrect associated identification tag*

As the aim of the study is to evaluate the aircraft identification A-SMGCS performance, it is assumed that the probability of aircraft detection (P_D) provided by MWS sensors is 100%.

It will be discussed the types of errors that can occur when recognizing the tail numbers of the aircraft type CRJ-200:

1. *Confusion of characters*: The optical sensor is unable to distinguish certain similar characters, such as zero and the letter O, the letters I and J, the number 5 and the letter S
2. *Problems in the processing of optical sensors*: Aircraft can be labeled with many characters in the fuselage, so the process of localization the tail number and translating the image into a reasonable string of characters may fail
3. *Problems with shadows*: On sunny days, from 11:00 UTC there is a shadow phenomenon affecting aircraft type CRJ-200 owned by Air Nostrum Airline that operate in apron T4. This phenomenon is that the engine of the aircraft projects a shadow over a part of the tail number, so that a portion of the tail number is in shade and some not

The Figs. 5 and 6 illustrate how shadows vary at different times of year.

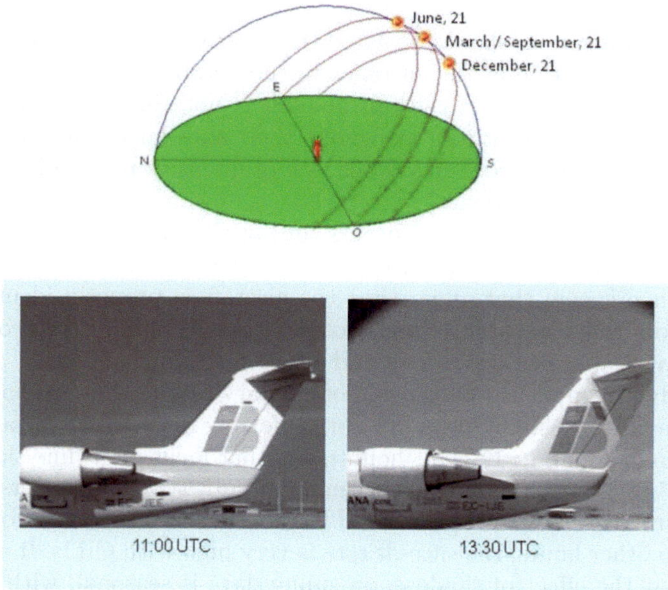

Fig. 5. Seasonal evolution of the shadows.

Fig. 6. Problems with shadows over the aircraft tail number.

6 Results

In terms of identification, should be noted that the system is only able to acquire the aircraft identification at the entrance to the apron, when the OIS reads the registration of the aircraft. If the tail number is not correctly recognized, the aircraft will lack identification throughout their stay on the apron. The tests were carried out in 2014, covering about 28 h of intensive monitoring with more than 700 movements of arrivals.

The identification errors often occurs during the movement of the aircraft, as they often result from errors in reading the tail number at the entry of the apron, which prevents that at no time the aircraft is properly identified. The types of errors that can occur when recognizing tail numbers are confusion of characters, optical sensor processing and shadow phenomenon.

The study accomplished by the end of January results that reading errors in aircraft type CRJ on sunny days are not uniformly distributed throughout the daylight hours (from 8 to 17 UTC hours), but that focus on a time-slot between 12 and 16 h UTC. It was found that when an aircraft type CRJ enters the apron T4 on a sunny day between 12 and 16 UTC hours most likely it is not correctly identified. In other hours, the success rate is very high with CRJs. It should also be noted that the effect of shadows on sunny days is seasonal, with maximum effect when the sun position is the lowest in December and a minimum effect when the sun position is the highest in June (Fig. 4). Thus, even on summer sunny days, this effect is negligible. The effect of shadows on the aircraft type CRJ-200 tail numbers disappears in cloudy days. Table 3 shows the summary of identification probabilities obtained in sunny days.

Table 3. Summary of Probabilities

Aircraft Identification	
P_ID	95.0
P_FID	2.1
P_no_ID	2.8

7 Conclusions

The goal of this paper was an independent system evaluation of the surveillance system used in the airport. This paper does not provide any technical details of the tail number recognition algorithm and it does not explain what machine vision methods were applied. Only a system level evaluation of a suveillance system for the recognition of Aircraft type CRJ-200 was studied. Thus, no one can re-produce the results without purchasing and evaluating the same system.

The values obtained allow affirming that the system provides a useful and reliable tool for supporting and managing the operation of flights in apron and stands areas. In order to provide a point of comparison, reference values for an A-SMGCS system specified by EUROCAE and ICAO [14] are provided:

- $P\_ID > 99.9\%$
- $P\_FID < 0.001\%$

However, at this point, it should be noted that according to EUROCAE and ICAO regulations [14], a surveillance system, as part of an A-SMGCS must integrate:

1. At least one non-cooperative sensor system
2. At least one cooperative sensor system

In addition, EUROCAE and ICAO do not define their indicators with the aim of covering the apron and parking areas, just runways and taxiways [15]. In the case of this study, it has been analyzed a single primary non-cooperative surveillance system. For this type of systems, EUROCAE sets a P_ID value higher than 99%. Although the results obtained for identification in our system are lower than the EUROCAE indicator for primary systems (95% vs. 99%), the values reported allow concluding that the system provides a useful and reliable tool in their supportive role in managing the operation of flights in apron and stand areas.

References

1. EUROCONTROL, ATM Airport Performance (ATMAP) Framework, A Framework for Measuring Airport Airside and Nearby Airspace Performance (2002)
2. EUROCAE, MOPS for Surface Movement Radar Sensor Systems for Use in A-SMGCS, EMMA Project. 074/ED-116 (2006)
3. Piazza, E.: A-SMGCS routing and guidance functions. navia aviation AS. IEEE AES Syst. Mag. (2000)
4. Piazza, E.: Increasing airport efficiency: injecting new technology, park air systems. IEEE Intell. Syst. (2002)
5. EUROCAE, A-SMGCS Implementation in Europe. International Federation of Air Traffic Controllers' Associations, December 2003
6. Del Corte Valiente, A.: Evaluation of a non-cooperative aircraft surveillance system based on short range continuous-wave radars. Prog. Electromagnet. Res. B **43**, 1–18 (2012)
7. EUROCONTROL, State of the Art in A-SMGCS, EMMA Project (2005)
8. ICAO, Advanced Surface Movement Guidance and Control Systems, International Civil Aviation Organization (2004)
9. Galati, G., Naldi, M.: Airport surface surveillance with a network of miniradars. IEEE Trans. Aerosp. Electron. Syst. **35**, 579–593 (1999)
10. Human Factors HMI Requirements, EMMA Project (2006)
11. Bombardier CRJ-200 Fact Sheet, http://www.bombardier.com/en
12. Indicators and Metrics for A-SMGCS, EMMA Project (2006)
13. Long-term Measurements of A-SMGCS Performance–CDG Case Study, EMMA Project (2006)

14. EUROCAE, Implement Advanced Surface Movement Guidance and Control System (A-SMGCS) Level I, ICAO (International Civil Aviation Organization), Doc AOP04. Eurocae (2005)
15. Functional Specification for A-SMGCS Implementation Level I, European Organization for the Safety of Air Navigation (2003)

Teaching Math in Elementary Schools by LabVIEW and Lego Robots

Pedro Ponce[1(✉)], Arturo Molina[1], Laura Hernández[2], Eloisa Acha[2], Berenice Morales[3], and Carlos Huitron[1]

[1] Instituto Tecnológico de Monterrey Campus Ciudad de México, Mexico City, Mexico
pedro.ponce@itesm.mx
[2] National Instruments, Austin, TX, USA
[3] Benemérita Escuela Normal Veracruzana Enrique C. Rebsamen, Xalapa, Mexico

Abstract. This paper shows a novel platform based on fuzzy logic system for developing a math tutor using LEGO® robots and LabVIEW programs to teach math in elementary schools. The platform is divided into two main parts; the first part is the LEGO® robots, which have preloaded programs for different math topics (exercises), and the second part runs LabVIEW programs for evaluating the progress of each child. The platform is used as a regular play in the classrooms. The robots are personal tutors for each child during the class. After the students finished to review the math topic with the robot, they have to solve a math exam that includes questions about the reviewed topic with LEGO® robot. The main program, which runs the complete math topic in the robot and the exam, was developed in LabVIEW and has an artificial intelligence method for searching the best set of questions and sending feedback regarding the exam results to the child. This platform was validated in two elementary schools located in Xalapa Veracruz, Mexico. The results confirm that the platform helps to increase the kid's motivation about math and it opens new possibilities for teaching mathematics. Moreover, the kids learn at the same time different topics and skills such as robotics, computer science, mechanical systems, teamwork, and leadership.

Keywords: Education · Math · Robotics · NI LabVIEW · Artificial intelligence · Platform

1 Introduction

In conventional math courses for elementary schools, the teacher gives explanations about the math topic and provides some examples. Sometimes those exercises are not good enough because the kid does not see visual results or practical examples. Absent hands-on exercises, though, many kids will not have an opportunity to design, construct an understanding of the process of measurement or a concept of measurement unit which can frequently result in mechanical and inappropriate applications of measuring knowledge and tools, this could be expanded to different concepts specially a math concepts. For instance, a lot of kids confuse basic geometric shapes because

I. Chang et al. (eds.), *Advances in Automation and Robotics Research in Latin America*, Lecture Notes in Networks and Systems 13, DOI 10.1007/978-3-319-54377-2_22

they do not keep in mind the shape at first sight; some common types of errors that are made by these kids when using a ruler they do not have clearly the difference between inches and centimeters and also they are not able to get an approximate distance value in centimeters using their pencils as references. This is consequence of having only abstract knowledge but hands on exercises are very powerful tools for giving the other part of the math knowledge.

The first attempts to use computers interactively for learning were made by Seymour Papert [1] who developed the LOGO programming language to teach students mathematical skills and concepts related to algorithms. Earlier attempts include Patrick Suppes drill and practice use of computers to teach mathematics [2]. His research inspired Alan Kay to develop the Dynabook concept and eventually the graphical user interface (GUI). However, in the 1960s and early 1970s computers were unsuitable for classroom use. Only with the development of the personal computer, schools were able to use computers in the classroom for the first time. Unfortunately, using computers remained difficult, as GUIs were not yet available and there was a paucity of teachers trained to use these first computers effectively [3]. On the other hand the use of Lego® robots shows good results in education in elementary schools. Tanja Karp proposed for attracting a large number of elementary schools and engaging teachers easily, an outreach program, which provides a large degree of flexibility to accommodate different schools' needs, situations, and ideas for the implementation. It also should excite students through hands-on engineering challenges and encourage teamwork, critical thinking, and trouble shooting skills while being relatively inexpensive and easy to implement. They decided in favor of robotic [4]. Taking into account all of those aspects is a better option to combine a LEGO® platform and NI LabVIEW programs for generating an integral educational system which covers the most important aspects for teaching math.

2 Education System in Mexico [6]

The Mexican education system is organized into four levels: preschool (K1–K3), compulsory basic education (grades 1–9), upper secondary education (grades 10–12), and higher education. The government is only officially responsible for providing compulsory basic education, although it is also involved at the other three levels through public provision of preschool and upper secondary as well as public funding of higher education in most states. Public schools serve 87 percent of all students in the country. Governance is centralized as the national level with the Secretaria de Educación (SEP). One of the main SEP efforts is to apply technology in elementary school classes, thus the Enciclomedia project digitalizes the school curriculum into CD-ROMs, consequently, the students can learn interactively with the aid of computers. In this regards, almost all the elementary schools in Mexico have computers for using this kind of resources. SEP's current school calendar consists of 200 days beginning in August and ending in June of each year. Primary schools can meet in three shifts: morning, afternoon and evening. All shifts last four hours, during which the four primary subjects are taught: Spanish, mathematics, natural sciences, and social sciences. There is a little room for other activities such as teamwork, robotics and technology, even though some schools do make time for those subjects. There is not enough time so the students have to be highly motivated for working without the teachers' assistance.

3 Empowerment Through Technology Planet NI

Planet NI helps engineers and students in developing countries work toward economic prosperity and sustainable development by providing access to technology and education. This initiative nurtures local innovation by making engineering tools affordable, accessible, and relevant to academic organizations, entrepreneurs, and small and medium enterprises. National Instruments offices around the world define and support their programs locally by collaborating with groups and individuals that share the Planet NI mission, which is to improve the world through technology. Considering this the generation of technology for education is a task that could be supported and designed trough National Instruments and different entities as research centers in universities from those developing countries in order to boost the elementary education up. The work presented in this paper runs under the support of Planet NI program [7].

4 Theoretical Background

There is an ongoing trend in the math courses towards practical exercises instead of only theoretical concepts.

There is an open possibility for implementing in regular classes in elementary school so the combination of robots and programs running under computers are good complement for a completed learning process, as it is shown in Fig. 1.

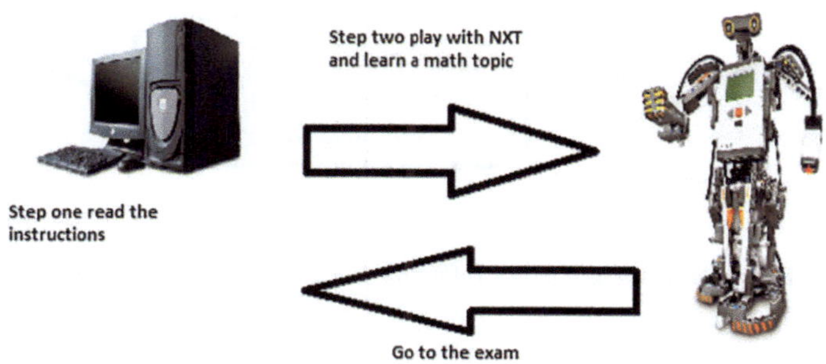

Fig. 1. Basic parts of the platform

Artificial intelligence method was based on a fuzzy logic system, the Mamdani inference is applied as an inference process to define what type of new exercises and the complexity of them is needed by the students in order to enhance the math knowledge regarding specific math topic. The Mamdani inference has the advantage that could include information from an expert, inside of the FAM (fuzzy associative memory). Thus, the information could come from the teacher using his experiences when they teach kids. Hence, the fuzzy logic system allows including the experience of a lot of teacher in a set of linguistic rules (IF-THEN). The basic parts of a fuzzy system are shown in Fig. 2. It could be observed how the input signals are mapped like fuzzy signals after that a set of rules are fired for getting the output signals [5].

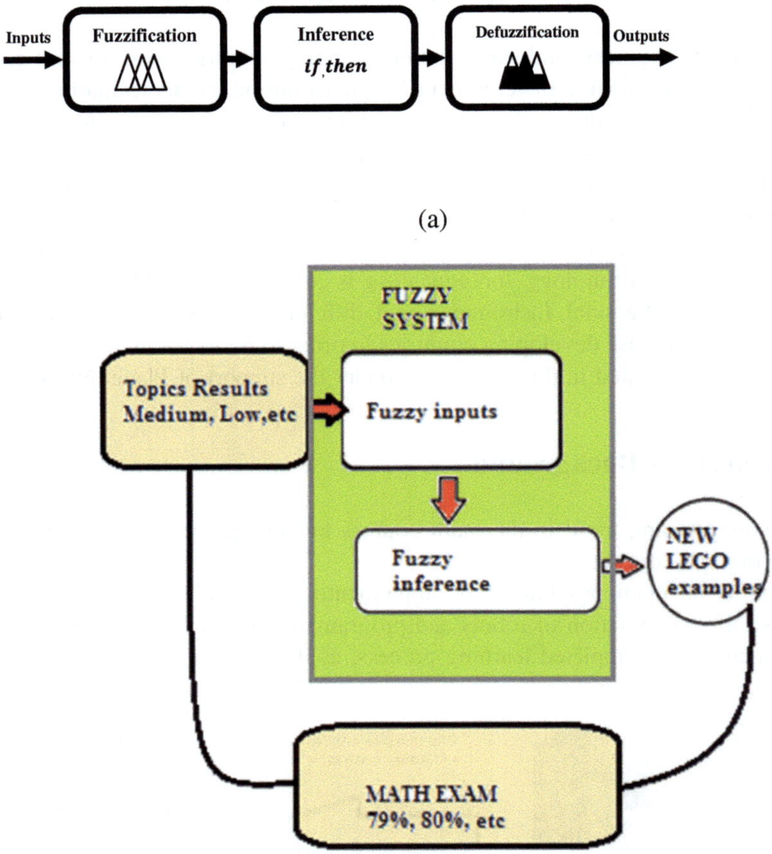

(a)

(b)

Fig. 2. Basic elements in a fuzzy inference (a) and fuzzy topology generated (b)

In the case of fuzzy system for educational purpose the fuzzy system is a very useful tool because the teacher could implement a set of linguistic rules in linguistic form if-then for defining how the feedback and the level of complexity in the questions is defined. Since Fuzzy logic was presented, it has been used for representing the information by fuzzy values instead of using crisp values for defining problems. To assign a crisp grade to the student is not always the best option, because you have many kids with the same grades, as result you do not know more information about the kid and the learning process. The fuzzy logic system proposed for this application allows sending feedback about the topic that requires the student for improving his math knowledge and a new set of questions and exercises are presented again. A LabVIEW program for downloading Lego® file was created, so the exercises and programs can be updated or deleted from the Lego® Brick (see Fig. 3).

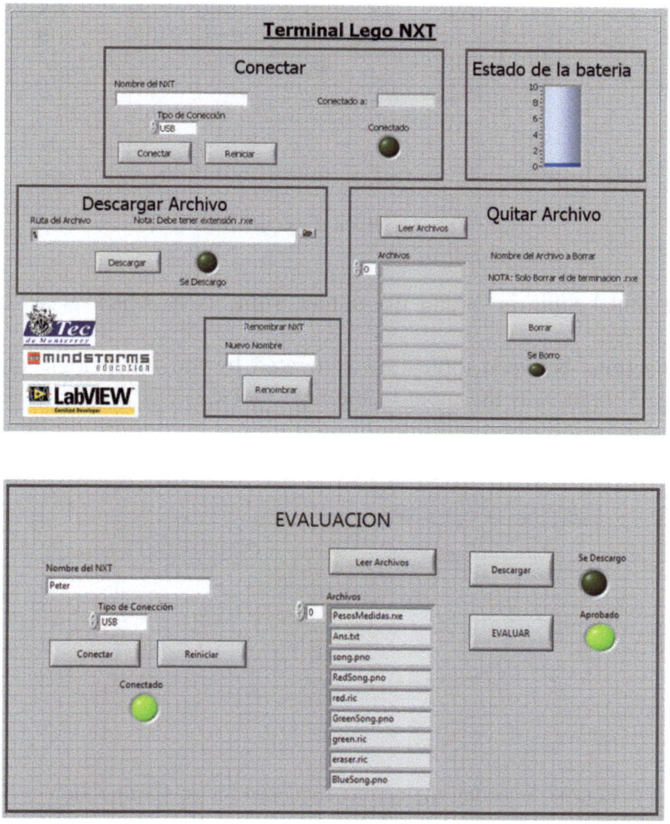

Fig. 3. Frontal panel for downloading programs to the LEGO® brick.

5 Learning Sequence

The learning sequence is not defined, so a specific learning sequence is designed by the artificial intelligence method for each student. This allows to move from one topic to another. Moreover, the sequence could be set to have a pre-defined sequence. The main goal is to have a sequence that follows the kid requirements; sometimes a specific sequence is not ideal for all the kids, so the computer program decides the sequence for each kid according to the previous information.

The next figure shows the inputs to the artificial intelligence method. One of the inputs is the information from the NXT Lego, the second input is the exam results, thus the Artificial Intelligence method sends a new topic. This topic is not following a serial sequence (A, B, C…etc.), it follows a sequence designed for a specific kid (A, X, W…etc.),

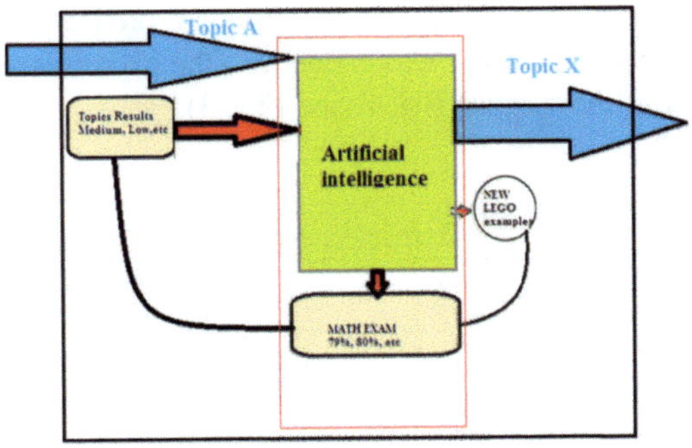

Fig. 4. Block diagram AI.

in Fig. 4 the initial topic is A and the output Topic is X which is set by the kid's requirements.

The connections between topics are generated by the fuzzy logic algorithm. As shown in Fig. 5 input X is the input to the next block.

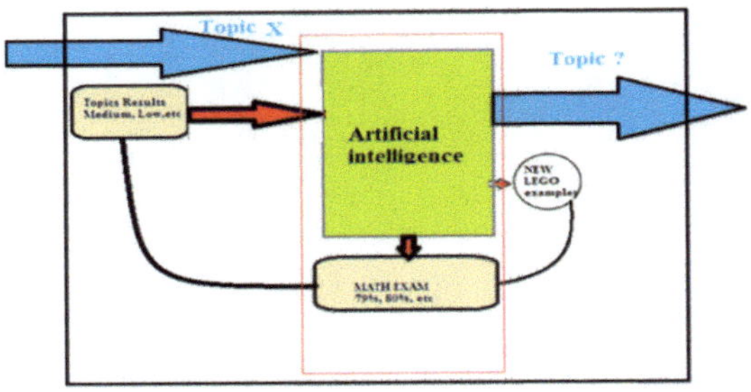

Fig. 5. Selecting the correct sequence for an specific kid.

At the end, all the math topics are covered but in different order according with each kind's requirements. In addition, the set of questions are not the same for all the kids, those questions change according to the kid's necessities.

6 Team Work

When the kids finished the exercises, the robots communicate with other robots (NXT robots send information about the kid progress and select the topic that will teach this kid to other kids in small groups). The robots send information about the learning progress for each kid and select the best kid for teaching a specific topic, so the kid is a team leader for this topic. For instance if the kid A learns very well topic A and the kid B learns very well topic X, the kid A teaches topic A and the kid B teaches topic X.

This teamwork generates different skills such as leadership, teamwork, team organization and writing. Figure 6 shows the communication of robots when they are using Bluetooth.

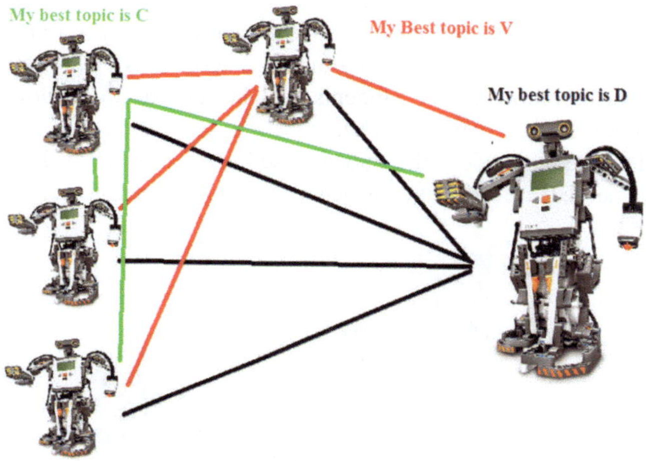

Fig. 6. Bluetooth communication by NXT

In this way, the robots propose a team leader who is the kid who get the best grades in each math topic as it is shown in Fig. 7.

Fig. 7. Students group (leader and members).

7 Implementation Issues

The validation of this platform was tested in an elementary school in Xalapa Ver. Mexico where the students have not worked with robots before. In Xalapa there are around 202 elementary schools. The first set of exercises with robots is for getting familiar with it, after that the exercises from the platform are in serial flow. It means that the topics have to be studied one by one in a serial way. The first step is to get familiar with the platform, so the students will be able to work alone without the assistance of teachers (see Figs. 8 and 9).

After reading the instructions the students use Lego® robots, this kind of play is an educational game because they have to learn an specific math topic. The platform is

Fig. 8. Getting familiar with Lego® robots

Fig. 9. Reading the instructions

Fig. 10. Kid as a mentor (teamwork)

quiet flexible and friendly. Thus, students can learn without the teacher assistance. Figure 10 shows a kid who is teaching to other kids, in this particular exercise they are learning geometry issues focus on angles.

The Lego® robots show information in their display during the learning process, this information is changing according to the process and result of each exercise. Figure 11 depicts the information presented in the display.

Other important issue is the classroom. The conventional classroom (see Fig. 12) becomes to a different place, the teachers and students are free to decide the place for learn. Hence, the classroom is adjusted according with the kid's comfort.

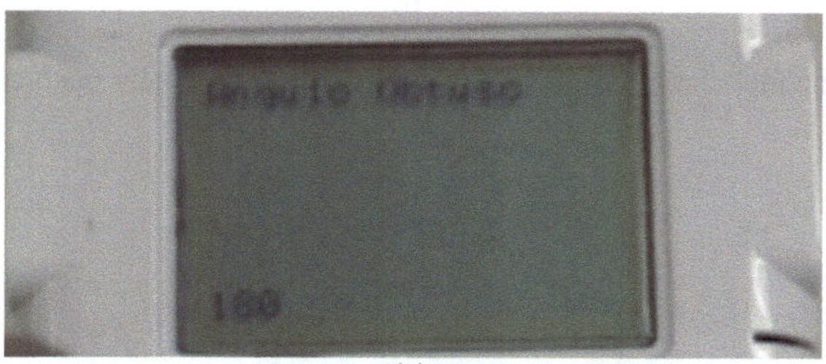

(a)

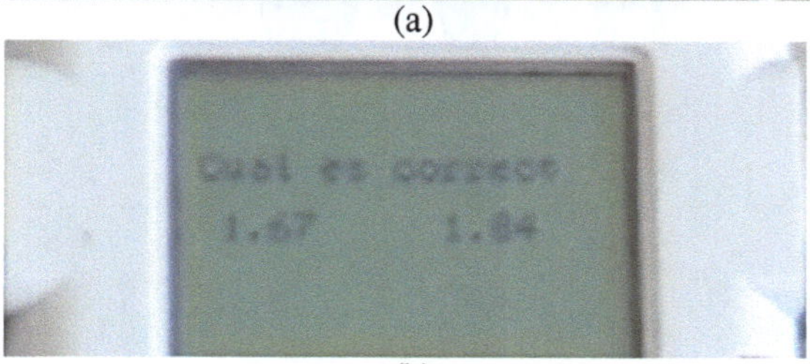

(b)

Fig. 11. (a) and (b) Lego® Display

a)

b)

Fig. 12. A Conventional classroom picture taken from [8] and a new math classroom using the platform.

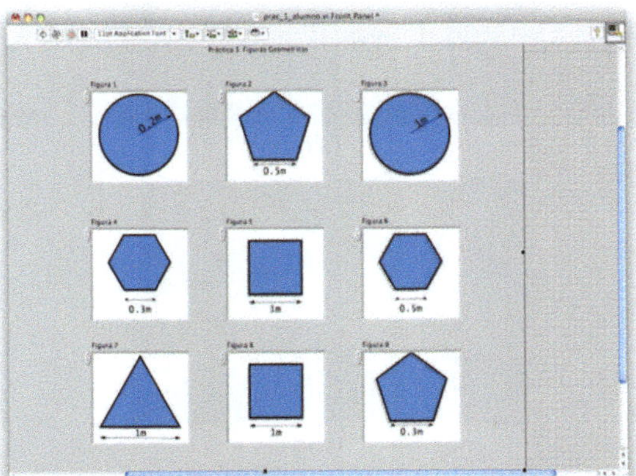

Fig. 13. An exam running on NI LabVIEW

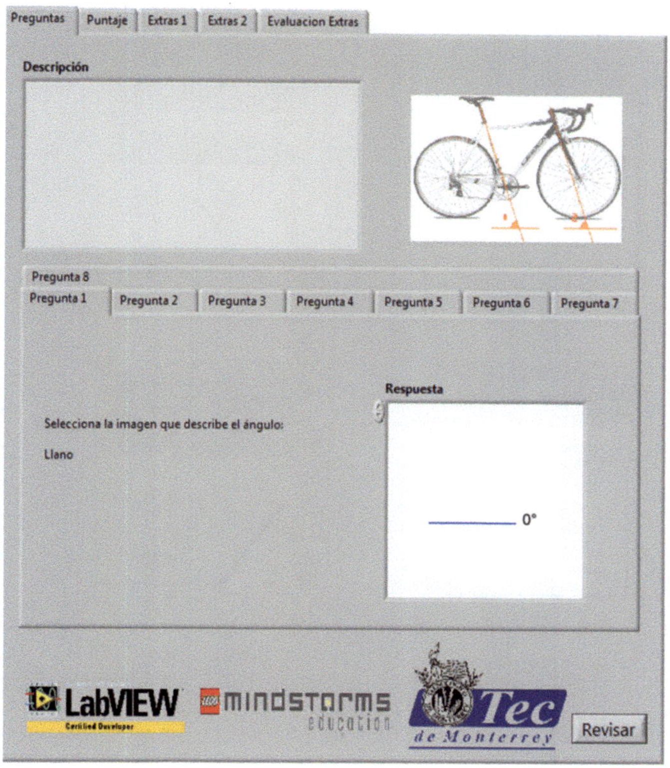

a)

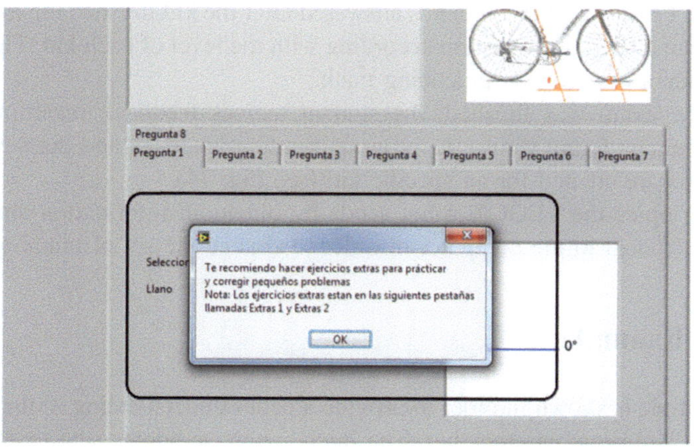

b)

Fig. 14. Feedback, the exam frontal panel (a), Information about the new exercises black rectangle (b) and proposed exercises (c).

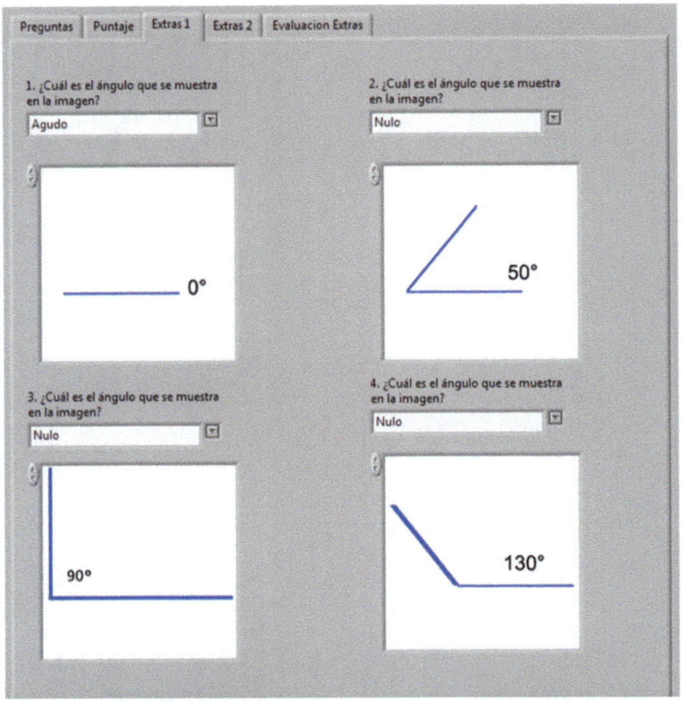

c)

Fig. 14. (continued)

The Fig. 13 shows an exam that is answered after the kid finished the robot activity. This exam can change the question according with the level of each kid. The main goal is to give kids a new way of practicing math.

When the exam was finished, the student recives feedback regarding the math progress. The next figure shows the program sending feedback and the next exercises. The questions are created for an specific kid (see Fig. 14).

In some topics the LEGO® robot sends by bluetooth information about the progress of the student to the computer in order to change the set of initial questions.

8 Topic Example

A basic example is shown in order to provide a better understanding of the potential of using robots and programs together. The inicial exercise begins setting an inicial and final distance using the robot and kid´s hand. The robots has an ultrasonic sensor that detects the distance from the robot to the kids hand, after that different questions are shown into the Lego display, the main topics are distance, fractions and conversion (cm-inches). When the answer is correct the robot plays a song and moves. A wrong answer does not generate any sound only the robot moves in a different way. When the robot decides that the kid is ready for begin with fractions new kind of questions are set, after that the robot moves to conversions. The last part of this topic is to use the wheel parameters diameter, perimeter to calculate the distance that the robot can reach using the number of wheel turns (see Fig. 15). In this sense, the number of concepts are increased but the kid is in the same topic. Additional topics could be covered as sensors, mechanical transmission and electric motors, by the same topic.

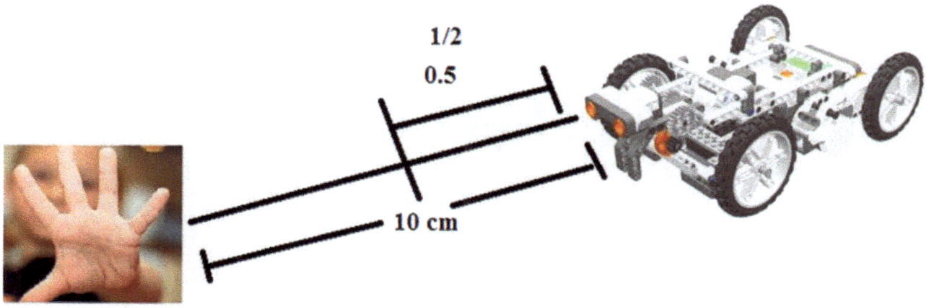

Fig. 15. Math topic (as a play for kids)

9 General Implementation Results

The results of this platform are presented in the following charts; these results were taken from an experimental validation of the platform in the two elementary schools in Xalapa, this information is a sample of four classrooms. The test covers 3 elementary grades. (80 students and 5 teachers took the survey). In addition, the response of the kid is based on the time that they are willing to spend with the robot and the improving that they can get in math. Figure 16 shows the comments of teachers and they agree on using the platform increased the math skills.

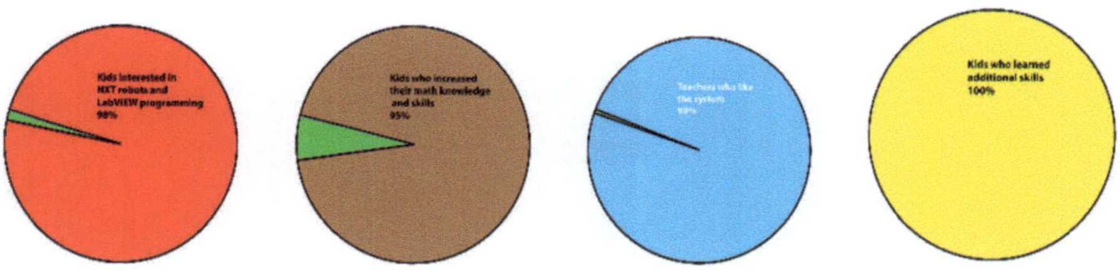

Fig. 16. General results of kids and comments of teachers about the platform.

The next results shows a national evaluation called enlace. Enlace is a test that defines the skills of a kid for understanding and solving problems in different areas and math is one of these topics which are evaluated. Figure 17 presents the results for different years, the arrow shows when the platform started (year 2009). The colors mean the following math label: Blue-low label, Green- minimum label, Yellow-good label, Red-excelent. It can be observed that the math level incresed. However, the robotic platform is only a tool in the classroom, so there are several factors that could affect the learning process. Nevertheless, robots and computers are very attractive for kids and they can help in this learning process (see Fig. 17).

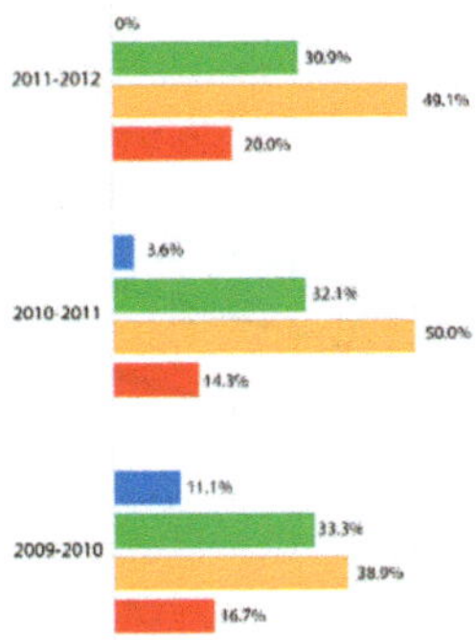

Fig. 17. Enlace test for 6 grade of elementary school that uses the platform

Figure 18 presents results about the 5th and 4th grade regarding Enlace national evaluation. The starting point was 2009, and the same color meaning is used in this chart.

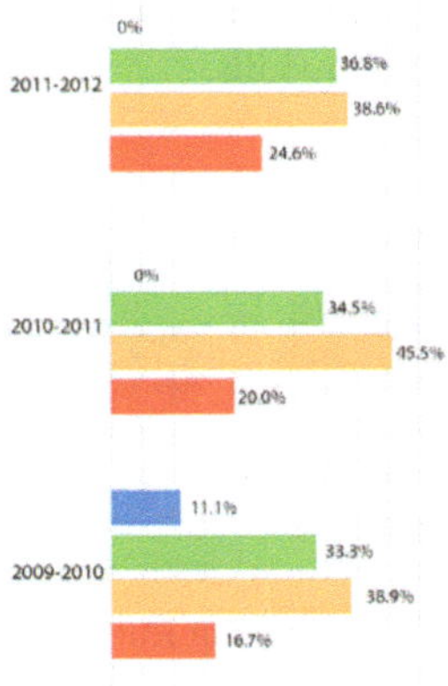

(a)

Fig. 18. Enlace test for 5th (a) and 4th (b) grade of elementary school

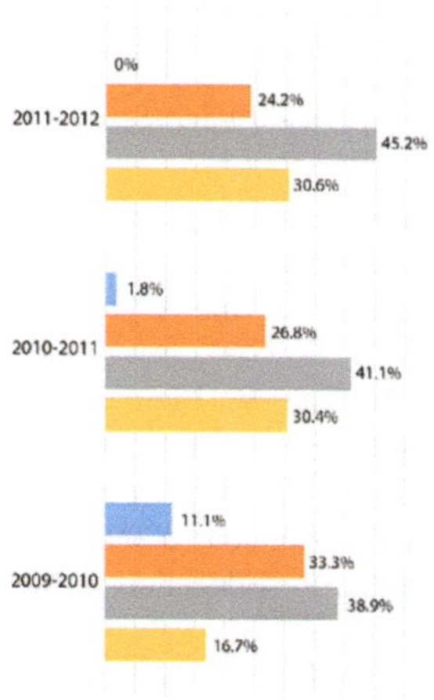

(b)

Fig. 18. (continued)

10 Conclusions

The proposed platform shows excellent results for motivating kids in learning math and teachers like using this platform because they can get more information about the students. These new concept of learning math using robots and computers allows the students to learn at their own pace, the robot and the computer can adjust the set of questions and exercises according to the students requirements. In addition, the use of artificial intelligence inside the program keeps some teacher knowledge regarding the evaluation form. Thus, the evaluation can be preserved inside the system. The motivation for learning math increased spectacularly when the kids regularly use the platform. In addition, this platform was developed for covering all the math topics assigned in conventional elementary school in Mexico, so it can be used in several schools or kids that do not attend classes. On the other hand, it is important to mention that all the teacher were willing to use the platform and the final version of the platform was accepted by 99% of them. In Mexico, there are many places that use robots for teaching robotics systems but there are few efforts for using robots in math classes in elementary schools. Thus, this work demonstrates that robots are flexible to teach kids several topics in elementary schools.

Acknowledgements. This paper was elaborated under the support of Tecnologico de Monterrey Escuela de Ingenieria, National Instruments and Lego. We like to express our gratitude to all the teachers of Acela Servin, Escuela Anexa a la Normal Veracruzana and Escuela Normal Veracruzana for the support and numerous discussions, which improved our knowledge about Elementary Education in Mexico.

References

1. Papert, S.: A Computer Laboratory for Elementary Schools, Computers and Automation (1972)
2. Suppes, P.: The uses of computers in education. Sci. Am. **215**, 206–220 (1966)
3. Fröhlich, T., Feinber, S.: Challenges to user testing E-learning games with children in elementary schools. In: Proceedings of the Eighth IEEE International Symposium on Multimedia (ISM 2006) (2006)
4. Karp, T., Gale, R., Lowe, L.A., Medina, V., Beutlich, E.: Generation NXT: building young engineers with LEGOs. IEEE Trans. Educ. **53**(1), 80–87 (2010)
5. Ponce-Cruz, P., Ramírez-Figueroa, F.D.: Intelligent Control Systems with LabVIEW™. Springer Science & Business Media, London (2009)
6. Santibañez, L., Vernez, G., Razquin, P.: Education in Mexico (2005)
7. http://www.ni.com/planetni/
8. http://www.thejakartapost.com/news/2009/05/22/character-building-the-missing-link-indonesia%E2%80%99s-public-school-curriculum.html

Learning Multivariable Controller Design: a Hands-on Approach with a Lego Robotic Arm

Victoria Serrano[1,2(✉)], Michael Thompson[3], and Konstantinos Tsakalis[2]

[1] Electrical Engineering Department, Technological University of Panama, David-Chiriqui, Panama
victoria.serrano@utp.ac.pa
[2] School of Electrical, Computer, and Energy Engineering, Arizona State University, Tempe, AZ 85281, USA
{vmserran,tsakalis}@asu.edu
[3] School for Engineering of Matter, Transport and Energy, Arizona State University, Tempe, AZ 85281, USA
mjthomp3@asu.edu

Abstract. In this paper we present the multivariable controller design of a Lego Mindstorms NXT robotic arm for educational purposes. System identification was performed to estimate the parameters of a coupled plant. Uncertainties and Bode plots of the coupled system were evaluated to determine that the system could be decoupled at the DC gain value. A Proportional-Integral-Derivative (PID) controller was designed for the system after it was decoupled. These steps conform one of the basic techniques to design the controller for a multivariable system. Therefore, it is a valuable model to learn multivariable controller design for undergraduate students. This provides a hands-on approach experience in controller design where some aspects of advanced controller theory are blended with implementation details.

1 Introduction

Multivariable systems are widely used in industry; therefore, a demand for more efficient and robust controllers are still needed. Since multivariable plants are capable of handling systems that have multiple inputs and multiple outputs (MIMO), many of these controllers are designed using extended versions of single input and single output approaches. But despite of the many advances in controller design, Proportional-Integral-Derivative (PID) controllers still prevails as the most used type of controllers used in industry. One of the most desired properties of PID controllers is their integral action that eliminates set-point errors and disturbance offsets. Also, it offers a phase lead to tune for crossover properties such as phase margin, resulting in good closed-loop damping.

Additionally, since PID controllers are relatively simple, it allows a straightforward implementation on hands-on experiments. Furthermore, their extension to multivariable systems makes them suitable in a myriad of applications.

In spite of the fact that MIMO systems are usually present in industrial applications, the interactions between loops is minimized by applying static or

I. Chang et al. (eds.), *Advances in Automation and Robotics Research in Latin America*,
Lecture Notes in Networks and Systems 13, DOI 10.1007/978-3-319-54377-2_23

dynamic techniques. That way the system becomes diagonal dominant, making it simpler and easier to control. Katebi [5] studied multivariable systems and provided a comparison of the different methods in terms of stability and robustness performance.

They used a 2×2 distillation column system to separate methanol from water and suggested some techniques that are more appropriated for simple designs. Two groups of methods emerged from this study. The first group of methods is classified as parametric techniques. They are based on the fact that a plant model is available. The second group of methods is known as non-parametric techniques. They are based on the assumption that a detailed model of the plant is not available. However, they should fulfill some requirements such as: the plant needs to be linear and time invariant, the uncontrolled plant must be stable and square, the controlled variables can be measured, the classes of input disturbance and reference input have to be known and the system must be controllable by a diagonal PID controller. Although this study provided a great contribution for the understanding of multivariable systems, the transfer of technology to audience of lower level expertise still remains an open issue.

One of the main problems that many engineering courses face are the lack of hands-on experiences. The predominant reason for this issue is the high cost of equipment to perform experiments in classes such as control systems [1]. That is why our proposal includes the use of Lego Mindstorms kits to develop control theory classes that not only spans from concepts such as system identification, controller design and discretization methods, but also tackles coupled multivariable systems. This brings a meaningful learning experience to undergraduate students since although many control system concepts have been taught using Lego Mindstorms kits, multivariable controller design has not been addressed to a satisfactory extent.

Cruz-Martin et al. [4] performed experiments with undergraduate students to teach data acquisition, control theory and real-time applications. Moreover, Kim [7] taught control theory classes through a Lego Mindstorms NXT motor. Valera et al. [13] and Pinto et al. [10] used the Lego Mindstorms NXT kits to teach concepts of Kalman filters and extended Kalman filters, respectively. In addition, Behrens et al. [3] performed experiments with Lego Mindstorms NXT together with Matlab in an introductory course for engineering students. Tse [12] presented the use of NXT-based robots controlled through Labview. Similarly, Kim and Jeon [6] used visual programming to control Lego Mindstorms robots while Be et al. [2] used voices to control them wirelessly. Furthermore, in an effort to understand multivariable systems, Serrano et al. [11] initiated a study of a decoupled plant. However, although a great contribution has been done in understanding control systems, all the previous studies lack in learning the hands-on approach of coupled multivariable systems for undergraduate students.

This paper provides a benchmark to design a multivariable controller for a Lego Mindstorms NXT robotic arm. System identification was performed to determine the plant model of the coupled motors. By having two motors physically coupled, a greater torque can be achieved in a system. A PID controller was designed for each motor after applying a decoupling technique and analyzing the Bode plot and uncertainties of the coupled plant. The results showed that the system although it follows closely the trajectory, it presents a small mismatch at different times due to the inherent backslash nature when two motors run in different directions.

2 Experimental Setup

A Lego Mindstorms NXT set, version 9797 was used to perform the experiment. This set contains multiple Lego pieces for mechanical construction, a brick to program the robot with inputs and outputs to connect up to three motors and different sensors such as light, sound, touch and ultrasound. The Lego brick is a 32-bit ARM7 microprocessor that has 64 kbytes of RAM memory and 256 kbytes of FLASH. It has Bluetooth capabilities to send/receive data. The cost of the kit was about $225 USD.

Additionally, Matlab and Simulink were used for data collection, analyses of results and motor control [9]. A 2014b version was required to run the Lego Mindstorms NXT toolbox [8]. A computer equipped with Bluetooth allowed the communication with the Lego brick to collect data while the USB cable was required to download the Simulink models from the computer to the brick.

3 Methodology

A set of two motors were physically coupled in the system motivated by the fact that their mechanical power can handle a higher payload. Two independent pseudo random binary sequence (PRBS) signals were introduced to each motor to analyze their coupling level. The system is shown in Fig. 1.

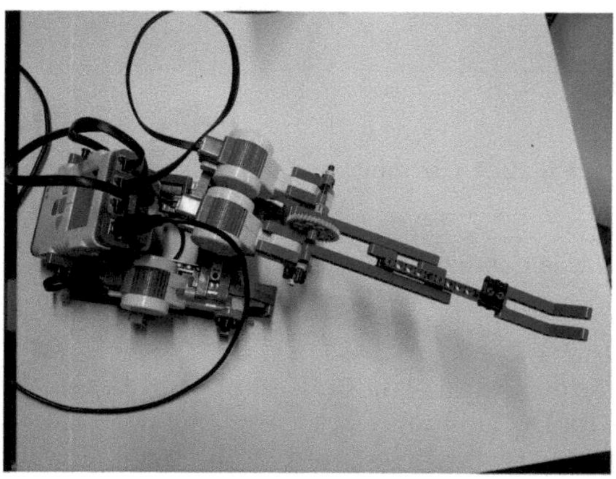

Fig. 1. Coupled Motors

This allowed to decouple the system to minimize the interaction between different loops. Figure 2 helps us to observe that the decoupling is achievable. Figures 3 and 4 show the Bode magnitude and the step response of the coupled system, respectively. Both figures show the interaction between different inputs and outputs. Input 1 (In1) and Input 2 (In2) are given by the power applied to motors B and C, respectively. Similarly, Output 1 (Out1) and Output (Out2) represent the angular position of motor B and C, respectively. Since our interest is to decouple the plant at the DC gain, the Bode magnitude plot allows us to determine the level of coupling between inputs and outputs at DC values. It is observed that the DC gain is approximately 25 dB for each input-output interaction, which makes the system coupled. Additionally, the step response of the coupled plant depicts a significant value different from zero between different inputs and outputs, which validates the coupling nature of the system observed in the Bode magnitude plot. Therefore, a decoupling technique was performed to obtain a diagonal dominant plant at least at the DC gain.

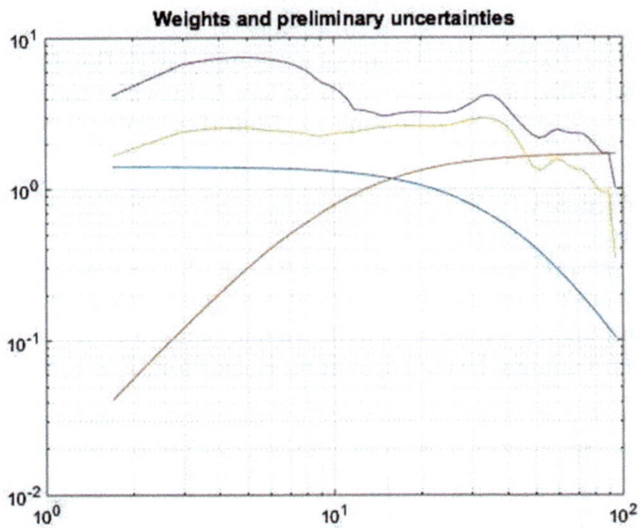

Fig. 2. Uncertainties of the coupled system

The new plant was defined by Equation (1).

$$P_d(s) = P(s)P(0)^{-1} \tag{1}$$

Here $P_d(s)$ is the decoupled plant (at least at DC), $P(s)$ is the original coupled plant and $P(0)$ is its DC gain.

Figure 5 illustrates the step response of the multivariable system when it has been decoupled. The system is diagonally dominant; however, the step response of the off-diagonal elements is not negligible. Therefore, this characteristic affected the performance of the multivariable system.

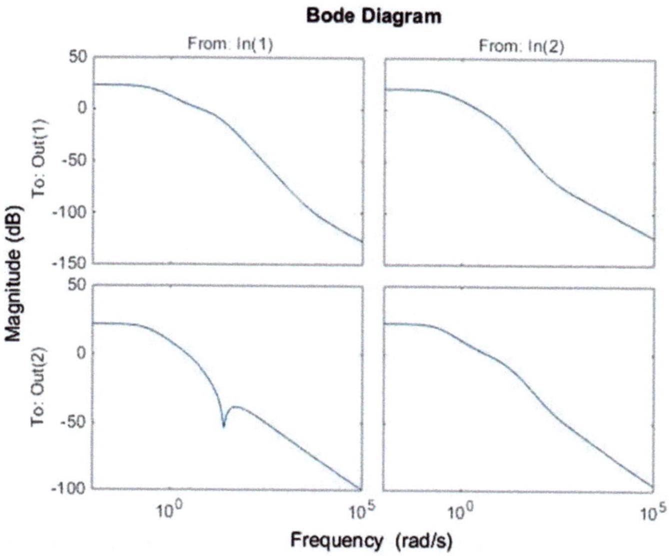

Fig. 3. Bode magnitude of the coupled system

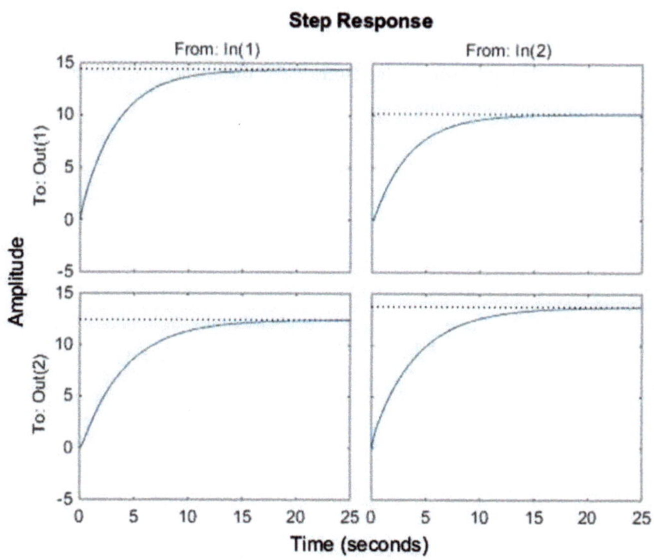

Fig. 4. Step response of the coupled system

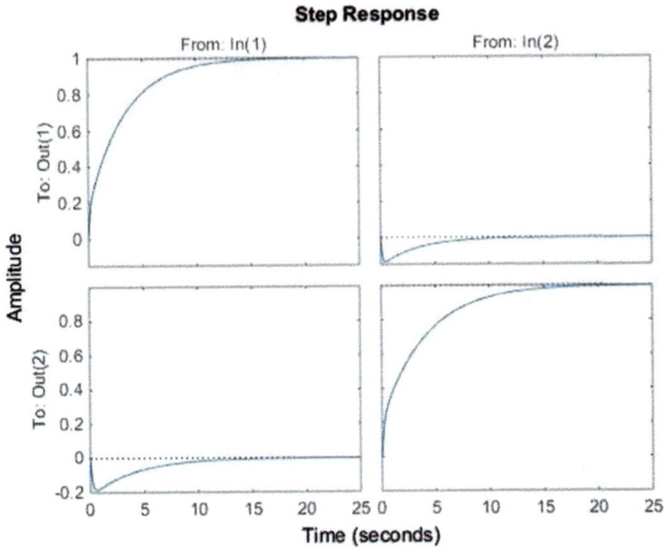

Fig. 5. Step response of the multivariable system after applying a decoupling technique

4 Results and Discussion

Once the PRBS signal allowed the system to be modeled and decoupled, a PID controller was designed and implemented for each of the motors of the

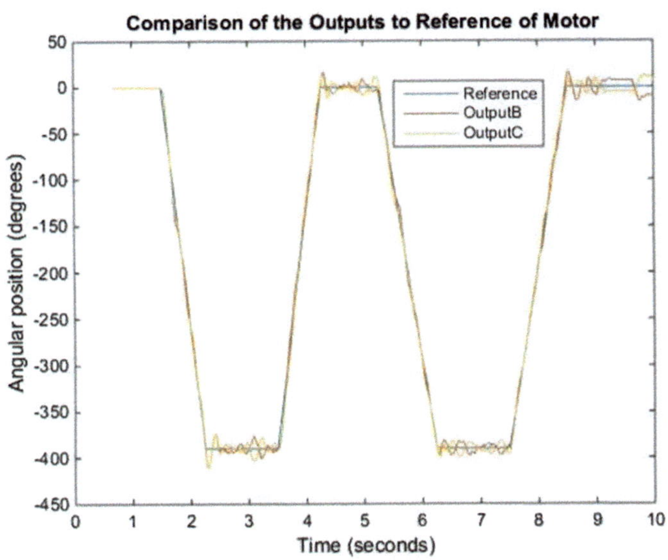

Fig. 6. Tracking reference of motors B and C

multivariable system. As shown in Fig. 5, the system can be now considered diagonal dominant. Thus, for controller design purposes, each motor can be treated independent from each other despite their physical coupling. The controller was discretized using a sampling time of 0.1 s and the Forward Euer method for simplicity.

Figure 6 depicts the tracking reference for motor B and motor C for a system which has been decoupled. The system did not follow the trajectory as well as the system with no physical coupling studied in [11]. This response is probably due to the coupling between motors and the decoupling technique applied to the system. Additionally, when the two motors rotate in opposite directions, a gear backlash is present in the system; therefore, these effects are reflected in the system by having a limited amplitude oscillation.

5 Conclusion

In this paper we developed the steps to design a multivariable controller on a Lego Mindstorms NXT robotic arm. The process started with the system identification performed to two motors that were physically coupled. Then, an analysis showed that the system could be decoupled at the DC gain value. Once the system was decoupled, two PID controllers were designed and implemented to each motor. The tracking reference showed that each motor did not always followed closely the trajectory which was probably due the coupling between motors and the decoupling technique used prior designing the controller. However, the approach followed in this paper is simple enough to be used with an undergraduate audience and can be applied to more complicated scenarios. Yet, a study of more challenging multivariable systems can be addressed in future studies that may include switching control or gain scheduling.

References

1. Atanasijevic-Kunc, M., Karba, R., Logar, V., Zupancic, B., Papic, M., Bester, J.: Remote laboratory for e-learning of multivariable control design. In: 30th International Conference on Information Technology Interfaces ITI 2008, Cavtat, Croatia, pp. 465–470 (2008)
2. Be, D., Escalante, M.: Wireless control LEGO NXT robot using voice commands. Int. J. Comput. Sci. Eng. **3**(8), 2926–2934 (2011)
3. Behrens, A., Atorf, L., Schwann, R., Neumann, B., Schnitzler, R., Ballè, J., Herold, T., Telle, A., Noll, T.G., Hameyer, K., Aach, T.: MATLAB meets LEGO mindstorms - a freshman introduction course into practical engineering. IEEE Trans. Educ. **53**(2), 306–317 (2010)
4. Cruz-Martín, A., Fernández-Madrigal, J.A., Galindo, C., González-Jiménez, J., Stockmans-Daou, C., Blanco-Claraco, J.L.: A LEGO mindstorms NXT approach for teaching at data acquisition, control systems engineering and real-time systems undergraduate courses. Comput. Educ. **59**(3), 974–988 (2012)
5. Katebi, R.: Robust multivariable tuning methods. In: Vilanova, R., Visioli, A. (eds.) PID Control in the Third Millennium. Advances in Industrial Control, pp. 255–280. Springer, Heidelberg (2012). Chap. 9

6. Kim, S.H., Jeon, J.W.: Programming LEGO mindstorms NXT with visual programming. Int. Conf. Control Autom Syst. **2007**, 2468–2472 (2007)
7. Kim, Y.: Control systems lab using a LEGO mindstorms NXT motor system. IEEE Trans. Educ. **54**(3), 452–461 (2011)
8. Mathworks. LEGO MINDSTORMS NXT Support from Simulink (2014). http://www.mathworks.com/hardware-support/lego-mindstorms-simulink.html?refresh=true. Accessed 2 Feb 2015
9. Mathworks. MATLAB & Simulink Student Version 2014b (2014)
10. Pinto, M., Moreira, A.P., Matos, A.: Localization of mobile robots using an extended Kalman filter in a LEGO NXT. IEEE Trans. Educ. **55**(1), 135–144 (2012)
11. Serrano, V., Thompson, M., Tsakalis, K.: Multivariable controller design of a lego mindstorm NXT robotic arm. In: Proceedings of the 2nd International Conference of Control, Dynamic Systems and Robotics, Ottawa, ON, no. 186, pp. 1–9 (2015)
12. Tse, S.B.: Mindstroms Controls Toolkit: Hands-On, Project-Based Learning of Controls. Master's thesis, Tufts University (2009)
13. Valera, A., Vallés, M., Marín, L., Albertos, P.: Design and implementation of Kalman filters applied to LEGO NXT based robots. In: 18th IFAC World Congress, Milano, Italy, pp. 9830–9835 (2011)

Towards Image Mosaicking with Aerial Images for Monitoring Rice Crops

Juan Rojas[1(✉)], Carol Martinez[2(✉)], Ivan Mondragon[2(✉)], and Julian Colorado[1(✉)]

[1] Department of Electronic Engineering, School of Engineering, Pontificia Universidad Javeriana, Bogotá, Colombia
{rojas-jp,carolmartinez}@javeriana.edu.co

[2] Department of Industrial Engineering, School of Engineering, Pontificia Universidad Javeriana, Bogotá, Colombia
{coloradoj,imondragon}@javeriana.edu.co

Abstract. Around 8 to 10 million Ton of rice are required in the following years to be able to supply the demand of the overall population. Analysis and monitoring of rice crops becomes nowadays very important issue for farmers, for ensuring a rice production level to cope this demand. This paper presents simulation results of an algorithm that allows to plan and create 2D maps using the technique of image mosaicking with multiple geo-referenced aerial images (multispectral images in the scope of the project). The planning algorithm is called Image Capture algorithm. It takes into account the area the UAV has to cover, the camera configuration, and the state of the UAV in order to define where to take the pictures to build the mosaic. The algorithm presented in this paper was developed in ROS (Indigo) and simulated in Gazebo. The results of this first approach to the 2D mapping of a rice crop allows to see that using the proposed algorithm, it is possible to automate the process of acquiring the pictures for creating the mosaic, ensuring that all the area of interest is covered. By using this algorithm, pictures will be acquired only in specific areas. Therefore, keeping the storage capacity on-board, under control.

Keywords: Homography · Stitching · Unmanned Aerial Vehicles · Computer vision · Image mosaicking

1 Introduction

The rice is one of the most important cereals. It represents the 25% of the calories consumed by humans in a day, and it is one of most cultivated cereals in the world [1]. There are four types of rice cultivated around the world, the first and the most important is the Indica type. This type represents the 75% of the world trade. The second type is the India Basmati. This type is aromatic and represents the 13% of the global trade. In third position with the 10% of the

I. Chang et al. (eds.), *Advances in Automation and Robotics Research in Latin America*, Lecture Notes in Networks and Systems 13, DOI 10.1007/978-3-319-54377-2_24

global trade is the japonica type. Finally, the fourth type is the glutinous type which is cultivated in some parts of Asia [2].

The most cultivated type of rice in Colombia is the Paddy Verde [3]. This specie of rice represents the rice core trade of the country, due to its fast growth, and resistance to plague and environmental changes. Colombia has two methods of growing rice. The mechanized and the manual methods. The mechanized can be divided in irrigated and upland rice. The difference between them is the source of the water. In the irrigated method the water proceeds of water districts, whereas in the upland method the rice is irrigated with rainwater. The mechanized method is used in the 94% growing rice fields, but the 70% of rice production is related with the upland method.

Colombia has many areas where the rice can be cultivated, but due to the cost of crop care, chemist and workforce [4], Colombia has began to import rice. In other countries such as China and USA, these problems have been reduced thanks to the use of technologies to monitoring the state of the crop (using for example drones, sensor networks, and satellite photos).

Image mosaicking is a computer vision technique used in many applications to generate a wider view of the scene, than the one obtained with only one image taken with a normal camera. This technique consists on aligning multiple overlapping images captured from a moving camera. One of the advantages of using image mosaicking in precision agriculture is to be able to capture in one image all the information of the crop for further analysis. Therefore, by analysing only that image, it is possible to extract information about the state of the crop without damaging the crop in the field. Therefore, it is considered a non-invasive technique for analysing the state of crops.

In this paper, we present results of the COLCIENCIAS 120371551916 project: "Aerial sensing and monitoring of rice crop fields applying precision agriculture techniques". The scope of this project is to develop a testbed for evaluating rice varieties in the field, by using sensors located on-board an aerial vehicles that allows to monitor in a non destructive way rice crops.

From previous results [5] at Pontificia Universidad Javeriana, a multiespectral camera and the ASCTEC-Pelican were used to generate a multiespectral mosaic that describes a specific terrain. the Normalized Difference Vegetation Index NDVI was used to estimate the healthy of the plants and the density of the vegetation. However, in this paper we focus on the image capture problem for creating the mosaic. The planning algorithm is called Image Capture algorithm. It takes into account the area the UAV has to cover, the camera configuration, and the state of the UAV in order to define where to take the pictures to build the mosaic. We present simulation results of the algorithm that allows to plan and create 2D maps using the technique of image mosaicking with multiple geo-referenced aerial images (multispectral images in the scope of the project). In the paper, the pictures to build the mosaic are taken in defined intervals during flight, in a specific altitude, and with a predefined overlap, to ensure the quality of the mosaic (in this paper 30% overlap is tested). The above values can be measured assuming the system has a predefined trajectory, a specific linear velocity and feedback from the sensors on-board the UAV (GPS and IMU). The algorithm was developed in ROS (Indigo) and simulated in Gazebo.

The paper is organized as follows. First, Sect. 2 presents related work on precision agriculture and image mosaicking. In Sect. 3, the UAV system is presented. In that section, the simulated UAV and its relation to the real drone are discussed. Section 4 explains the image capture algorithm and the image mosaicking algorithm. Finally, Sect. 5 presents results of the image capture and image mosaicking algorithms; and Sect. 6 presents the conclusions and direction of future work.

2 Related Work

Precision farming has become an essential tool to improve agricultural production and productivity. It includes techniques for crop monitoring and management. The crop management essentially depends on the irrigation type, nitrogen levels, and chemists varieties used in it. In a specific site, crop management requires a fast diagnosis method that allows the farmer to have the measurement and analysis of the crop field in real time. This is to avoid crop losses, increase crop quality, and optimize agriculture supplies. The latter will impact the production cost (reducing it), and will also help protecting the environment. Remote sensing and monitoring applications based on unmanned aerial platforms, equipped with a set of specific sensors and instruments have been widely developed on the last few years, as a rapid deployment tool for field applications [6].

Field studies in Colombian rice zones have demonstrated that the biomass dynamics is related to a higher performance of the crop, when the conditions of the field are the best [7]. It is important to notice that the temperature and low solar radiation also affect the biomass dynamics.

To take the best behavior of the crop, many techniques and technologies have been developed that give a measurement of the field and allow the farmer to have solid information of the crop. An example of the developed technologies can be seen in [8], where a depth camera, a Laser, and a UAV are used to estimate the status of the soil. In [9], a sensor network is implemented to measure and classify the percentage of arsenic, ammonium, and other chemical substances in the field, with the aim of finding their origin and reduce their impact to the field. In [10], a satellite IKONOS is used to take several multispectral images of a grape crop, with the aim of analyzing the state of the crop; and to provide the farmers information to improve the crop care. To address this problem, the Normalized Difference Vegetation Index NDVI is used to classify segments of the crop. The size of the leafs is also measured to estimate their growth. On the other hand, in [11] a method to estimate crop area coverage is presented. This method uses a UAV and a spatial sampling method, based on Moderate Spatial Resolution (MSR) image classification results, to estimate the area.

It is important to mention that the use of UAVs in combination with image processing is a low cost tool compared with the analysis of satellite images or the implementation of network sensors. Additionally, UAVs allow to have a full view of the field quickly, speeding up the process of measurement and analysis of the crop variables. In the work presented in [12], authors present a method to classify rice quality, based on high resolution images taken by a UAV flying at low altitudes.

Digital image mosaicking is a technique that has increased importance over time in many fields of the industry (military intelligence [13], image processing [14,15] and biomedical approaches [16], etc.), especially in precision agriculture [17–19]. There are different approaches that allow to generate an image mosaic. Some are focused on improving the motion estimation problem, and others focused on post-processing the mosaic for solving problems related to drift.

In the work of Tom Botterill et al. [20] an aerial vehicle was used to capture multiple images of a specific area. The main idea was to provide the operator with an image that has a larger field-of-view. For feature extraction and matching they use Bag-of-Words, which allow to find, in an efficient way, the matching of points with wide-baseline. A seam-placement algorithm to de-noise and rendering the mosaic was also implemented.

Moussa [21] proposed an algorithm that allows to manipulate several images in an efficient way. SIFT (Scale-invariant Feature Transform) features were used; and the correlation between images was used for feature matching. Then, the algorithm estimates the motion of the images, warps them, and generates the mosaic.

In [22], three algorithms for mosaic generation, were compared: controlled, uncontrolled and semi-controlled mosaics. The images are taken from two different test flies and has several geo-referenced points, based on GIS (Geographic Information System) and GCPs (Ground Control Points), that allow the algorithm to generated the stitching between images using these points as references.

As a general conclusion from the woks presented above is that the most important step in image stitching is the selection of the feature extraction and matching algorithm. In this sense, to detect growing areas it is possible to implement different techniques of image processing that segment the blocks of crops and other elements, as shown in [23].

For UAV-based image mosaicking, a mission planning algorithm has to be carefully designed in order to ensure the coverage of the terrain, taking into account the UAV autonomy. The research presented in [24] presents a solution to the Coverage Path Planing (CPP) problem. The proposed algorithm allows to reduce the consumption of energy; and to ensure the image resolution, the speed and acceleration of the drone. In the work of [25], a new path planning method is proposed. It allows to optimize the drone task, path length, and coverage lost. The algorithm is based on three planers, Route Planer (RP), path planer (PT), and coverage planer (CP). All the called planers are based on the modified versions of Genetic Algorithms and Dijkstra Algorithm ($A^{\star}$).

3 UAV System

The drone that is going to be used in the project is the ASCTEC-Pelican from Ascending Technologies, shown in Fig. 1a. The system is composed by a ground station (a standard PC), and the drone which is equipped with a Mastermind board (for image processing and data acquisition); the ASCTEC autopilot; and cameras: an RGB camera SONY FCB-EH6500, and a multispectral camera ADC Lite from Tetracam. The multispectral camera captures visible light wavelengths longer than 520 nm and near-infrared wavelengths up to 920 nm. It is located looking-downwards in the bottom of the drone, and aligned with the center of mass.

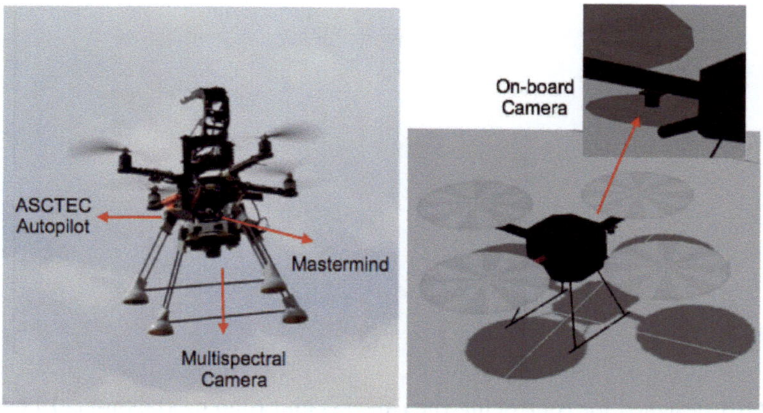

(a) ASCTEC-Pelican UAV used for rice crop monitoring (b) Hector quadrotor for simulation tests.

Fig. 1. UAVs used for rice crop monitoring. Figure 1a shows the ASCTEC-Pelican test-bed used in the project. It is equipped with a multispectral camera, the ASCTEC autopilot, and a Mastermind (on-board computer). Figure 1b shows the simulated drone used for software development.

The ASCTEC-Pelican comes with a GPS waypoint navigation strategy with ±1 m of accuracy. Data from sensors are sent via 2.4 GHz XBee link to the ground station. In the ground station there are two computers, one for running the Graphical User Interface that comes with the ASCTEC-Pelican, that allows to control the drone (autonomous take-off and landing; and waypoint navigation) and provides information about the state of the drone.

The second computer runs a customized GUI, design in Qt with Ubuntu 14.04, shown in Fig. 2. This GUI allows to start the different processes that run in the Mastermind board, and as shown in the figure, allows to have visual information about the different cameras on-board the UAV.

On the other hand, the Mastermind computer on-board the UAV, runs Ubuntu 14.04 and the framework used to control the autopilot from the mastermind was

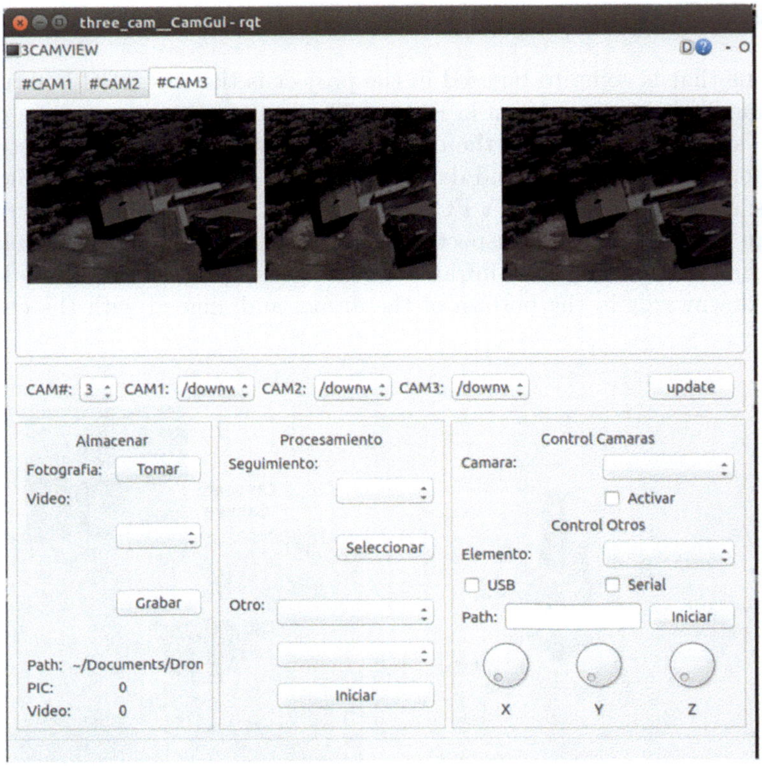

Fig. 2. GUI designed for controlling the cameras and running the on-board computer vision algorithms

ROS (Robot Operating System). Under this framework, the trajectory planning algorithm, the position and linear velocity control are independent nodes, and all of them run on-board, but are monitored from the ground station.

3.1 Simulation Environment

In order to develop the different algorithms required by this project, a simulated environment, based on ROS, was created. The drone model used was the Hector Quadrotor, a ROS package that allows to simulated and control a UAV of type quadrotor [26]. The model of this drone runs in the Gazebosim [27], a simulation environment that allows to integrate multiple sensors like IMUs, cameras, pressure sensors, thermal cameras, among other sensors. The drone model communicates with ROS under specific topics related to the multiple elements that are placed on-board the simulated drone (e.g. sensors). For the case of this project, an RGB camera that points to the ground, an IMU, and a GPS were used. Figure 1b shows the Hector quadrotor with the hardware that was configured.

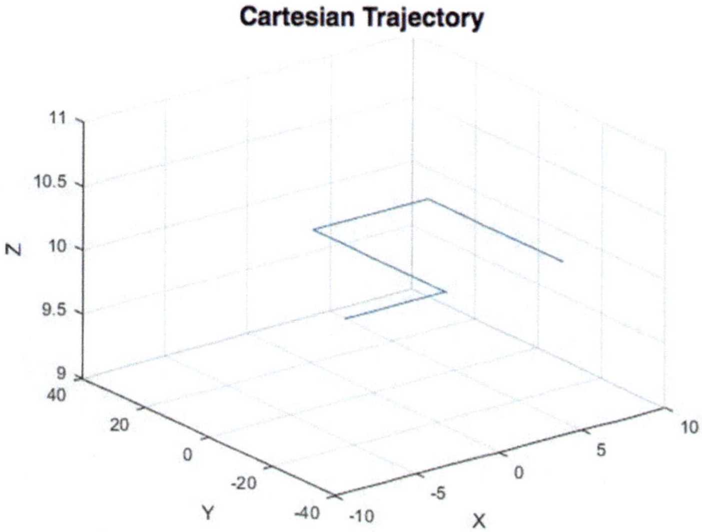

Fig. 3. Example of the generated trajectory points, when using the trapezoidal move profile, used in the simulation tests.

The predefined trajectory model can be seen in Fig. 3. This trajectory was generated using a trapezoidal move profile. This model gives the cartesian points, the linear velocities, and the acceleration that the drone needs to follow, in every moment, to cover the working area.

To simulate the environment a satellite image of a rice crop field was taken from Google Earth, and a digital elevation model was taken from Google Sketchup. All the hardware place on-board the simulated UAV, and the physics variables are scaled and related with the world definition file from Gazebo. Once the world model is created in gazebosim, the drone is loaded, and the simulated system is ready to work. Figure 4 presents a picture of the environment and the trajectory followed by the UAV.

4 Image Mosaicking

The main objective of the COLCIENCIAS project is to analyze the state of the rice field. Assuming the area of the rice crop is known, the UAV should be able to plan the proper trajectory to follow, in order to take the pictures required to create an image mosaic of the field. This image mosaic will be the one used to analyze the state of the field.

The Image Capture algorithm presented in this paper will take into account the characteristics of the camera, the field area, and information from the state of the UAV, to define the places where the images from the crop will be acquired. This algorithm will ensure that the captured images will cover the flight area with the enough overlap to create the mosaic of the crop area. Once image are

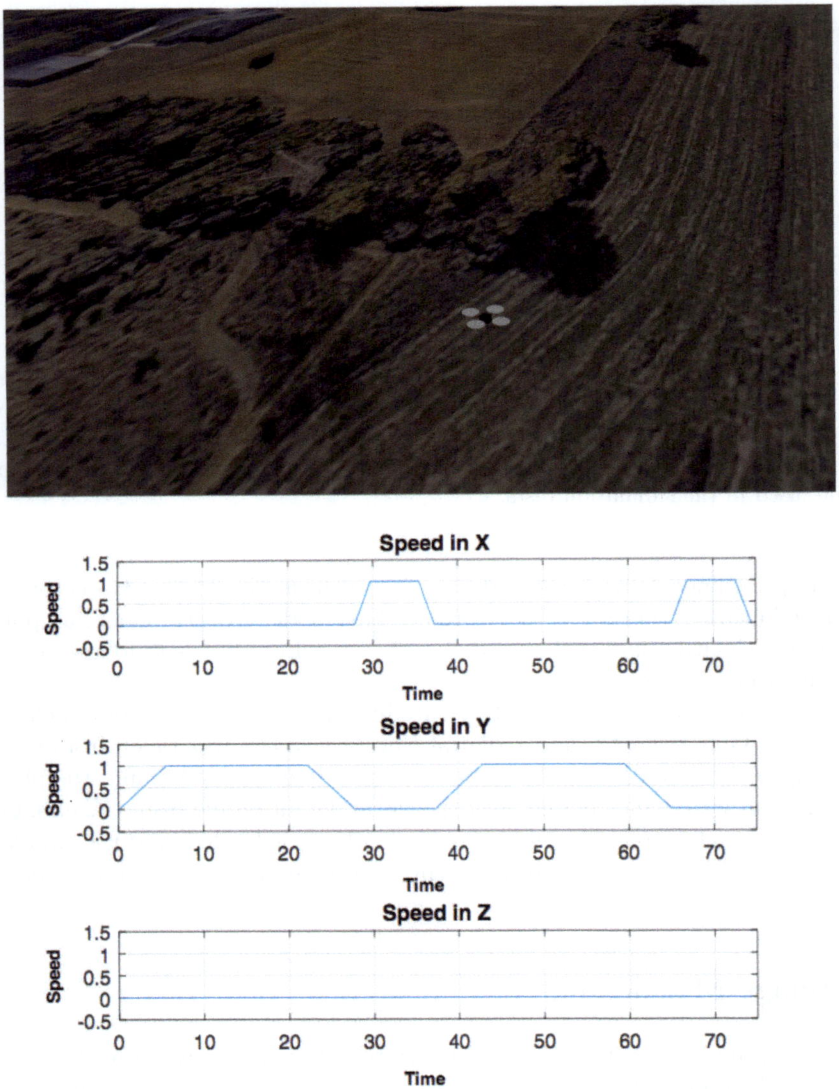

Fig. 4. Simulation environment. A satellite image of a rice crop field was taken from Google Earth and a digital elevation model was taken from Google Sketchup. Once the world model is created in gazebosim, the drone is loaded, and the simulated system is ready to work.

captured, the image mosaicking function (stitching) from OpenCV is used to generated the mosaic.

4.1 Image Capture Algorithm

The implemented algorithm assumes that the intrinsic camera parameters (the camera calibration matrix **K** Eq. 1), the size of the camera sensor s_x and s_y, and the altitude of the drone, are known. This information is used by the algorithm to estimate the area covered by the camera.

$$\mathbf{K} = \begin{bmatrix} f_x & 0 & c_x \\ 0 & f_y & c_y \\ 0 & 0 & 1 \end{bmatrix} \tag{1}$$

where f_x and f_y are the focal length, in pixel units, in the x and y axes; and c_x and c_y are the coordinates of the optical center.

From the focal length (f_x and f_y), the size of the camera sensor (s_x and s_y), and the number of pixels per unit distance (m_x, m_y), it is possible to estimate the field of view (FOV) of the camera, as shown in Eq. 2; where θ_x and θ_y correspond to the FOV in each axis.

$$\theta_x = 2 * \tan(\frac{s_x/2}{f_x/m_x}); \; \theta_y = 2 * \tan(\frac{s_y/2}{f_y/m_y}) \tag{2}$$

To estimate the distance covered by the camera in the x and y axes ($\mathbf{d_{cam}} = \{d_{cam_x}, d_{cam_y}\}$), Eq. 3 is used. This equation relates the height of the UAV (h_{UAV}) and the FOV of the lens calculated in previous equation (θ_x and θ_y), to estimate the covered area, as shown in Fig. 5.

$$\mathbf{d_{cam}} = 2 * h_{UAV} * \begin{bmatrix} \tan(\theta_x) \\ \tan(\theta_y) \end{bmatrix} \tag{3}$$

The distance $\mathbf{d_{cam}}$ gives the measurement in meters; and the area covered by the camera is given by:

$$a_{cam} = d_{cam_x} * d_{cam_y} \tag{4}$$

In order to calculate the number of pictures required to cover the rice field; and the point where those pictures has to be taken, it is required to know the percentage of the desired overlap between images (p_{ov}); the distance covered by the camera, given by Eq. 3; and the area of the crop field, given by the width (w_{crop}) and height (h_{crop}) of the crop field. With these values, Eq. 5 estimates the number of pictures in x and y axes that has to be taken in order to cover the crop area.

$$n_x = \frac{w_{crop}}{d_{cam_x} - (d_{cam_x} * p_{ov})} \; n_y = \frac{h_{crop}}{d_{cam_y} - (d_{cam_y} * p_{ov})} \tag{5}$$

In this application, the percentage of overlapped p_{ov} tested was 30% in order to ensure that consecutive images have enough common features to create the mosaic.

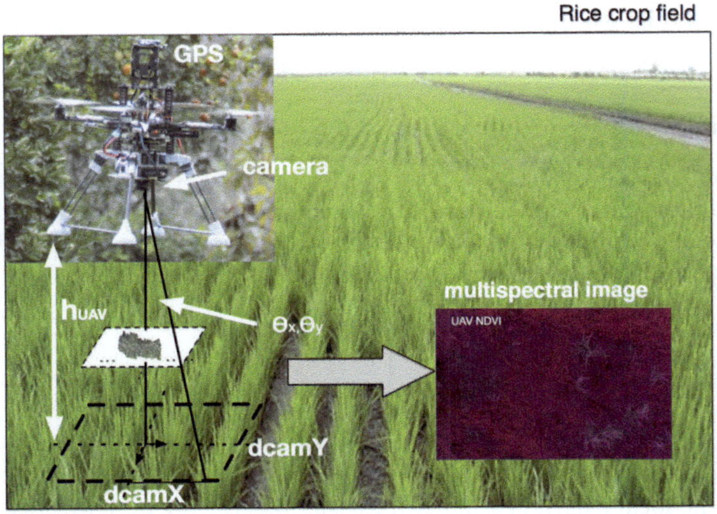

Fig. 5. Area covered by the camera. The height of the UAV and the lens field of view is used to estimate d_{cam_x} and d_{cam_y}, see Eqs. 3 and 4

To calculate the places where the images have to be taken ($\mathbf{x}_{\mathbf{pic}}^{\mathbf{T}} = [x_{pic}, y_{pic}]$), an iterative model is used. This model takes into account that the origin of the crop coordinate system is at the center of the crop area (the working area); and that the UAV starts in the lower right corner of the working area and moves towards the lower right corner (UAV initial position $\mathbf{x_{UAV}}\ (0) = [\ w_{crop}/2, -h_{crop}/2])$.

The initial iteration is:

$$\mathbf{x}_{\mathbf{pic}}^{\mathbf{T}}(0) = \mathbf{x}_{UAV}^{T}(0) + \mathbf{d_{cam}} \begin{bmatrix} -1 + p_{ov} & 0 \\ 0 & 1 - p_{ov} \end{bmatrix} \tag{6}$$

and when iterating, the x_{pic} or y_{pic} position is updated as follows:

$$\begin{aligned} x_{pic}(i+1) &= x_{pic}(i) + d_{cam_x} * (-1 + p_{ov}) \\ y_{pic}(j+1) &= y_{pic}(j) + d_{cam_y} * (-1 + p_{ov}) \end{aligned} \tag{7}$$

Once Eq. 7 estimates the coordinates where pictures have to be taken. The UAV is ready to fly and to acquire the images required to generate the mosaic.

During the flight, the only communication with the UAV is to request GPS points and IMU data to integrate, and estimate the position of the UAV related to the starting point of the algorithm. When the UAV is close to the places where the pictures have to be acquired, a flag is activated and the image is acquired.

When the fly ends a node with the OpenCV stitching algorithm is launched and begin the process of generating the mosaic (feature extraction, feature matching, motion estimation and image warpping).

4.2 Mosaic Generation

To generate the mosaic, a function that uses the stitching class of OpenCV is used. This class contains different functionalities that allows the user to estimate the parameters that relate multiple images. In this particular case, default parameters were used to create the mosaic.

The stitching function resizes the images, finds features, matches them, and finally returns the blend of the images to generate a unique images (the mosaic), that contains all the given images.

The stitching class of OpenCV uses SIFT (Scale-invariant feature transform) algorithm for feature extraction and matching. Feature matching is improve with Bayes's rule. On the other hand, RANSAC is used for adjusting the best homography that aligns the images, and the model of Burt and Adelson [28] is used for blending the mosaic. OpenCV class allows to select different algorithm for the different stages required to create the mosaic.

4.3 Algorithm

The developed system for mosaic generation can be summarized in the algorithm shown in Algorithm 1. The algorithm is divided in three sections, two offline and one online.

In the first part (offline), the Image Capture algorithm estimates the positions where the images have to be captured. This is based on information about the crop size, the camera, and the planned UAV's height. The second part occurs online. The UAV moves following a predefined trajectory and when it reaches the positions were images have to be captured, the image is stored. Finally, after the flight has finished, the OpenCV image stitching algorithm is used to generate the mosaic, using the stored images.

5 Results

In order to test the image capture algorithm and the mosaic generation for rice crops, different tests were conducted, using the simulation environment create in Gazebo and the simulated drone: Hector quadrotor.

5.1 Image Capture Algorithm

The Image Capture algorithm allows estimate how many images have to be taken with some percentage of overlap, in a specific area; and where they have to be taken in order to cover the working area. To test the algorithm, the simulated quadrotor was commanded to fly an area of 50 m^2, at two different heights 10 m and 15 m. This was done with the aim of analyzing how the image capture algorithm covers the working area.

The camera parameters (see Eq. 1) used in the tests were: a focal length of 40; c_x and c_y correspond to the center of the image (320×240); $s_x = 39.3$ and

Offline Image Capture planner (Before Flying);
input : Crop size (w_{crop} and h_{crop}), Camera Sensor Size (s_x and s_y), Drone Height (h_{UAV})
output: Array of cartesian world points

1. **Compute** Area covered by the camera $\mathbf{d_{cam}}$, Eq. 3
2. **Compute** Number of photos **n** in each axis, Eq. 5
3. **Compute** Initial pose $\mathbf{x_{pic}^T}(0) = [x_{pic}(0), y_{pic}(0)]$, Eq. 6

```
for i ← 1 to n_x do
    for j ← 1 to n_y do
        Update y_pic position, Eq. 7;
        Save x_pic^T = [x_pic(i), y_pic(j)];
    end
    Update position x_pic position, Eq. 7;
    Save x_pic^T = [x_pic(i), y_pic(j)];
end
```

Online Image Capture (During Flight)
input : initialize motors and sensors reading. Read file with $\mathbf{x_{pic}^T}$ coordinates
output: Stored mages

```
Take off;
for i ← 1 to n_x * n_y do
    Move to next pose;
    Read current UAV position x_uav^T ;
    if if x and y coordinates of x_uav^T ≈ x_pic^T(i) then
        Store image;
    else
        Save current pose x_uav^T
    end
end
```

Offline Mosaic Generation;
input : Stored images
output: Image Mosaic

```
Initialize nPic = n_x * n_y;
for i ← 1 to nPic do
    Read stored image;
    Append images in an array;
end
Call OpenCV stitching function;
```

Algorithm 1. General Algorithm. In the first stage (offline), the places where images have to be captured are estimated. The second stage (online) focuses on capturing and storing the images; and the third stage (offline), deals with the creation of the mosaic.

$s_y = 44.5$; and $m_x = 16$ and $m_y = 10$. With these parameters, when the UAV flies at 10 m, the algorithm has to take 5 pictures in the Y axis, and 10 in the X axis, in order to cover the working area (see Fig. 6). However, when the UAV flies at $h = 15$ m, it requires to take less pictures to cover the working area: 3

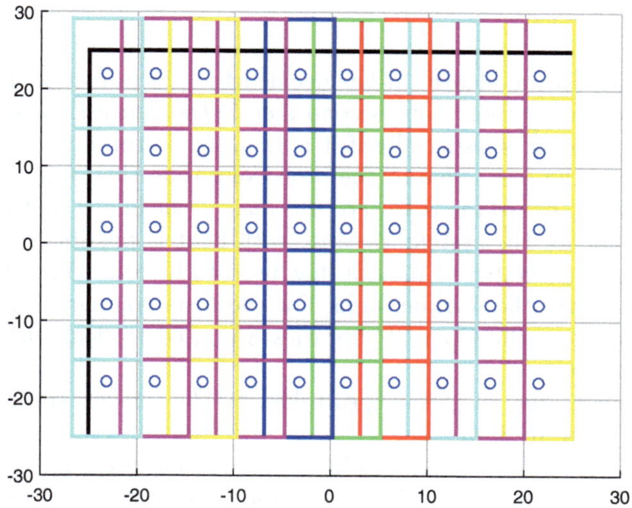

Fig. 6. Mission planing test at 10 m. Working area 50 m^2

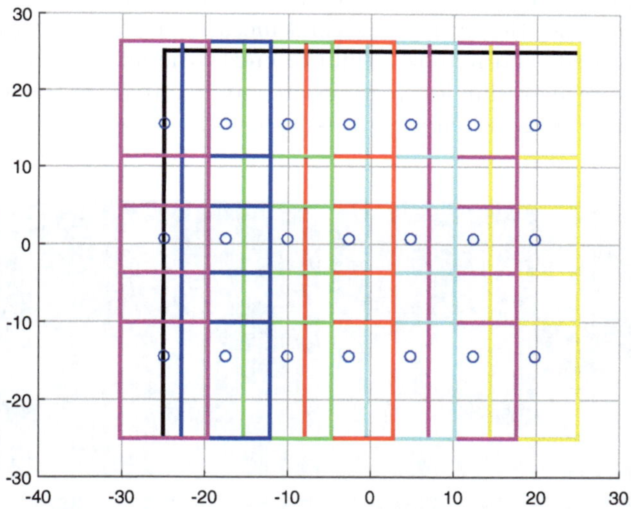

Fig. 7. Mission planing test at 15 m. Working area 50 m^2

in the Y axis and 7 in the X axis (when being higher, the field of view of the camera covers more area).

Figures 6 and 7 also show how the percentage of overlap defined (30%) is preserved, and it is possible to see the different grids created by the algorithm. The circles represent the coordinates where images have to be captured, which are used during flight for capturing the different images required to build the mosaic.

5.2 Stitching

Once the image capture algorithm has decided the places to take pictures, and the UAV has acquired them. The second step is the generation of the mosaic. This is a critical process, especially if it is required to run online on-board the UAV (this is not the case), due to the processing time (high) and the required storage capacity on-board the UAV. For this reason, the image capture algorithm is very important for the system, because with it, it is possible to reduce, in a significant way, the number of images that need to be taken in order to ensure a mosaic that covers the working area. For the tests conducted in this paper, the Stitching class of the OpenCV libraries was used to deal with the mosaic generation problem.

Figure 8 shows some of the images taken by the image capture algorithm, when the UAV flew at 10 m.

Figure 9 compares two mosaics. One created using images acquired by the image capture algorithm, see Fig. 9a; and the other one created using all the images captured while flying, see Fig. 9b. In the figures, it is possible to see that both mosaics look similarly. However, when using the image capture algorithm, the mosaic shown in Fig. 9a was created using only 6 images and it took 5 min to create the mosaic. However, the second mosaic was created with 19 images (manually captured) and it took 30 min to create the mosaic. Therefore, using the image capture algorithm less pictures are required to cover the working area, and therefore the image stitching process is faster.

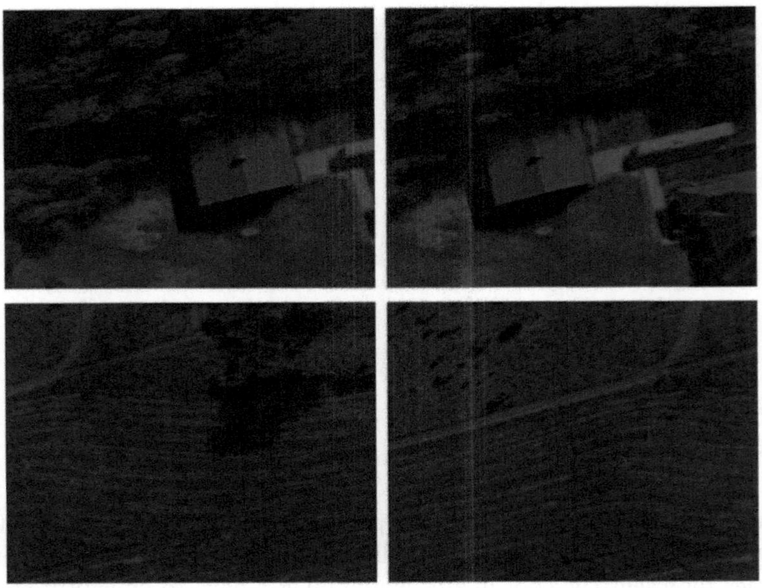

Fig. 8. Representative images taken using the mission planning algorithm, when flying at 10 m.

(a) With mission planning algorithm. Images 6, processing time 5 min.

(b) Without mission planning algorithm. Images 19, processing time 30 min.

Fig. 9. Comparison of two different mosaics created. Using the mission planning algorithm less pictures are required to cover the working area, and therefore the image stitching process is faster.

Finally, Fig. 10 shows another mosaic created using the OpenCV stitching function and the simulation environment created for the project. For this mosaic, the UAV took 10 min to fly around the area, and the stitching algorithm took 85 min to create the mosaic. From this figure, it is possible to see the quality of the mosaics that can be generated with the proposed strategy.

It is important to mention that in this process the GPU was not used; and that depending on the resolution of the images, it is possible to require or not the use of aditional filters to eliminate some noise due to the overlapping process.

Fig. 10. Mosaic generated after using the image capture algorithm, when flying at 10 m. Processing time 85 min.

6 Conclusions

In this paper we have presented an image capture algorithm for ensuring the generation of image mosaics of rice crop fields.

Different tests were conducted using a simulation environment created in ROS. From the tests we have shown the importance of using the image capture algorithm, which allows to reduce, significatively, the number of images required to cover the area of interest, and to reduce the processing time required to generate the mosaic. In the area of UAVs, this is a very important feature taking into account the limited payload capacity on-board these vehicles. It has also been shown that by reducing the amount of images required to cover a specific area, the quality of the results is not affected.

It is important to highlight that with the proposed algorithm, it is possible to generate a mosaic that describes the environment with a small amount of images, but ensuring the resolution and overlap between them required by the image stitching algorithm.

The tests have also shown that with 30% of overlap the stability of the image stitching algorithm was preserved (with this value there were common features in consecutive images to ensure the feature matching process). Nevertheless, current work in the project is focused on conducting real-flight tests to analyze the behavior of the algorithm using multispectral images, and to define with more exhaustive tests the different parameters required by the algorithm, such as the required flight height and the appropriate percentage of overlap. Future work will focus on analyzing the state of the rice crop, based on the multispectral mosaic.

Acknowledgements. This work has been funded and sponsored by the research project: Desarrollo de una herramienta para la agricultura de precision en los cultivos de arroz: sensado del estado de crecimiento y de nutricion de las plantas usando un drone autonomo (Aerial sensing and monitoring of rice crop fields applying precision agriculture techniques). Funded by: COLCIENCIAS. ID 120371551916.

References

1. C. for the Common Organisation of Agricultural Markets, "Rice market," European Commission; AGRICULTURE AND RURAL DEVELOPMENT, Technical report, November 2016
2. Childs, N., Burdett, A.: The U.S. rice export market; rice situation and outlook. Economic Research Service - USDA, Technical report (2000)
3. Chauhan, B., Jabran, K., Mahajan, G.: Rice Production Worldwide, 1st edn. Springer, New York (2016)
4. Mazuera, C.A. Neira, J.D.: ANALISIS DE LOS COSTOS DE PRODUCCION DE ARROZ, Oriza sativa L., EN EL MUNICIPIO DE SALDAÑA, TOLIMA. METODO PULVER VS METODO TRADICIONAL DE MANEJO. Universidad De La Salle Facultad De Administracion De Empresas Agropecuarias, p. 44 (2009)
5. Patino, D., Colorado, J., Navia, J., Mondragon, I.: Multispectral mapping in agriculture: terrain mosaic using an autonomous quadcopter UAV. In: International Conference on Unmanned Aircraft Systems, ICUAS, pp. 1351–1358 (2016)
6. Pajares, G.: Overview and current status of remote sensing applications based on unmanned aerial vehicles (UAVs). Photogram. Eng. Rem. Sens. **81**(4), 281–329 (2015). http://www.sciencedirect.com/science/article/pii/S0099111215300793
7. Takai, T., Matsuura, S., Nishio, T., Ohsumi, A., Shiraiwa, T., Horie, T.: Rice yield potential is closely related to crop growth rate during late reproductive period. Field Crops Res. **96**(2–3), 328–335 (2006)
8. Tripicchio, P., Satler, M., Dabisias, G., Ruffaldi, E., Avizzano, C.A.: Towards smart farming and sustainable agriculture with drones. In: Proceedings of the International Conference on Intelligent Environments, IE, pp. 140–143 (2015)
9. Ramanathan, N., Balzano, L., Burt, M., Estrin, D., Harmon, T., Harvey, C., Jay, J., Koheler, E., Rothenberg, S., Sirvastava, M.: Rapid deployment with confidence: calibration and fault detection in environmental sensor networks. Commun. ACM **47**(6), 34–40 (2006). http://ieeexplore.ieee.org/xpls/absall.jsp?arnumber=1180542
10. Johnson, L.F., Roczen, D.E., Youkhana, S.K., Nemani, R.R., Bosch, D.F.: Mapping vineyard leaf area with multispectral satellite imagery. Comput. Electron. Agric. **38**(1), 33–44 (2003)
11. Shen, K., Li, W., Pei, Z., Fei, W., Sun, G., Zhang, X., Chen, X., Ma, S.: Crop area estimation from UAV transect and MSR image data using spatial sampling method. Procedia Environ. Sci. **26**, 95–100 (2015). http://www.sciencedirect.com/science/article/pii/S1878029615001747
12. Tri, N.C., Hoai, T., Duong, H.N., Trong, N.T., Vinh, V., Snasel, V.: A novel framework based on deep learning and unmanned aerial vehicles to assess the quality of rice fields. In: Akagi, M., Nguyen, T.-T., Vu, D.-T., Phung, T.-N., Huynh, V.-N. (eds.) ICTA 2016. AISC, vol. 538, pp. 84–93. Springer, Cham (2017). doi:10.1007/978-3-319-49073-1_11
13. Goodman, R., I.U. Ltd.: Too many sensors not enough receptors. The Ranger J. Defence Surveyor Assoc. **1**(1) (2011)
14. Kang, Z., Zhang, L., Zlatanova, S.: An automatic mosaicking method for building facade texture mapping. ISPRS, vol. XXXVI (2014)
15. Shum, H.Y., Szeliski, R.: Construction of panoramic image mosaics with global and local alignment. IJVC **36**, 101–130 (1998)
16. Ulman, V.: Mosaicking of high resolution biomedical images. Ph.D. dissertation, Masaryk University, Brno (2005)

17. Guo, T., Kujirai, T., Watanabe, T.: Mapping crop status from an unmanned aerial vehicle for precision agriculture applications. In: ISPRS - International Archives of the Photogrammetry, Remote Sensing and Spatial Information Sciences, vol. XXXIX-B1, August 2012
18. Valente, J., Guillen, A., Manrique, O., Arenal, M.: On the acquisition of high resolution maps with open source software and commercials off the shelf quadrotors. In: Gonzalez, P., Ribeiro, A. (eds.) Second International Conference on Robotics and Associated High Technologies and Equipment for Agriculture and Forestry New Trends in Mobile Robotics Perception and Actuation for Agriculture and Forestry (2014)
19. Primicerio, J., Gennaro, S.F.D., Fiorillo, E., Genesio, L., Lugato, E., Matese, A., Vaccari, F.P.: A flexible unmanned aerial vehicle for precision agriculture. Prec. Agric. **13**(10), 517–523 (2012)
20. Botterill, T., Mills, S., Green, R.: Real-time aerial image mosaicing. In: International Conference on Image and Vision Computing, New Zealand (2010)
21. Moussa, A., Sheimy, N.E.: A fast approach for stitching of aerial images. In: ISPRS, vol. XLIB3, July 2016
22. Hussain, R.Y.: Production of digital mosaics from aerial images. J. Eng. Devel. **17**(4) (2013)
23. Claudia, A., Porto, S.M.: Classification of crop-shelter coverage by RGB aerial images a compendium of experiences and findings. J. Agric. Eng. **41**(3), 1–11 (2010)
24. Di Franco, C., Buttazzo, G.: Coverage path planning for UAVs photogrammetry with energy and resolution constraints. J. Intell. Rob. Syst. Theory Appl. **83**(3–4), 445–462 (2016)
25. Hawary, A.F., Chipperfield, A.J.: Routeing strategy for coverage path planning in agricultural monitoring activity using UAV. In: International Congress on Recent Development in Engineering and Technology, RDET (2016)
26. Meyer, J., Sendobry, A., Kohlbrecher, S., Klingauf, U., Stryk, O.: Comprehensive simulation of quadrotor UAVs using ROS and Gazebo. In: Noda, I., Ando, N., Brugali, D., Kuffner, J.J. (eds.) SIMPAR 2012. LNCS (LNAI), vol. 7628, pp. 400–411. Springer, Heidelberg (2012). doi:10.1007/978-3-642-34327-8_36
27. Gazebo, "Gazebosim org." http://gazebosim.org/. Accessed: 15 Sep 2016
28. Adelson, E.H., Burt, P.J.: A multiresolution spline with application to image mosaics. ACM Trans. Graph. **2**(4), 217–236 (1983). http://dl.acm.org/citation.cfm?id=245.247

Mechanical Design of a Robotic Exoskeleton for Upper Limb Rehabilitation

Jorge García Montaño[1(✉)], Cecilia E. García Cena[2], Luis J. Monge Chamorro[1], Marie André Destarac[2], and Roque Saltarén Pazmiño[3]

[1] Aura Innovative Robotics and Escuela Técnica Superior de Ingenieros Industriales of the Universidad Politécnica de Madrid, C/ José Gutiérrez Abascal, 2, 28006 Madrid, Spain
george_14_7@hotmail.com, luisj.mchamorro@gmail.com

[2] Aura Innovative Robotics and the Centre for Automation and Robotics UPM-CSIC, C/ José Gutiérrez Abascal, 2, 28006 Madrid, Spain
cgarcia@upm.es, manded@gmail.com

[3] Centre for Automation and Robotics UPM-CSIC, C/ José Gutiérrez Abascal, 2, 28006 Madrid, Spain
roquejacinto.saltaren@upm.es

Abstract. This work presents the conceptual design, construction and testing of an upper limb rehabilitation system consisting of an exoskeleton of 6 degrees of freedom and a Human Machine Interface (HMI) to configure and control the exoskeleton. We present the study of the kinematics of the mechanical device followed by a three-dimensional modeling. Before the design of the exoskeleton was finished, we proceeded to the construction of a prototype in order to validate the model, using the Fused Filament Fabrication (FFF) technology and a three RX-64 Dynamixel motors from Robotis. LabVIEW® software from National Instruments was chosed to develop the HMI to communicate the exoskeleton with a computer for control the device and obtain relevant data, as the position or movement velocity. The results show the feasibility of the assistance of the robotic device in a rehabilitation processes.

1 Introduction

Nowadays we can find many applications of robotics in the medical field. However, we found few devices developed specifically to assist during the rehabilitation processes of musculoskeletal injuries of upper limb [1–3].

Rehabilitation therapies currently carried out by techniques that are imprecise. They are generic therapies that produce long-term results and require great personal effort by a specialist.

There has been an increase in musculoskeletal injuries of 45% between 1990 and 2010 in the worldwide population [4]. For these reasons and according to literature [5–7], a robotic exoskeleton could be a tool for rehabilitation specialists and also offer the patient a personalized and more intense process [8–10] with repeatability [11] and better short-term results [12, 13].

I. Chang et al. (eds.), *Advances in Automation and Robotics Research in Latin America*, Lecture Notes in Networks and Systems 13, DOI 10.1007/978-3-319-54377-2_25

Medical literature differs in the number of degrees-of-freedom (DOF) that has the human upper limb. It is estimated that exist between 13 and 16 DOF. This quantity of movements is excessive to be mechanically addressed therefore, we simplify the design choosing the relevant DOF for a rehabilitation treatment of the human shoulder.

Complementing the other studies, we analysed the different technics and movements that take place during upper limb rehabilitation treatment. These are the movements that a healthy patient makes along the day, in order to cleaning and feeding itself.

We based our development on a previous works which presents a musculoskeletal model of the upper limb that allows some movements such as the elbow flexion/extension and shoulder abduction/adduction [14, 15], two of the basic movements of arm rehabilitation. The reliability of this musculoskeletal model has been proven through simulations, as shown in [16, 17]. In [15] can be found the results of the simulations of both movements for the case of a healthy subject and one with upper brachial plexus injury. The software platform used was the version 2.2 of the Musculoskeletal Modeling Software (MSMS). MSMS is a free software developed by the University of Southern California [18, 19], which incorporate different tools to perform animations and simulations of the biomechanical behavior of musculoskeletal models.

The aim of this work is to develop a system for upper limb rehabilitation that includes a mechanical device controlled by a software system, which has an integrated musculoskeletal model. The ultimate purpose is to provide a tool for the rehabilitation staff, which allows a more precise and controlled therapy, and create a record of the evolution of the patient. With this the patient rehabilitation process would be optimized.

2 Requirements Engineering

This section presents the device requirements imposed by the medical team based on their experience. They have placed special emphasis on work space, ergonomics, safety and adaptability.

Contrasting the information obtained from the position of the DOF with the movements analyzed in the rehabilitation treatment, we observed which are the most decisive joints. We consider as determinants, those joints that allowed us to reach most of the different rehabilitation movements. The result was that we reduced the DOF to six. The exoskeleton's shoulder has 3 DOF: flexion-extension, abduction-adduction and scapula elevation-depression. Three others DOF can be found in the arm: forearm pronation-supination, elbow flexion-extension and humeral rotation. It is necessary to have one actuator for each DOF of the exoskeleton, therefore the kinematic of this system will be the same as the human shoulder.

Before starting the development of the mechanical part of the exoskeleton, we need to consider some restrictions for the use of the device in a medical environment. These recommendations were provided by physicians of the Infanta Sofia Hospital in Madrid. As a first step, the exoskeleton does not be able to overtake the operating space area of a human shoulder, because this would cause serious injuries to the patient. Secondly, the device has to be mechanically adaptable to the different human anatomies. This fact determines the target group that could use the exoskeleton. In our development, we are considering to make a rehabilitation system for adults.

In addition, the exoskeleton has to be comfortable, because patients will use the device many hours per week. If the device causes some kind of discomfort in the patient, they will reject their use and the therapy will not be effective.

We also have to consider the materials to use, the geometry or the presence of annoying noises. The chosen material was aluminum, a commonly material used in medical devices because it provides a good mechanical behaviour with a low weight. In addition, aluminum does not cause interference with other medical devices.

Other important point is that the exoskeleton should have suitable dimensions for its use in hospitals, being better to have a reduced design.

3 Mechanical Design

For the development of the exoskeleton, we estimate the torques needed in each joint. To obtain the torques we calculate the weight of a human arm of a 100 kg person, based on [20]. The results of this analysis were that each joint has a different torque. However, we could appreciate that the 3 DOF of the shoulder need a considerably higher torque than the rest, so in order to reduce the final application cost, we considered to use two types of motor size.

We chose motors with a higher torque for the shoulder abduction-adduction, flexion-extension and the scapula elevation-depression. These motors were located in a tower behind the patient and thus reducing the weight of the exoskeleton and simplify the structure. In addition, the position of the motors shift the centre of gravity of the structure providing a greater stability. Another advantage is that the motor noises and vibrations are reduced, and this point is important to not affect the concentration of the patient. However, to obtain these results we have to design a transmission system to communicate the motors with the exoskeleton joints.

The motors with a lower torque was attached in the robotic arm directly and they enable the elbow flexion-extension, arm internal-external rotation and forearm pronation-supination.

To carry out this work, we had three Dynamixel RX-64 servomotors, which characteristics are specified in [21], therefore we decided to actuate the three more important joints: the elbow flexion-extension, shoulder abduction-adduction and shoulder flexion-extension. We also decided to exclude the transmission system, because it would complicate the design stage, and with three motors the global weight wouldn't be high. These joints are the most relevant because they allow us to move the arm in a tridimensional space. The other three movements are executed in a passive mode, in order to show the exoskeleton workspace.

The next step was to design a virtual modelling of the mechanical system. We used the software Autodesk Inventor Professional 2016® to develop the three-dimensional (3D) model, taking into account the medical restrictions previously mentioned.

The device adaptability is achieved with movable parts in the forearm and in the arm of the exoskeleton. The exploded view of the mechanism is showed in Fig. 1. It has different position point that are manually fixed with a screw, indicated with parts A and B of Fig. 1. The adjustments of the lengths must be in the range of the patient's dimensions. In addition, we use the motor of the scapula to adjust the height of the patient, as is shown in part C of Fig. 1.

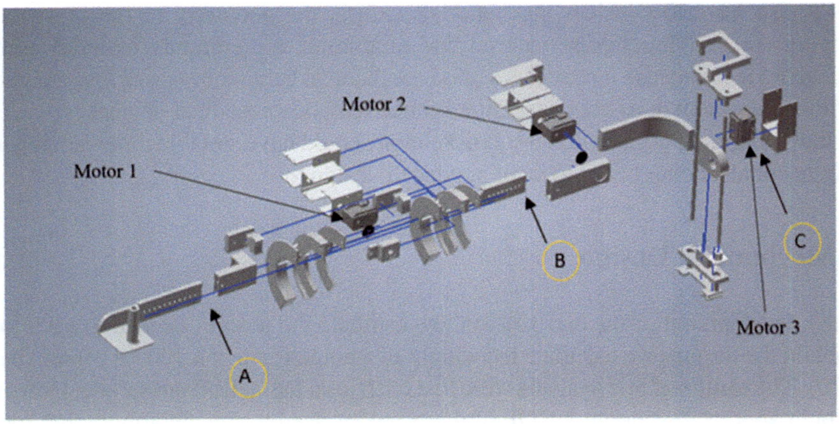

Fig. 1. Exoskeleton exploded view. Part A: adjustment of the forearm. Part B: adjustment of the arm. Part C: adjustment of the shoulder height

Furthermore, in Fig. 1 it is represented the three motors position that will move the actuated parts.

Other important aspects can be seen in Fig. 2. The final effector has a convenient form for the hand grip, the main idea is to have a support which can accurately guide the rotation during the pronation-supination movement. In addition, we developed another hand grip showed in Fig. 3. This element has an elliptical form, to connect the

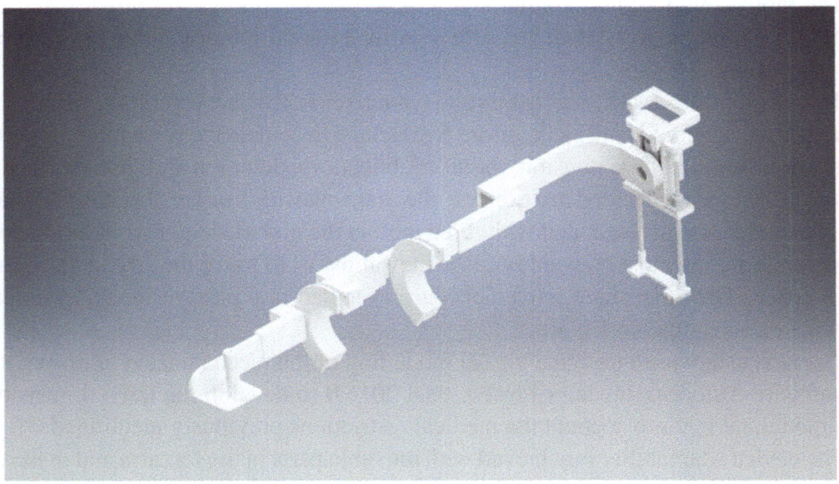

Fig. 2. Three-dimensional model of the exoskeleton

patient's hand with the exoskeleton. This union allows us to control the hand position in the space and leave it free in order to endow the patients with hand mobility. The advantage of this system is to give the patient the ability to interact with objects in the environment. This increases the available exercises in rehabilitation therapies.

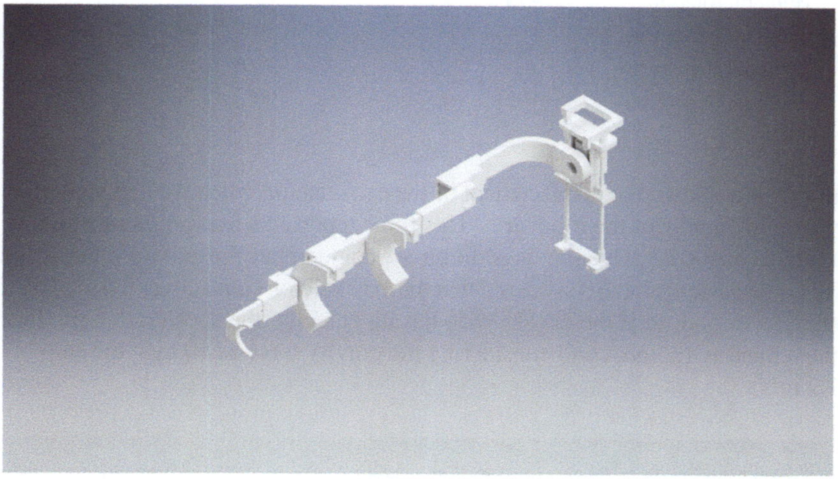

Fig. 3. Three-dimensional model of the exoskeleton with alternative hand grip

The following remarkable element was one of the most difficult to modelling and reproduce: the forearm rotation and the humeral rotation. This element was created due to the fact that rotational axes of pronation-supination and humeral rotation are coincident with the axes of the bones. This fact determine that we have to place the motors outside the axes. In order to solve this problem, we developed a semicircular piece with an internal track. This element is situated around the arm and the forearm and place the rotation motors axes in the joint arm axes. Finally, we approximate the scapula elevation as a shoulder rotation combined with a vertical displacement and this fact, allows us to simulate the vertical movement with two vertical guides. These will serve to adjust the height of the device to different anatomies.

Then we made simulations of the different rehabilitation therapy movements with the 3D model created in order to obtain the angular limitations. With these simulations, we can demonstrate that the model could make one of the most difficult movement to achieve with an exoskeleton: take the patient's hand to the back, in internal rotation. To reach this complex movement is necessary to rotate all the joints of the device, which is part of the last phase of the rehabilitation therapy. The exoskeleton can achieve the range of motions showed in Table 1.

The 3D model was mechanically analyzed. We made a static simulation to calculate the structure with the finite element method (FEM). This test considered the material used as aluminum 6061, and this will be the material used in a next stage of the exoskeleton.

Table 1. Range of motion

Joint	Minimum angle (degrees)	Maximum angle (degrees)
Forearm prone-supination	−75°	75°
Elbow flexion	0°	135°
Humeral rotation	−75°	75°
Shoulder extension	0°	50°
Shoulder flexion	0°	180°
Shoulder abduction-adduction	0°	90°

In Fig. 4 is shown the displacement of the exoskeleton when we apply one force of 60 Nm in the middle of the robot arm. This force represents the weight of a human arm for a subject of 100 kg of weight. In addition, we contemplate the gravity value as 11.2 m/s^2. This is because we have considered that the exoskeleton could cover 0.6 m of distance during one second, so if the exoskeleton has no velocity, the acceleration produced in the movement is 1.2 m/s^2, and considering the gravity force as 10 m/s^2 the sum of both is 11.2 m/s^2.

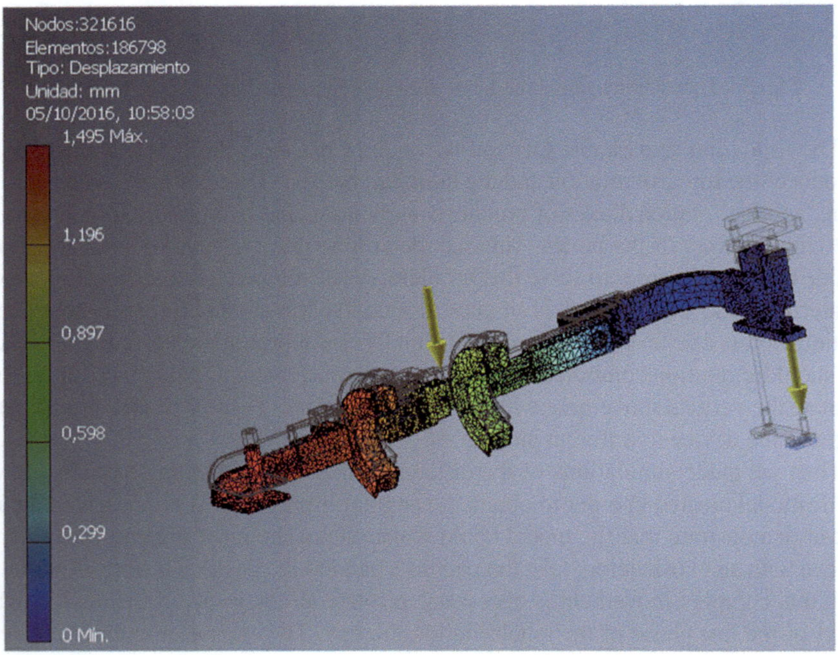

Fig. 4. Analysis of the 3D model by FEM method

The chosen position of the exoskeleton is the worst position from the point of view of the efforts, because the generated torque is higher. With this analysis, also we got the security coefficient of the structure, which is higher than 2 in the worst condition.

4 Experimental Results

4.1 Software Control Architecture

A software architecture was developed to interrelate the different elements of the system: the exoskeleton, a musculoskeletal model, the control system and the human machine interface (HMI). A diagram of the software system and the relationship between its elements it is shown in Fig. 5. The main objective was to obtain a robust, fast and modular system in order to obtain optimal results in the integration and a proper operation of the device.

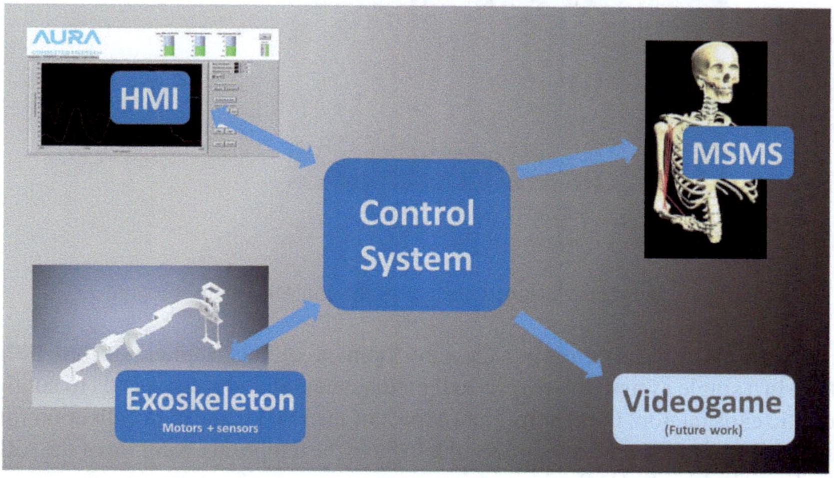

Fig. 5. Software architecture to control the exoskeleton

For the development of the control software and HMI, we chose NI LabVIEW®, and the reasons was based on the capacity and robustness of this platform, as well as powerful tools offered in relation with the user interface.

Communication with servomotors that integrates the exoskeleton has been carried out by serial RS-485 port, allowing bidirectional communication to read the angular position data of the motors, applied load torque exerted, errors and send set points of motion and configurations parameters.

Communication with the musculoskeletal model of the upper limb developed with MSMS [14, 15], is performed by User Datagram Protocol (UDP) communication. The model can reproduce the movement patterns collected from the patient or a therapy movement introduced by the rehabilitation specialist.

The control system governing the robotic exoskeleton is designed as a first stage in a simple and functional way, with the idea of implementing more functionality in the future, taking into account the feedback from medical specialists when they use this prototype.

The user interface has been made taking into account the characteristics of the end user, which are a medical staff with biomedical training. We consulted many professionals in the medical field and defined the main features and functionalities they need for this type of device, in order to facilitate their incorporation in hospitals and rehabilitation clinics. In Fig. 6 we can observe the appearance of the user interface.

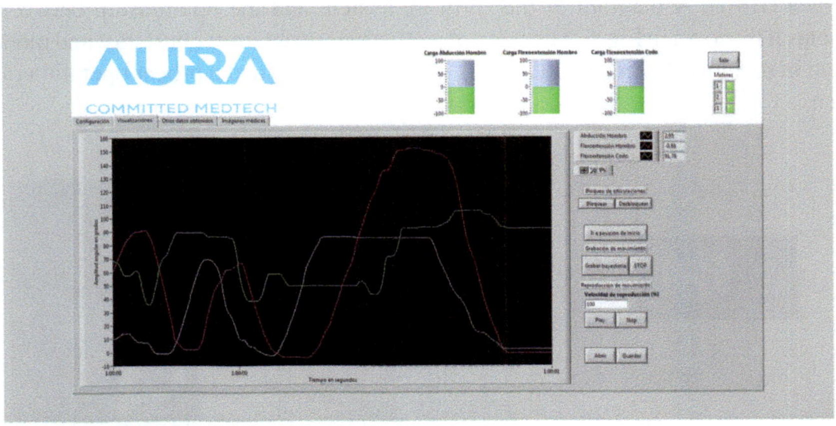

Fig. 6. Human Machine Interface of the control system software

The principal objective during the development of this application has been to combine in a single system the control and management that the user needs to configure the robotic exoskeleton. Other important aspect of the HMI is the data collection for the posterior interpretation and storage in order to have the temporal evolution of the patient.

We have designed a HMI with a main screen session configuration in which they are declared the gender, mass and height of the patient, we can establish joint limits for patients with limited mobility and can also generate a rehabilitation routine with the desired configuration. On a second tab, we can see a graph with the joint positions, as well as tools for managing the exoskeleton as the beginning of therapy, stop, recording movements, load or save a data file, and select between active and passive mode functioning, as is shown in Fig. 6.

Once completed the mechanical development of the device and the associated software system, we proceed to the construction of a first prototype to validate it.

4.2 Practical Results

The prototype showed in this work was manufactured with the Fused Filament Fabrication (FFF) technology. The reasons are that is a low-cost technology, with a fast implementation and it allows to make complex geometries. The chosen material was Acrylonitrile butadiene styrene (ABS) due to the fact that has a better behavior with flexion efforts. Regarding the motorization of the device, we have decided to integrate the Dynamixel RX-64 servomotors of Robotis manufacturer. The reliability of these servomotors has been proven repeatedly through simulations, as shown in [22, 23].

Once completed the construction of the prototype, we proceed to integrate the software application to perform validation testing system. These operations consist of recording the movements of rehabilitation therapy performed by a subject, generating a file with all the data produced by the dynamics of the movement. Moreover, we can load a data file in order to reproduced defined exercise. In Fig. 7 we can observe the system reproducing a previous defined movement.

Fig. 7. Testing process of the prototype of the rehabilitation system

The results of the FEM analysis show that the require torque in every joint of the exoskeleton is higher than the torque generated by the Dynamixel motors. This fact determines that the motors cannot be used in the final device, because the system request others with a better torque characteristics.

Making the same analysis without the arm force and changing the material to ABS, we calculate the torque needed in the prototype. The conclusion of this analysis is that the Dynamixel can be used to move the prototype, but not the human arm combined with the exoskeleton.

Once the tests with different subjects was carry out, we found that the prototype reaches the required angular ranges, complying with medical restrictions. As we calculated, the prototype in ABS has had a good performance against the efforts required in the exercises. In addition, it has been proven to reproduce a path entered through a data file and has been shown to faithfully execute it according to the established control configuration, as is shown in Fig. 8.

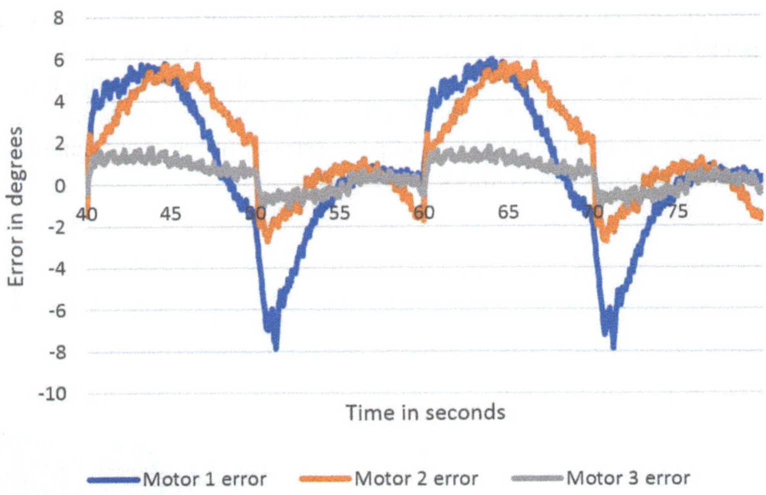

Fig. 8. Exoskeleton tracking error chart

5 Discussion

Having analyzed the results, we can say that the device reach the objectives. The mechanical system has properly responded to the expected movements as well as the software system gives accurate and necessary information, and provides the user with sufficient tools for handling the device. Therefore, the functionality of the rehabilitation system is demonstrated.

The prototype was tested with healthy human subjects. In this experiment, we confirmed the simulation results. The prototype could move itself without a human arm but it can't move both.

Among all the results we have obtained in the course of this work, we can highlight the confirmation that the material of the structure must be more resistant than ABS. Also, it has been found that the mechanical structure allows the user a natural and comfortable movement in all GDL, beside placing the driving shaft of the exoskeleton perfectly aligned with the upper limb joints, all with a reduced visual impact. The software system includes a stable communication and important user information is obtained. It also allows to save the record of the evolution of the patient and load a previously designed therapy.

In order to actuate the passive joint and starting working in the transmission system, we decided to design a transmission that could be yoke to the passive joints. The idea was to use the bike wires to reproduces the motor movement in the exoskeleton joints.

This wires has a semi rigid plastic jacket, this allow us to modify the length of the exoskeleton without losing wire strain. The problem was that the exoskeleton has a big quantity of movement. This fact force the wire jacket to bend, and make more difficult to move the exoskeleton joints, in other words, we are increasing the torque needed. For this reason, we decided not to use the transmission system.

We left three joints as passive and this issue would be treated in future works, transforming them in active joints.

6 Conclusion and Future Work

Once built a first prototype and made the relevant tests, it is demonstrated the feasibility of the assistance of a robotic exoskeleton in rehabilitation processes. The repetitive movements performed by the physiotherapist can be emulated by a robotic mechanism. This system provides a valuable and accurate information that allows to particularize the subject rehabilitation therapy, and create a record of the evolution.

As future work the system update is pending after receiving feedback from health workers and implement an adaptive control of the system. Also, the next prototype will be manufactured in aluminum and all of the joints will be actuated.

As a result of the work we have created an exoskeleton prototype that can be used to obtain data, to design a new version, like quantity of movements, mechanical and communication problems, kinematics limits, physicians feedback. In addition, the prototype can be used to reproduces some experiments in which we could get data of subjects, combined with different muscles measure techniques like electromyography.

References

1. Colombo, R., Pisano, F., Micera, S., et al.: Robotic techniques for upper limb evaluation and rehabilitation of stroke patients. IEEE Trans. Neural Syst. Rehabil. Eng. **13**(3), 311–324 (2005)
2. Perry, J., Rosen, J., Burns, S.: Upper-limb powered exoskeleton design. IEEE/ASME Trans. Mechatron. **12**, 408–417 (2007)
3. Loureiro, R., Harwin, W., Nagai, K., et al.: Advances in upper limb stroke rehabilitation: a technology push. Med. Bio. Eng. Comput. **49**(10), 1103–1118 (2011)
4. Vos, T., Flaxman, A.D., Naghavi, M., et al.: Years lived with disability (YLDs) for 1160 sequelae of 289 diseases and injuries 1990–2010: a systematic analysis for the global burden of disease study 2010. Lancet **380**, 2163–2196 (2012)
5. Barreca, S., Wolf, S.L., Fasoli, S., et al.: Treatment interventions for the paretic upper limb of stroke survivors: a critical review. Neurorehab. Neural Repair **17**, 220–226 (2003)
6. Platz, T.: Evidence-based arm rehabilitation – a systematic review of the literature. Der. Nervenarzt. **74**, 841–849 (2003)
7. Dobkin, B.H.: Strategies for stroke rehabilitation. Lancet Neurol. **3**, 528–536 (2004)
8. Ottenbacher, K.J., Jannell, S.: The results of clinical trials in stroke rehabilitation research. Arch. Neurol. **50**, 37–44 (1993)
9. Kwakkel, G., Wagenaar, R.C., Koelman, T.W., et al.: Effects of intensity of rehabilitation after stroke a research synthesis. Stroke **28**, 1550–1556 (1997)
10. Nelles, G.: Cortical reorganization-effects of intensive therapy. Restorat. Neurol. Neurosci. **22**, 239–244 (2004)
11. Bütefisch, C., Hummelsheim, H., Denzler, P., et al.: Repetitive training of isolated movements improves the outcome of motor rehabilitation of the centrally paretic hand. J. Neurol. Sci. **130**, 59–68 (1995)

12. Sunderland, A., Tinson, D.J., Bradley, E.L., et al.: Enhanced physical therapy improves recovery of arm function after stroke. A randomised controlled trial. J. Neurol. Neurosurg. Psychiatr. **55**, 530–535 (1992)
13. Kwakkel, G., Kollen, B.J., Wagenaar, R.C.: Long term effects of intensity of upper and lower limb training after stroke: a randomised trial. J. Neurol. Neurosurg. Psychiatr. **72**, 473–479 (2002)
14. García Cena, C.E., Saltarén Pazmiño, R., Destarac, M.A., Vega, E.L., Gomez, R.E., Santonja, R.A.: Skeletal modeling, analysis and simulation of upper limb of human shoulder under brachial plexus injury. In: Armada, M.A., Sanfeliu, A., Ferre, M. (eds.) ROBOT 2013: First Iberian Robotics Conference. AISC, vol. 252, pp. 195–207. Springer, Heidelberg (2014). doi: 10.1007/978-3-319-03413-3_14
15. Destarac, M.A., García, C.E., Saltarén, R.J., et al.: Modeling and simulation of upper brachial plexus injury. IEEE Syst. J. **10**(3), 912–921 (2016)
16. Destarac, M.A., García, C.E., Saltarén, R.J.: Simulation of the length change in muscles during the arm rotation for the upper brachial plexus injury. In: Ibáñez, J., Gonzalez-Vargas, J., Azorín, J.M., et al. (eds.) Converging Clinical and Engineering Research on Neurorehabilitation II. BB, vol. 15, pp. 1263–1268. Springer, Heidelberg (2016). doi: 10.1007/978-3-319-46669-9_206
17. Destarac, M.A., García, C.E., Saltarén, R.J., et al.: Kinematic and kinetic simulation of upper brachial plexus injury in the arm rotation. In: Fernández, R., Montes, H. (eds.) Open Conference on Future Trends in Robotics, Consejo Superior de Investigaciones Científicas, pp. 11–18 (2016). ISBN: 978-84-608-8452-1
18. Davoodi, R., Todorov, U., Loeb, G.E.: Development of clinician-friendly software for musculoskeletal modeling and control. In: Proceedings of the 26th Annual International Conference of the IEEE EMBS, vol. 2, pp. 4622–4625 (2004)
19. Khachani, M., Davoodi, R., Loeb, G.E.: Musculo-skeletal modeling software (msms) for biomechanics and virtual rehabilitation. In: Proceedings of American Society of Biomechanics Conference (2007)
20. Zatsiorsky, V., Seluyanov, V.: Estimation of the mass and inertia characteristics of the human body by means of the best predictive regression equations. In: Winter, D.A., Norman, R.W., Wells, R.P., et al. (eds.) Biomechanics IX-B, pp. 233–239. Human Kinetics, Champaign (1983)
21. Dynamixel RX-64 Product Manual. Robotis Inc.
22. Monge, L.J., Destarac, M.A., García, C.E., et al.: Modelling and simulation of servomotors for a rehabilitation exoskeleton. In: Fernández, R., Montes, H. (eds.) Open Conference on Future Trends in Robotics. Consejo Superior de Investigaciones Científicas, pp. 29–36 (2016). ISBN: 978-84-608-8452-1
23. Monge, L.J., García, C.E., Destarac, M.A., Saltarén, R.J.: Simulation of rehabilitation therapies for brachial plexus injury under the influence of external actuators. In: Ibáñez, J., Gonzalez-Vargas, J., Azorín, J.M., et al. (eds.) Converging Clinical and Engineering Research on Neurorehabilitation II. BB, vol. 15, pp. 1043–1047. Springer, Heidelberg (2016). doi: 10.1007/978-3-319-46669-9_169

Author Index

I. Chang et al. (eds.), *Advances in Automation and Robotics Research in Latin America*,
Lecture Notes in Networks and Systems 13, DOI 10.1007/978-3-319-54377-2

MIX
Papier aus verantwortungsvollen Quellen
Paper from responsible sources
FSC® C105338

If you have any concerns about our products,
you can contact us on
ProductSafety@springernature.com

In case Publisher is established outside the EU,
the EU authorized representative is:
Springer Nature Customer Service Center GmbH
Europaplatz 3, 69115 Heidelberg, Germany

Printed by Libri Plureos GmbH
in Hamburg, Germany